# RESEARCH MONOGRAPHS ON HUMAN POPULATION BIOLOGY

*General Editor*: G. A. HARRISON

RESEARCH MONOGRAPHS ON HUMAN
POPULATION BIOLOGY

R. L. Kirk: *Aboriginal man adapting*
A. E. Mourant, Ada C. Kopeć, and Kazimiera Domaniewska-Sobczak:
*The genetics of the Jews*
G. T. Nurse, J. S. Weiner, and Trefor Jenkins: *The peoples of Southern Africa and their affinities*

# THE PEOPLES OF SOUTHERN AFRICA AND THEIR AFFINITIES

G. T. NURSE, J. S. WEINER, and
TREFOR JENKINS

CLARENDON PRESS · OXFORD
1985

Oxford University Press, Walton Street, Oxford OX2 6DP

Oxford New York Toronto
Delhi Bombay Calcutta Madras Karachi
Kuala Lumpur Singapore Hong Kong Tokyo
Nairobi Dar es Salaam Cape Town
Melbourne Auckland

and associated companies in
Beirut Berlin Ibadan Nicosia

Oxford is a trade mark of Oxford University Press

Published in the United States
by Oxford University Press, New York

© G. T. Nurse, J. S. Weiner, and Trefor Jenkins, 1985

British Library Cataloguing in Publication Data
Nurse, G. T.
The peoples of Southern Africa and their
affinities
1. Human population genetics—Africa, Southern
I. Title    II. Weiner, J. S.
III. Jenkins, Trefor
573.2'15'0968    QH431
ISBN 0-19-857541-6

Library of Congress Cataloging in Publication Data
Nurse, George T.
The peoples of southern Africa and their affinities.
(Research monographs on human population biology)
Bibliography: p.
Includes index.
1. Anthropometry—Africa, Southern.    2. Human population
genetics—Africa, Southern.    3. Ethnology—Africa,
Southern.    4. Africa, Southern—Social life and
Customs.    I. Weiner, J. S. (Joseph Sidney), 1915–
II. Jenkins, Trefor.    III. Title.    IV. Series.
GN58.A356N87    1984    573'.0968    84–1138
ISBN 0-19-857541-6

Typeset by Joshua Associates, Oxford
Printed in Great Britain
at the University Press, Oxford
by David Stanford
Printer to the University

# PREFACE

Southern Africa is a region of especial interest to the human biologist. It was the home of early hominids, perhaps on the threshold of true humanity; it is the present home of the Khoisan, who may or may not constitute the smallest and most distinctive of the major races of man; it has received two large immigrations, of Negroes during two millennia and of Caucasoids in the course of the past three and a half centuries, from populations which have evolved in distant ecosystems. The descendants of both have continued to evolve unmixed *in situ*, and in addition have thrown up new hybrid populations by miscegenation with modern or ancient neighbours. In Southern Africa the study of the effects of racial admixture on a variety of traits in a range of contrasting ecosystems is eminently possible. It is possible, too, to make some assessment of the biological and technological accommodations which early man as well as later arrivals had to make to an area in which some of the process of hominization may have taken place. It is also disturbingly easy to prognosticate about the long-term effects of the types of resource utilization, and of the present high technology, on the human and other assets of a fragile ecology already subject to extensive changes.

We have attempted to take all this into consideration in this book, though our larger objective has been to describe the various peoples of Southern Africa in their biological settings up to and during a particular period of time, the time when we have been able to study them at first hand. A few years ago we presented in a book called *Health and the hunter-gatherer* (published by Karger of Basel in 1977) an account, rather smaller in scope, of a numerically small but biologically important group of these peoples. Some of the data from that book have been incorporated into this one; much has been added; but the emphases of the two books intentionally diverge. The earlier work gave little latitude for the exploration of interactions among human populations; and that is one of the principal themes of the present one.

Some previous work on the human biology of Southern Africa has tended towards the creation of false categories by assuming sharp demarcations and convenient uniformities. We have preferred to consider the Khoisan, the Negroes and the Caucasoids not only in their unity but in their diversity, though the more bewildering diversity of the hybrid populations has come close to defeating us. We hope that we have helped to show that even in the Southern African context, where human categorization sometimes attains a high level of absurdity, the biologically useful concept of race nevertheless represents only shifting and arbitrary boundaries in a single continuum of humanity.

We have ventured on rather more extensive interpretations and conclusions than we were able to draw in *Health and the hunter-gatherer*, and are conscious that several are arguable. Where in the earlier work we covered a narrower range both of peoples and eco-systems, in this one we have been led to try to find some explanations for our own and other workers' observations, and to unify them relative to their settings. Our shortcomings in this direction may be the consequence of our not having had the advantage during the final phase of writing of the wisdom and guidance of our senior member. The book was originally designed as the collaborative production of a physiologist–palaeoanthropologist, a human geneticist well grounded in clinical epidemiology, and a medically qualified anthropologist with experience of practice in the tropics and some knowledge of linguistics. Since all three of us had some knowledge in the fields of the others, we did not feel any need for any section to be written by any one of us exclusively. The preliminary drafts of the chapters were to be prepared variously by individuals or combinations among us, and then alterations and the final form agreed by all of us jointly. It was hoped that by that means not only a consensus, but a unifying style, would emerge.

These best-laid plans were set agley by the untimely death of Joseph Sidney Weiner. That indefatigable man worked on drafts of his sections of the book, as on the other of his productions which have appeared posthumously, and on the remarkably wide and committed teaching tasks he had set himself, until shortly before the end. The approach of death, expected for months, did not provoke in him any slackening, nor more desire for rest than his condition forced on him. He accomplished a good deal of what was asked of him, but the greatest boon of all, the benefit of his expertise in the making of books and the intellectual power he would have brought to the coordination and unification of the sections, has been denied to us. We have tried to preserve what he wrote, as far as possible as he wrote it.

We have benefited greatly from consultations and discussions with a wide range of our colleagues. James F. Murray, former Director of The South African Institute for Medical Research, read through the penultimate typescript and corrected some of our worst mistakes; to the Directors who preceded and followed him, James Gear and Jack Metz, we are likewise grateful not only for the facilities they, like James Murray, put at our disposal, but also for continuing friendly advice. The Librarians and Libraries of the Institute, and of the University of the Witwatersrand, the London School of Hygiene and Tropical Medicine, and the School of Oriental and African Studies of the University of London, have given us a great deal of assistance; less often consulted, but deserving acknowledgement, are the Librarians and Libraries of St. George's Hospital, Tooting, the Medical School of the University of Papua New Guinea, Boroko, the Centre Georges Pompidou, Paris, and the Australian National Library, Canberra. Among our major discussants were Patricia Draper, Allie Dubb, Revil Mason, Taya Mollison, Arthur Steinberg, Phillip Tobias, and Hertha de Villiers, but helpful comments on various aspects of our work also came from V. Balakrishnan, Megan Biesle, Rocky Chasko, Julia and Walter Bodmer, Peter Booth, M. C. Botha, Liz Cashdan, Luca Cavalli-Sforza, Desmond and Betty Clark, Ken Collins, Renki Eibl-Eibesfeldt, Maurice Godelier, Michael Greenacre, Henry Harpending, Robin Harvey, Bob Hitchock, Hoppy Hopkinson, Bob Kirk, Gerhard Kubik, John Marshall, Lorna Marshall, Hermann Lehmann, Pierre Lemonnier, Derek Roberts, A. E. Mourant, Jim Neel, Patrick and Rita Pearson, Skip Rappaport, Abe Rosman, Paula Rubel, Wolf Schiefenhövel, George Silberbauer, Eric Sunderland, L. D. Sanghvi, Jiro Tanaka, Tony Traill, Polly Wiesner, Ed Wilmsen, Jim Wood, and Alwyn Zoutendyk. Trixie Serjeaunt was a tower of strength in the arrangement of text and typing at many intermediate stages in London, as Shirley Alper, Christa Marschall and Stella Woods were in Johannesburg. The Papua New Guinea Institute for Medical Research permitted some typing to be done there by Kay Beynon, Miriam Korarome and Berenice Williams. Final versions of the maps were drawn by the Map Unit, and photographed by the Photographic Unit, of the SAIMR in Johannesburg. David Dann took over much of the compilation of the Tables, including last minute revisions. Material support was provided by amongst others, the Medical Research Councils of the United Kingdom and of South Africa, and the University of the Witwatersrand. Finally, we are deeply grateful to Geoffrey Ainsworth Harrison for his encouragement, guidance and patience towards authors whom he must at times have hoped would remain even further flung.

*London*                                                    G. T. N.
*Johannesburg*                                              T. J.

# CONTENTS

# FIGURES

# TABLES

# 1
# SETTING AND SUBSISTENCE

The two continents which bestraddle the equator have a number of characteristics in common. Both Africa and South America have large river basins, stretching almost right across them. These rivers run from east to west in the former and in the opposite direction in the latter, in both at about the level of the equator. In both cases the basins are thickly afforested. Moreover, Africa and South America are both connected to adjoining continents by narrow isthmuses. Inland from both western littorals, indigenous civilizations once flowered, while on the northern and eastern coasts of the two continents occurred the earliest traceable contacts with foreign peoples. In both continents the population is and apparently always has been unevenly distributed and, on the whole, sparse. Each is broad at the northern end, and tapers southwards; but where the northern coast of South America faces on to a warm unconfined sea, the northern coast of Africa is, of course, Mediterranean and temperate, and even at its southernmost Africa does not extend nearly so far towards the Antarctic as does South America.

## Africa in early recorded history

The historical parallels between the two continents are similarly curious. Africa could be thought of as standing in somewhat the same relationship to the ancient familiar world as South America did to the world of the Renaissance. Egypt was known to Greece and Rome, but its hinterland merged with the realms of myth. Herodotus garnered and recorded what he could, and some of the rest was assimilated by fashion and literature. Traders and mercenary soldiers, less fanciful than poets, less inquisitive than historians, travelled up the Nile, carved their names like *conquistadores* on the colossi of

Abu Simbel, and bargained or fought with the black Nubians who had earlier given Egypt its twenty-fifth dynasty. Pharaoh Necho hired Phoenician sailors and sent them down the Red Sea. At his orders they may in three years have circumnavigated Africa, an achievement no less remarkable than that of Magellan, though they left no record of the inhabitants to the south (Cary and Warmington 1929).

Refugees from Aegean droughts founded cities in Libya which were perhaps the first instances of European colonies in Africa. It was from them that Europe first became aware of the Sahara desert and of the peoples who lived near and in it. These people generated one large and powerful state, Numidia, perhaps in response to outside challenges, and settlers from the eastern Mediterranean set up Carthage, the most nearly successful of all the colonies. Both of these fell to the rising Romans, but not before the Carthaginians had expanded down the west coast of Africa as the Phoenicians had down the east. After the Roman conquest, though, this was abandoned, and subsequent generations preferred to learn what they could of the hinterland through Egypt. The Romans made contact with Meroe and Axum as well as establishing the existence of Lake Tsana and the Blue Nile. Ptolemy tells of one Diogenes, blown down the east coast probably as far as Bagamoyo, who travelled into the interior, discovered two large lakes, and widened the mythography of Africa with the tale of the snow-capped Mountains of the Moon. There their recorded explorations ended.

The first known crossing of the Sahara is described by Herodotus. Five young Nasamonians, members of a tribe to the west of Cyrene, set out and came at length to a country of black dwarves and a river which flowed from west to east, almost certainly the Niger. They returned home safely, and so, several centuries later, did Julius Maternus, enlisted by the Garamantes for an expedition against the 'Ethopians'. The Garamantes of what is now southern Tunisia plainly had, from Ptolemy's account, a good knowledge of suitable routes across the desert; but they themselves had left no account either of these or of the peoples with whom they traded.

Early written history, in short, tells us tantalizingly little of the peoples of Africa south not just of the Sahara but even of the Mediterranean seaboard. The curiosities of Egypt did not exclude the Egyptians from the comity of civilized nations. Greeks had settled the coasts next to the west, the Phoenician founders of Carthage and the Numidians beyond them were heirs of familiar cultures, and the scholarly king of Maurusia, Juba, himself an explorer and discoverer of the Canary Islands, was respectfully plundered by the elder Pliny. But inland, apart from the myths and monsters, only the cave-dwellers, whose colour is nowhere mentioned, and the dark Ethiopians and Nubians, emerge as possible progenitors of the modern peoples of sub-Saharan Africa.

## A brief geography

Most of the African land mass lies in the northern hemisphere. Southwards from the latitude of the Bight of Benin it narrows un-evenly to its southernmost point, Cape Agulhas at latitude 34° 50' S, hardly distinguishable as a projection from the bluntly rounded coast. Topographically the whole continent is roughly symmetrical; everywhere the land rises from a coastal plain, on average only thirty to thirty-five kilometres wide, to the intricate plateau system which occupies most of the interior. In eastern and southern Africa this great plateau complex lies along an axis running roughly south-west and north-east, and is transected across the axis by the valleys or gorges of large but not, in general, particularly long rivers such as the Rufiji, the Rovuma, the Zambesi, largest of them all, and the Limpopo, emptying into the Indian Ocean, and the Orange, Kunene, and Kwanza which discharge into the Atlantic. The Congo Basin is too distinctive and unique to be said to be associated with these plateaux at all except as their north-western boundary, but certain of the tributaries of the Congo, such as the Luangwa and Luapula, play major parts in diversifying their relief.

Africa has the shortest coastline for its area of any of the con-tinents (Hance 1964), and is singularly lacking in well-protected harbours. In the whole of Southern Africa only Maputo and Durban give adequate shelter against all prevailing winds. River mouths are frequently blocked by sand bars, and some of the harbours which might appear most promising lie adjacent to unproductive hinter-lands. This factor, combined with the harshness of the western coast and the warlike nature of the peoples of the southern and eastern coasts, as well as the notoriously unpredictable seas and winds around the Cape of Good Hope, must have contributed to the long period of relative isolation which Southern Africa enjoyed subse-quently even to the voyages of Bartolomeu Dias and Vasco da Gama.

## The limits of Southern Africa

Southern Africa is rather an imprecise entity, lacking political or ethnological definition and capable of being allotted any one of several geographical boundaries. The northern border of the Bantu-speaking peoples, stretching from the Niger–Benue confluence in the

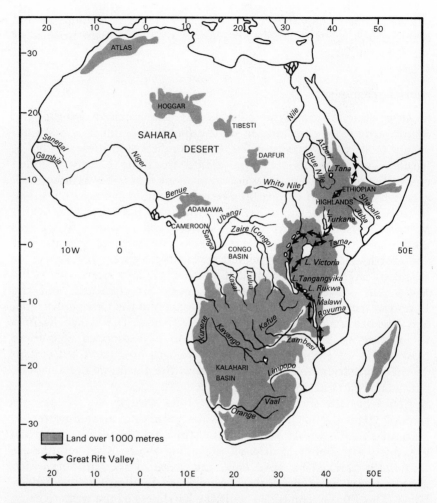

Fig. 1.1. Africa, showing the principal topographical features and the main drainage systems. Note that though the main system extending from north of Lake Victoria to south and west is deeply incised by large rivers, none except the Zambesi system would be difficult for migrants to pass across or avoid. Note also that the ill-defined southern end of the Great Rift Valley lies in a system of coastal lowlands continuous with the valley of the Limpopo, which in its upper reaches passes near Taung.

west to the Juba river in Somalia in the east, might at first sight seem to constitute an acceptable limit; but the Bantu languages may be simply part of a rather more widespread language family (Greenberg 1955), and this boundary is made dubious by the presence of non-Bantu languages to the south of it which are obviously related to

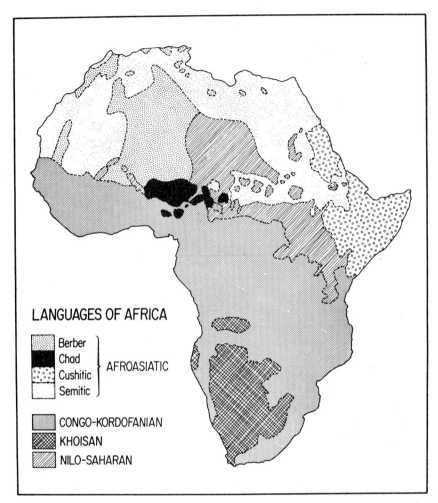

Fig. 1.2. The linguistic divisions of the African peoples, according to Greenberg (1963).

those to the north. Besides, many of the groups inhabiting this large area are politically or culturally closer to their northern or western neighbours than to peoples further south. A geographical feature which could be used as a line of demarcation is the Congo River; but the Congo is a navigable waterway, and eases rather than limits contacts among the peoples along its banks. The southern edge of the Congo Basin, again, might do; but that is a watershed hardly noticed by the people who live along it, and it would not be easy to draw an acceptable boundary to the east. If it were drawn to the Rovuma, a border between peoples as well as between states, one or other of the peoples of Malaŵi would live on both sides of it. If the line were drawn from the headwaters of the Luangwa to the nearest point on the Zambesi, the fact that the banks of the Luangwa unite rather than divide peoples would again be ignored. The Zambesi itself is a preferable alternative. For most of its length it does indeed separate peoples; its passage is almost everywhere difficult and is a familiar obstacle in traditional histories. But to the west of its head-waters, again, there is no obvious projection to the Atlantic; the axes along which the Angolan peoples are distributed run in the main north and south, and even the upper reaches of the Zambesi itself are inhabited by similar peoples on the right and left banks.

The next obvious boundary to the south has similar defects of in-completeness, but is reinforced by coinciding through most of its length with fairly rigidly observed political borders and changes of terrain. The Kunene River runs through a dramatic but by no means impassable gorge forming the western sector of the border between Namibia and Angola. The border leaves the river at the point where it begins to run westwards and travels in a straight line eastward to a point near where the southward-running Kavango bends to the east, after which it follows that river until it turns southward again. From this point the boundary inclines in a straight line approximately north of east to intersect with and subsequently coincide with the Linyati or Chobe River, a tributary of the Zambesi. From there it proceeds in a roughly south-easterly direction along the border between Botswana and Zimbabwe, the last portion following the Shashi River to its confluence with the Limpopo and thereafter run-ning along the Limpopo to the sea.

The principal objections to this boundary are, that Herero-speaking peoples live on both sides of the Kunene, and the Kavango peoples on both sides of the Kavango River; that the line joining the rivers effectively bisects the Ambo; that the Chobe divides Eastern Caprivi from the rest of Namibia; that the Botswana/Zimbabwe border is an artificial colonial construct; and that the lower Limpopo separates not distinct peoples but two divisions of the one people, the Tsonga or Tswa. The first two of these objections are virtually unanswerable, but the Chobe and the Botswana/Zimbabwe border do or did coin-cide approximately with ethnic boundaries, and the peoples on either side of the Limpopo are culturally more different than their close-ness of speech might suggest. It will nonetheless be impossible to deal fully with the peoples of Southern Africa without extensive refer-ence to conditions to the north of them. Not only do the peoples extend across the boundary in places, but there is even today a fairly constant flow across even the most stringently patrolled of the frontiers. Not all of this movement is in one direction, or politically inspired. Local population pressures fluctuate, kinship ties are strong, labour migration, despite all that is said against it, is persistently attractive, and in times of food shortage people will drift towards where the food is whether international tensions exist or not.

For present purposes, therefore, we may take Southern Africa to consist of the whole of the Republics of South Africa and Botswana and the Kingdoms of Lesotho and Swaziland, all of Namibia except Eastern Caprivi, and that part of Moçambique which lies south of the Limpopo river.

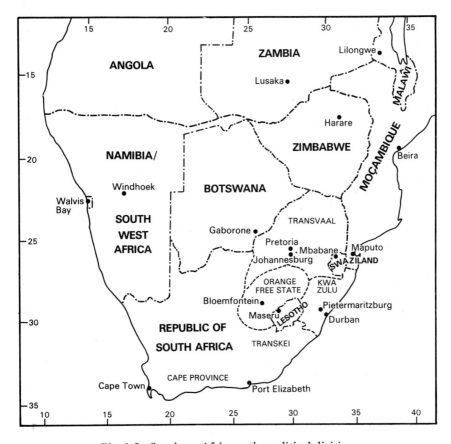

Fig. 1.3. Southern Africa—the political divisions

## The topography of Southern Africa

In the relatively small compass of Southern Africa we find almost all the typical topographical contrasts of the rest of the continent.

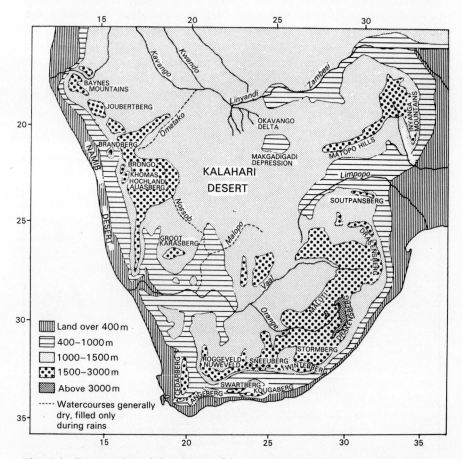

Fig. 1.4. Topography of Southern Africa. Note the narrowness of the coastal plain, the restricted range of elevation of the plateau system, and the virtual absence of southward drainage. Note also that the branching Okavango Delta receives the waters of the Kavango and Kwando rivers, with a small escapement to the Zambesi provided by the Linyati. Only after exceptionally heavy rains does the delta overflow into the Makgadikgadi Depression.

There is a sandy desert, the Namib, and a desert mainly scrub, the Kalahari, as well as a largely man-made and unavowed desert, the Great Karroo. There are cold high mountains and torrid plains, rocky shores in the south and east and pebbly and sandy shores in the west, deep gorges and gentle rolling hills. While it is possible to distinguish a great variety of landform patterns, the overall configuration is one of a narrow coastal lowland belt, widest in the east, rising abruptly inland to the plateau system of the interior. The plateaux vary in elevation between 1000 and 2000 metres, and are bounded by chains of mountains which form escarpments oriented to face towards the coasts. These ranges are highest along their eastern front, where the Drakensberg rises in places to above 3000 metres. The highest mountain of the range is Thabana Ntlenyana in Lesotho, the summit of which is 3482 metres above sea level. It forms one of a series of high peaks running roughly along the border between Lesotho and Natal. Further to the north the Drakensberg becomes less imposing, sending out an eastward outlying range as the Biggarsberg into northern Natal and becoming less regularly a range as it stretches towards the Limpopo. Below this northern continuation of the Drakensberg, and its westward outlier the Zoutpansberg, which divides the plateau from the Limpopo Valley, lies the lowveld. This is generally at an altitude below 600 metres, but where the range is less well demarcated it may extend up the slopes to constitute a middle veld. South of Lesotho and Natal the Drakensberg divides into a series of parallel folded ranges, which approach the coast and are separated by alluvial valleys of varying breadth, those to the north being higher and dryer than those closer to the coast.

The plateau surface of the interior is on the whole an area of plainland interrupted in places by broken country and by steep table-topped masses. It has three major sectors: an area of stepped terraces representing intermontane transition in the south, the Cape middle velds with elevations between 700 and 1400 metres and generally relatively smooth-surfaced; the Highveld itself, mostly above 1400 metres except for sections in the south-east, and in places above 2000 metres, from which rise a few lines of hills and numerous *koppies* or small inselbergs; and the Transvaal plateau basin. This last is in effect a projection of the highest sector of the highveld but tilted slightly towards the west. It is thus continuous with the great Kalahari Basin which fills the north-western part of the plateau and is bounded by the mountains which run north and south through central Namibia and culminate in the escarpment overlooking the Namib Desert.

The Orange and Vaal rivers rise in the Drakensberg and flow westward across the plateau of the highveld, eventually uniting and cutting through a gorge between the Namibian uplands and those of the northern Cape Province of South Africa to debouch into the Atlantic Ocean. Low limestone hills form an indefinite boundary between the northern part of the Orange Valley below the confluence, and the Kalahari Basin. The river courses in the southern part of the basin are generally dry except after heavy or continuous rains, and even then they peter out in the red sands of the south-west of the Kalahari Desert, though an old course can be made out which once drained the united waters of the Nosob and Auob, the Molopo and the Kuruman towards, if not perhaps actually into, the Orange. Below its junction with the Vaal the Orange gains only one large tributary, the Hartebeest, which collects the catchment from most of the plateau south of the Orange but which is nevertheless not perennial. The escarpment to the south of the Orange Gorge is irregular and not very abrupt in the north, but as it continues southwards becomes more impressive and eventually continuous with the majestic ranges of the western Cape Province.

There is no clear transition from the Kalahari Desert to the coastal plain in the southern part of Namibia, where the escarpment is lower than in the north. Here the plain represents the southernmost part of the Namib Desert, and its narrowest. The Great and Little Karas ranges project eastwards across the south of the country, and in the flat dry Namaland plain the extinct volcanic massif of Brukaros provides some relief. As the coastal lowlands widen northwards the escarpment becomes more impressive and throws out more eastern outliers culminating in the Auas Mountains and the Khomas Hochland. North of these and the Erongo range the escarpment is more gradual, interrupted only by the great isolated Brandberg system, until along the Kunene Gorge it rises again into the Joubert and Baines Mountains.

The Kalahari Basin contains virtually the whole of Botswana with the exception of the low hills in the east, which form the watershed of the upper Limpopo. It lies generally at a lesser altitude than the highveld sectors of the plateau, mostly between 700 and 1400 metres, and drains a considerable area. From the south and east a number of usually dry watercourses carry water after heavy rains to, or at least towards, Lake Xau and the Makgadikgadi Pans. The better-watered north drains into the swampy delta of the Kavango, which is connected with Lake Ngami. There is also a spillway towards Makgadikgadi, and the lake occasionally receives or donates a contribution itself along a spillway connecting it with the Chobe. None of the waters poured into the Kalahari has any outlet to an ocean; evaporation disposes of the greater part, and the rest goes to swell underground water reserves. In fact the bulk of the water flowing into the Kalahari is derived not from precipitation occurring over Southern Africa, but from the much more rainy *planalto* of Angola. The Kavango River is perennial and fast-flowing, collects its principal tributaries from central and south-eastern Angola, and loses comparatively little to evaporation until it dissipates itself as a delta.

The watershed between the northern Kalahari and the Zambesi is imperceptible, but it is noteworthy that in this stretch neither the latter nor the Chobe receives a regular tributary from the south. The Kavango, on the other hand, gets the ultimate overflow of the seasonal Omuramba Omatako, the main drainage channel of Namibia north of the Erongo Mountains. In northern Namibia the Etosha drainage system is analogous to that of the Kavango; the Etosha Pan receives the outflow of a number of small seasonal rivers running from the north out of Angola through the gap between the Kunene and Kavango valleys. None of these, however, flows strongly enough to provide a constant supply to the Pan; most sink into the sand before reaching it.

## African climatic regions

The transequatorial position of Africa has resulted in an almost symmetrical distribution of climate to north and south. The bulk of the northern portion of the continent, however, secures a certain uniformity for the zones there, while to the south of the equator the greater variability of elevation leads to the existence of greater climatic contrasts in a smaller compass. The extension of the plateau system into the north-eastern quarter, interrupting the equatorial belt of tropical rain forest, nevertheless ensures that some of the mainly southern subequatorial zones continue further northwards than any of the individual northern zones do southwards.

Systems of classification of African climates abound; in all of them the limits of the various regions are quite arbitrarily demarcated, except where the presence of strongly marked topographical features impose distinctions. Elsewhere each region tends to grade imperceptibly into the next; there are seasonal and cyclical fluctuations in the boundaries, and even ongoing processes of land utilization are quite certainly constantly changing them. Hance (1964) gives examples of different classifications in three maps which nevertheless show broad overall agreement. At the northernmost and southernmost ends of the continent are winter rainfall areas bounded by zones of semi-arid tropical and sub-tropical steppe beyond which lie deserts: the vast expanse of the Sahara in the north and the comparatively narrow tracts of the Kalahari, Namib, Moçâmedes and Karroo in the south. Equatorially to these lie other belts of steppe, bordering again on tropical savannah, which swings across the equator to the east of the equatorial rain forests if we agree with

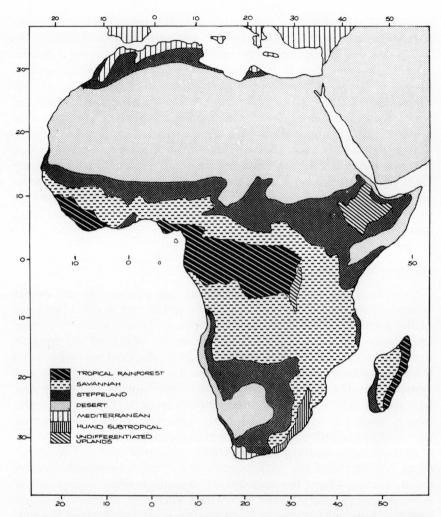

Fig. 1.5. Climatic regions of modern Africa. Note the rough symmetry, modified by the presence of high ground and the narrowing abruptly southward of the continent, to north and south of the Equator. There is, however, no zone north of the Equator corresponding to the humid sub-tropical south-east coastal plain.

Ady (1965), but in the opinion of Trewartha *et al.* (1957) is interrupted by a discontinuous belt of savannah and undifferentiated highlands, and according to Meigs (1952) by patches of tropical and temperate highland and, at lower elevations, by desert.

A large portion of Africa is relatively unproductive on account of these climatic patterns, which are responsible for the high relative aridity of the continent. Africa possesses about a third of all the arid lands of the world. Among the other continents only Australia has a higher percentage.

### The climates of Southern Africa

All of the climates of Africa with the exception of the equatorial rain forest are represented on a small scale within Southern Africa. It is even possible to find patches of rain forest with relatively high temperature along the lower Kavango valley and in parts of Natal, Swaziland, and the eastern Transvaal and Moçambique. The Mediterranean climate of the south-west extends over a comparatively small area, but is responsible for high productivity there. The winter rainfall spreads out to part of the somewhat drier upland interior and for some way eastwards along the coast into an intermediate area which receives rain throughout the year. Still further east the rain falls on the coastal plain mainly during the summer, and the decreasing latitudes bring an increase in mean temperature which renders the climate on the whole humid and sub-tropical. It is in this region, lying below the highest peaks of the Drakensberg, that the greatest mean annual precipitation, up to 180 cm and more, occurs.

Precipitation over the region as a whole varies considerably, from this maximum down to the approximately 2 cm annually which the Namib Desert derives mainly from the cold sea fogs which sweep periodically across it. The western and central areas are the driest; inland from the west coast precipitation ranges from 10 to 15 cm to north and south of the gorge of the Orange River, to about 30 cm in central Namibia and more than 50 cm in the northern parts of that country. The Kalahari Basin has a surprisingly high rainfall, ranging from 30 to 60 cm per year, but most of this either seeps at once through the porous surface or collects in large shallow pans and quickly evaporates. Seasonality of the rains, and their violence when they do come, contribute significantly to erosion from most of the plateaux and mountains. More than 85 per cent of the precipitation in the interior of Southern Africa occurs during the months from October to April, and in the intervening period the land surface becomes exsiccated and fractured by strong irradiation and winds, so that when the sudden thunderstorms arrive in late spring and early summer their runoff carries with it an appreciable portion of the topsoil. Less than 3 per cent of the region averages an annual rainfall greater than 100 cm, about 25 per cent receives between 60 and 100 cm and approximately a further 25 per cent less than 25 cm. Southern Africa is subject to recurrent severe droughts, exacerbated by the difficulties of conserving water in dams or lakes owing once again to the sharp seasonality of the rainfall and to the absence of cloud cover to minimize evaporation during the dry season.

The mean annual temperature over most of the region is remarkably uniform, due principally to the plateau structure of the interior. Temperature differences along the coasts, however, are the result of oceanic effects: the warm Moçambique current runs southwards down the east coast and the Benguela current carries sub-Antarctic waters northwards along the west coast, causing marked contrasts in mean annual temperature between the areas adjoining these shores. The similarity of the means elsewhere, too, tends to obscure the pronounced differences of annual temperature ranges, which attain greater amplitude in the interior. Over the plateaux frosts occur at least once in each winter, and high summer temperatures promote further evaporation and reduce the effectiveness of the rainfall. Where mountain formations are most marked, as in Lesotho and parts of the Cape Province, both temperature and precipitation may vary within quite circumscribed areas according to elevation and the aspect of the slope. The Maluti Mountains of Lesotho, running southwestwards from the Drakensberg and with them enfolding the upper basin of the Orange River are, like the Drakensberg, often covered with snow for at least part of the winter. Snowfalls tend to be heavier but less regular on the western Cape mountains, where although the precipitation is highest during the winter the ambient temperature does not fall as persistently low as it does in the Drakensberg. The Malutis and the western slopes of the Drakensberg, exposed as they are to these extremes of temperature and variations in the rainfall, are situated in possibly the most heavily eroded area in the whole of Southern Africa.

Though summer rainfall is universal except in the south-west its effect combines less harshly with the changes in temperature in the most northerly and easterly parts of the region, where the range north of the Tropic of Capricorn fluctuates considerably. The hotter savannah belt bordering the Kalahari Basin to the north and continuous with similar country in Angola and Zambia moderates the winters along the Kavango and Chobe, though not along the Kunene, which for the last part of its course runs between the Namib and Moçâmedes Deserts. Frost is not unknown in the Ambo country lying across the non-riparian section of the border, but occurs more commonly on the western than the eastern side. It has even been known to strike down into the eastern coastal plain, blasting the tropical and sub-tropical crops.

**The vegetation**

Indigenous hardwood forests once covered a much greater area than they do now. In the north-eastern part of Namibia, across the Caprivizipfel and northern Botswana into the northern Transvaal and adjacent parts of Moçambique, a dry forest cover, with in places little or no undergrowth, and dispersed among open woodlands, persists and is continuous with similar vegetation in Zambia, Zimbabwe and Angola. This is of mixed type, with the *mopani, Colophospermum mopani*, in greatest abundance, but including numerous hardwood trees. The grasses here are generally rank and sour except during their first few weeks of growth. Herbivorous animals are dependent on shrubs and immature saplings and what they can browse from the lower branches of the trees. Further south, the only indigenous high forests are confined to the area of constant rainfall and to the seaward slopes of the mountain ranges. The slow rate of growth of many species has meant that with the coming of man there has been little opportunity for forest regeneration, and much of the reafforestation which has occurred has been with exotic species recently introduced.

On the higher reaches of the plateau system areas of savannah, with moderately dense tree cover and long grasses of a mesophytic type adapted to dry conditions, are found particularly in the less heavily populated or hitherto unexploited parts. Where grazing and settlement have occurred a partial or complete transformation to steppeland has taken or is taking place, with the long grasses being ousted by shorter varieties. The grasses are the most important elements in the economy of pastoralists. The bushes and shrubs, which vary in density and predominance, may provide supplementary browsing for large stock, but the bulk of the nutriment of such stock must come from the grass cover. The grasses of the interior include both annual and perennial varieties. The former depend on favourable temperature and moisture conditions for the germination of their seeds, and, since in an inconstant climate these cannot always be relied upon, are not generally of much significance to the cattle-keeper, though their spring flush may be of value and they can provide some enrichment even of the later grazing. The latter retain water and nutritive substances more effectively, and have a spring flush whether the early rains come or not. The dry mass of both types of grass is related to rainfall rather than soil fertility, with the unexpected result that grass generally grows more profusely in the Kalahari than in some areas not classed as desert. Desert grasses, however, are usually deficient in mineral content and there are great variations in their protein component.

In the Karroo and adjacent areas small sub-desert shrubs with large roots and small leaves, and consequently adapted to retain water, are the predominant vegetation. They supply good browsing for wild ungulates and small domestic stock, and are fairly rich in protein and essential minerals. The bushes and trees of the savannah and steppe lands serve much the same purpose, but for the larger animals they cannot be anything more than a supplement to their grass fodder. Some larger game such as giraffe and elephant seem, unlike the ungulates, to be able to make better use of these than of the grasses; in addition they are able to cull fruits and nuts and to thrive better in the more heavily tree-covered areas. The bulk of such animals limits their efficient use as a source of food, since much of their flesh cannot be entirely consumed by migrant parties of the usual size before it becomes inedible or is stolen by scavenger animals or birds. Nevertheless, their different pattern of dispersion would make them an important alternative to the ungulates, once the appropriate techniques of hunting them had been mastered.

### The fauna, including man

It is unnecessary to list here the variety of wild animals and birds which originally swarmed over Southern Africa. They are comprehended in the so-called Ethiopian division of the world's fauna, which extended, and to some degree still extends from the Mediterranean to the Cape. The Ethiopian faunal system consisted originally of a balanced and self-replenishing association of antelopes and bovids with larger browsers and grazers, an abundance of ground-living birds, small mammals and simians, with scavengers and predators ranging from the crocodiles of the rivers through the canids to the major felines.

Though hunting and gathering societies subsist more considerably on vegetable than animal foods, man must be included among the predators. Domestication of animals for food could certainly be counted as a form of predation. From early on, however, domestic animals also provided hides, and the caprines and ovines could have supplied wool. Dogs were used for hunting from an early stage, though not, it seems, by the San. Archaeological evidence indicates that though the prehuman australopithecines may originally have been vegetarian, the later hominids were certainly omnivorous. With rare exceptions, man has continued so until today. But man has not only hunted and herded; throughout his evolution he has been, at one or another level, prey himself.

During recorded history man has been present in Southern Africa in a wide range of cultural and ethnic varieties. It is no longer fashionable to regard any form of man as feral, but that is how the earliest writers on the area saw the hunter–gatherers. The San or Bushmen, who constituted most of these, were viewed as oddly shaped, underclad, dirty people, without laws, barely human. The itinerant cattle-keeping Khoi or Hottentots were hardly more advanced, or so it seemed. They, like the San, spoke languages including sounds such as few Europeans thought could represent anything at all. The Negroes were rather more familiar, but it was only after the two former peoples had been encountered that any Negroes were found in the region. Late-comers, the Caucasoids of Europe, were the first literate settlers, and it is their impressions which have been preserved. It is in the cultural details which they recorded that these are of most value.

No known major division of the peoples of Southern Africa has ever been extinguished. The contemporary populations of Southern Africa include, in fact, one more than all the populations that could possibly have immigrated. There are still Khoi and San, though not spread as widely as they once were. The distribution of the Negroes, on the other hand, is probably wider now than it ever has been before. The peoples who entered the region by landing on the coasts, the Caucasoids of Europe and India and the Mongoloids of South-East Asia and China, persist. And in addition there are the hybrid peoples, including but not confined to the so-called 'Coloureds'. Not all of these peoples have been exposed in equal degree to the formative rigours of the Southern African environment; but all of them have had to some extent to respond to interactions with indigenous or introduced fauna, either in their own persons or through effects on livestock on which they depend for all or part of their subsistence.

### Macroscopic and microscopic predators as hazards to man

At some time during his evolution man developed the skills and the tools for combating the larger predators. They have continued into modern times as hazards to persons or, more frequently, to stock, but it does not seem likely that they constituted at any time a major threat to human movement and spread. Lion and leopard were common before the introduction of firearms, and crocodile still lurk in many rivers. Though poisonous snakes can hardly be accounted predators any more than scorpions and spiders can, all three of these posed and occasionally still pose danger to man and beast. In the deserts and on the highveld scorpions are very common, though their sting is rarely fatal to adults, and they are even a food source to the remaining hunter–gatherers. It has never been necessary for man or his cattle to undergo biological adaptation to cope with any of these. Such processes seem only to follow the assaults of very much smaller predators.

It is not known what bacteria and viruses were introduced into the region by human migrations and their associated domestic animals, or which of those already present found new hosts in the migrants. The devastation wrought in the human populations by the recorded epidemics of smallpox in the eighteenth century argue for low previous exposure and low resistance. The equally devastating epidemics of cholera which swept East Africa in the nineteenth century appear never to have extended beyond Lourenço Marques into Southern Africa, which experienced its first major outbreak in the 1970s. The rapid spread of tuberculosis, and its relative severity, in non-Caucasoids after contact with Caucasoids, suggests that it was unknown before. It is less possible to be dogmatic about leprosy, which is common all over Negro Africa and may have been brought in by early Negro immigrants. The prescence of brucellosis in both man and domestic animals in the region has long been recognized. It is more likely to have arrived in stock coming from Europe than to have been brought overland by early migrants. Treponematosis, in the form of yaws in the tropical moist areas and non-venereal syphilis in the Kalahari, could have been present long before the introduction of venereal syphilis which has to a great extent supplanted them (Murray *et al.* 1956). A more certainly native spirochaetosis is tick-borne relapsing fever, and other ticks carry the rickettsiae which are responsible for tick-bite fever. Ticks are also vectors of *Babesia*, which causes red-water fever in cattle but is not found in wild bovines and is almost certainly recently introduced. Identical or similar ticks possess a toxin in their saliva which can cause fatal paralysis in man or, more usually, in sheep.

The most important microscopic predators of man and beast rely for their transmission on insect vectors or intermediate hosts. Most survive in wild animal reservoirs. Plague is probably a fairly recent introduction; though it is holoenzoötic in wild rodents in several areas, the main known outbreaks in man have been in settled agricultural areas of Botswana and Namibia. Its vectors, the *Xenopsylla* fleas, are widespread in Africa. Rabies is similarly holoenzoötic in wild canids, and spread by their bites; it is probably also a relatively new disease in the region.

The most important transmissible disease of man in the region is historically probably malaria, endemic in the better watered, more tropical regions, almost unknown on the highveld, and tending to be epidemic after the rains in the Kalahari and adjoining parts. Any immigrants into the area from the north would have been compelled to pass through a belt where conditions for its spread, the presence of suitable breeding-grounds and temperatures for its anopheline mosquito vectors, existed. But since there appear to be no wild animal hosts of the human parasite, this would not matter unless those areas were already inhabited by infected individuals or one or more members of the party carried the parasite. This fact is probably responsible for the clear-cut distinctions in distribution among the various modern populations of those genetic polymorphisms which represent adaptive protection against the disease and will be discussed in later chapters. The other important protozoal parasite endemic in the same belt is the trypanosome, which has an ungulate reservoir and immunity against which in man is acquired rather than inborn. It is devastating to bovine cattle, but goats and asses seem not to be susceptible to it. Again, its presence and spread depend on conditions suitable for its vector, the tsetse fly. The tsetse prefers shady areas and settles for preference on darker skins; it is possible that the reddish cattle of the early pastoralists were helped through the belt by their colour.

Both malaria and trypanosomiasis can be fatal to man. The other protozoal diseases found in man in Southern Africa are merely incapacitating, but to a hunter–gatherer or a pastoralist even moderate incapacitation may lead to death. Leishmaniasis is rare, almost confined to Namibia, spread by sand-flies and has a reservoir in hyraxes (Grové and Ledger 1975). In man it has been described only in settled individuals; it is possible, however, that it may have occurred in earlier inhabitants, and it is uncertain when it first entered the area. Amoebiasis is probably a newcomer among diseases; it needs no vector or reservoir, is spread by contaminated food, and probably arrived with immigrants from India in the nineteenth century. It would not have affected the earlier inhabitants, though other dysenteric diseases, among them giardiasis, spread in the same way, certainly would have.

Schistosomiasis has spread gradually through Africa from a probable original focus in the Nile valley. The snail intermediate hosts are present in almost all the rivers and streams, often to quite high altitude, but we cannot be sure how early they first became infested with the responsible worms. It is unlikely to have been before the arrival of agriculturalists and their settlements along river banks, as the human host is almost certainly necessary for the cycle. Filariasis, whether of *Wuchereria* or *Onchocerca*, is uncommon in the region, though the mosquito and fly vectors are present. Far more important helminths are the ankylostomes, both *A. duodenale* and *N. americanus*, the latter certainly a relatively recent arrival, the former probably more anciently present. Both are responsible for severe iron-deficiency anaemia which is often a contributory cause of death. Less serious intestinal helminths are common, and in the sandy areas of the east coast larva migrans due to the recently introduced *Ankylostoma braziliense*, a parasite of cats and dogs, is often confused with a burrowing mite (? *Tetranychus* spp.) which may have been there for much longer.

Burrowing insects may once have been more potent causes of disease than they are now. Inexpert evacuation of the sac of *Tunga penetrans*, the jigger flea, can cause severe secondary infection. *T. penetrans* has been known since the seventeenth century in west Africa (Davies 1979), but its spread to the east coasts has only happened lately; it is impossible to be sure when it arrived in Southern Africa. Cutaneous myiasis due to the larvae of various flies affects man less dangerously than it does horses, sheep and cattle. Scabies probably first arrived with the Caucasoids, and even today is more of an urban than a rural affliction.

The main epizoötic diseases to appear in the region in recent times have been caused by viruses. Rinderpest or cattle plague is now eradicated, but devastated the herds of pastoralists and sedentary stock farmers during the nineteenth century. It also brought about a serious reduction in the numbers of wild ungulates (Davies 1979). Neither rinderpest, nor blue tongue disease of sheep, is transmissible to man; the latter, and the virus of African horse sickness, are carried by *Culicoides* midges. Horses were a late introduction to the domestic fauna of the region. Rift Valley fever, or enzoötic hepatitis, is transmitted to man and cattle, and particularly sheep, by the bites of mosquitoes. The first recorded outbreak in Southern Africa was quite recent (Alexander 1951) and it is unlikely to have had any earlier effect on the flocks and herds of the pastoralists. Foot-and-mouth disease may also affect man and wild ungulates. For how long it has been present in the region is unknown. Other important virus diseases of cattle are the petechial and ephemeral fevers, and malignant catarrh, to none of which is man susceptible and the dates of whose spread in Africa are obscure. The same holds for caprine pleuropneumonia and the orfe and *jaagsiekte* of sheep.

Diseases of domestic animals to which humans were less susceptible would have had a greater social than direct selective effect on human populations. There are, for instance, strong indications that trypanosomiasis was responsible for the reversion of previoulsy cattle-keeping

Negro peoples in Namibia and northern Botswana to hunting and gathering. Losses from rinderpest may have accelerated the urbanization of large sections of the rural population at the end of the nineteenth century. Before there were towns to absorb such people the loss of livelihood by pastoralists from stock diseases must have been even more catastrophic. We may presume that epizoötics among game could have produced similar catastrophes for hunter–gatherer parties.

### Animal husbandry and agriculture in Southern Africa

In each of the three patterns of subsistence followed by the inhabitants of Southern Africa before the coming of the Europeans, we can distinguish a tendency towards different food procurement roles between men and women. Among hunter–gatherers, the men hunt and the women gather; the women of the pastoralists continue to gather while the men look after their flocks and herds; in an agricultural community the women hoe, sow, weed and reap and the men appear to do little or nothing except during brief annual seasons of quite strenuous activity. At such times they clear bush, build and repair houses, and organize and maintain the defences of the villages against human and other threats. Settlement imposes a pattern of responsibilities more varied in type than any which devolve on peoples who are not tied to a single place, or who are able to leave it only at the price of some social disruption. The apparent idleness of the male agriculturalist in a traditional society cloaks a constant wariness and readiness to defend the established, and consequently exposed, community. Pastoralists, careful of their herds, will always, unless they believe themselves to be much stronger than their opponents, move away rather than fight, while hunter–gatherers have learned to elevate elusiveness into an art. But for the agriculturalist, once he and his womenfolk have invested time, energy and expertise in attaching themselves to a particular location, the stakes are far too high. He has to be prepared to use violence to keep his investment remunerative and his women to himself.

Seeds of food plants, and possibly poultry, with the expertise for their management, were the coin of this investment and the principal importations into the region by the agriculturalists. These, in their turn, imposed on the choice of the area of settlement more rigorous restrictions than those which pastoralists or hunter–gatherers would have to face. Good agricultural land is rare in Southern Africa, where the general poverty of the soils is reinforced by the uncertainty of rainfall. The climates most propitious for agriculture are to be found on the whole in areas possessing unsuitable landforms, while the best landform areas tend to suffer from capricious rainfall and a wider range of temperature. The most favourable of all, the Mediterranean climate below the Karroo escarpment in the extreme south-west, was hardly exploited at all before the coming of the Europeans. The humid sub-tropical area of the eastern coastal plain and the Limpopo Valley suffers from porousness and sandiness of the surface except in the river valleys themselves, which, except along the lower Limpopo, are not of any great extent. The good soils of the eastern Transvaal are limited by the folded mountainous ridges of the escarpment, and those of Swaziland by erosion depletion and irregular water supplies. The poor soils generally are difficult to manage and maintain. The humus content is low, and the derivation of many of the soils from mechanical weathering of ancient rocks makes them liable to compaction. Elevation and steepness of slope, combined with the violence and suddenness of the early summer storms, leads to a great deal of erosion. There is often a serious deficiency of phosphates in the soil (Wellington 1955, 1967).

The wild species from which the early food plants of the Southern African agriculturalists were derived were indigenous not to the region itself but to the sub-Saharan Sudan belt, and were first cultivated there. It seems probable that the seeds of a greater variety were carried from there than eventually became established further south, where only sorghum and some of the millets were successfully cultivated as staples. Maize, the present staple, is of comparatively recent introduction, and whatever the date of its introduction into West Africa, which may have been pre-Columbian (Jeffreys 1966), has even today not completely supplanted millet in some areas, though the latter is most commonly used for the fermentation of beer. African beer, however, is a by no means negligible food. Other modern food plants include pumpkins and other edible gourds, and beans and other pulses, including the Bambara ground-nut. The dates of introduction of these varieties are not certainly known, but the last-named, though not grown as generally as its valuable protein content and nitrogen-fixing properties might make desirable, is indigenous to Africa. Most of the other pulses appear to be of American or Asian origin.

The seasonal nature of these crops makes it necessary that some form of food storage should be devised. Granaries are a feature of the household of the Southern African agriculturalist, and inevitably played a part in the very earliest periods of agricultural settlement. Defence of the village would have meant defence of the granaries as much as of the inhabitants. On the whole agriculturalists do not hold large reserves of cattle, which may invade and damage the gardens. It would perhaps be fair to say of the men of agricultural villages that they retained and amplified hunting techniques while their women cultivated the crops. A certain amount of garden preparation was probably also carried out by the men, who would furthermore have tended such large stock as had been bartered from pastoralists or bred.

It is likely, in fact, that the margins between a pastoral and an agricultural way of life were less clear-cut than the foregoing might have suggested. A community of prudent cultivators would be aware of the wisdom of keeping cattle as a reserve food supply, while pastoralists who found good grazing of reliable duration would be well advised to encourage their womenfolk to make small temporary gardens to supplement the often arduous collection of wild plants. This is not, however, described by Engelbrecht (1936) for the Korana or !Ora, the Khoi group which retained its traditional way of life longest. On the other hand the Nguni peoples in the southeast of the region, and the Herero of Namibia, although Bantu-speaking Negroes, have and seemingly always have had cultures in which cattle are of paramount importance. They also make extensive use of wild plant foods, particularly at times when the harvests are not yet in.

As the agriculturalists brought seed into the region, so did the pastoralists introduce animals. Wild cattle existed in north Africa in the early Pleistocene, and it is not beyond the bounds of possibility that they could have been domesticated there. On the whole, though, African experiments in the domestication of animals came too late to compete with species being brought into the continent from Asia, and except for the ass, the cat and the guinea-fowl, no modern domesticated animals are certainly indigenous to Africa. Even these three seem not to have reached Southern Africa in their domesticated forms. The dogs, chickens, sheep, goats and at least some varieties of the cattle are all derived from types of Asian origin. The degree of Asian content in or influence on the culture of Southern African pastoralists has been the subject of recurrent controversy (cf. Jeffreys 1968).

## The nature and extent of cultural competition

The requirements of the pastoralists generally did not compete with those of the agriculturalists except where the latter ventured especially ambitiously into pastoral activities, or a shortage of grazing tempted pastoralists to encroach on cultivated land. Apart from the seizure of springs and lands from the Basters in the northern Cape by nomad Caucasoid pastoralists in the late eighteenth and early nineteenth centuries, we have no sure knowledge of violent contact between the proponents of those cultures before the long and bitter nineteenth-century wars of Nama and Herero in Namibia. The goals of the agriculturalist and the hunter–gatherer would be in the main mutually exclusive. Land considered desirable by the agriculturalist could be no more intensively exploited by the hunter–gatherer than other land which the agriculturalist would spurn. Cultivated crops would represent to the hunter–gatherer strange and potentially dangerous foods, not worth marauding when familiar alternatives were available. The hunting activities of agriculturists are generally of low intensity and pose no threat to the hunter–gatherer's food sources. In fact, in many parts of Africa, a symbiotic relationship has existed between agriculturalists and hunter–gatherers, with extensive bartering of goods and services (Maquet 1961).

On the other hand, animosity between pastoralists and hunter-gatherers could be expected to occur much more commonly. The hunter–gatherer is unfamiliar with the concept of property in animals and is inclined to regard any untended beast as fair prey. Moreover, stock-farming, or even the passage of large nomadic herds, reduces the quantity of game the land can support. When supplies of game are suddenly reduced the hunter–gatherer may have no alternative but to seek his food among domestic animals. Sometimes a compromise can be reached. Some pastoralists in historical times took hunter–gatherers either forcibly or by persuasion into their service, used their skills to augment their own food supplies and rewarded them with the artefacts of a superior technology. But pastoralists with no regular cycle of nomadism will be more tempted to graze an area to extinction than will settled agriculturalists who happen to possess cattle. They can easily exhaust the capacity of such an area to support game, and them move on to repeat the process elsewhere. In Southern Africa, however, there are large tracts over which cattle keeping is only possible where and when surface water is procurable, and until recent times the interactions among the contrasting cultures of the inhabitants must have been relatively infrequent owing to the sparseness of the population. Nomadism on the one hand and cyclical migration on the other, with the intercalation of permanent settlement where conditions were favourable, nevertheless must have made encounters among them more frequent than would have been the case if the only culture had been of settlement.

The arrival of European settlers added a totally new dimension to a process of interaction which had not yet attained a stable pattern. The pastoralists were substantially ousted from the best agricultural land in the region; those of them who remained in the western Cape were forced into roles marginal to and dependent on those of the newcomers. The first land occupied was put to uses more rewarding than grazing, and new technologies rendered it much more productive. As the Caucasoids spread out of the area the old way of life retreated, though it remained sufficiently well entrenched to be adopted by some of the less fortunate among the descendents of the immigrants. The most significant additions to the culture of the indigenous pastoralists were brought by such people. Firearms made hunting, and defence against the raids of the San, much easier, but it was probably the introduction of the wagon which most considerably changed and improved the lives of the Khoi. Up till then large cattle had been used as beasts of burden and as sources of milk. They were, incidentally, hardly ever eaten: there was too much flesh on an ox for consumption by a single household, and the preparation of *biltong*, strips of wind- and sun-dried meat, was to some extent restricted by the activities of scavenger animals and birds. The principal domesticated food animals were sheep and goats, which could be eaten at a sitting. The only vehicles were crude ox-drawn sledges, of use only where the land was relatively smooth or recognizable roads had been worn into the ground. With the wagon came superior facilities for the transport, and hence for the accumulation, of goods. Trade began to play an important part in the lives of the pastoralists as soon as they realised that their cattle could be bartered for items of clothing and equipment superior to any that had been available to them before.

These advantages were, however, counterbalanced to some degree by their exclusion from or social degradation on lands which had previously been theirs. Despite the demand for labour on the farms of the Caucasoids, a large proportion of the Khoi withdrew from the Cape rather than abandon their freedom to wander and work as they would. At the same time the remnant San, enemies of the Khoi, were either extirpated or driven away by campaigns in which Khoi and Caucasoid cooperated. The advance of the Negroes into South Africa was happening at a slower and probably more peaceful pace than that of the Caucasoids; in what may have been a thousand or more years they had only reached the eastern marches of what is now the Eastern Province. Nevertheless, they blocked any attempt of the retreating Khoi and San to occupy the productive lands further east, and the refugees were forced back into the less favourable interior. At first this was of little moment; but the Caucasoid expansion, supported by the immensely more effective Caucasoid technology, continued during the next hundred and fifty years and culminated during and after the so-called 'Great Trek'. Competition for land had even before this led to clashes between Caucasoids and Negroes, and the path along the east coast was as effectively closed to Caucasoids as it had been to Khoi and San. Penetration of the interior by large and well-armed parties bent on the ownership rather than the simple use of the land not only disrupted the remaining Khoi polities and decimated the San, but resulted in considerable changes in the way of life of the settled agricultural peoples who were already making use of a great proportion of the arable land.

## Man-made changes in the Southern African environment

Traditionally, it is war which is held to be the human activity which makes the most drastic alterations in man's environment. To direct the economic resources of a nation towards non-productive ends will deplete them. Not all post-war reparative phases are always adequate. On the other hand, the exigencies of war, particularly in modern times, lead to technological advances which can be carried over and become more socially beneficial in peacetime. The small aggressions of autonomous agriculturists, however, rarely have either severely detrimental or markedly beneficial results, and it is unlikely that warfare in Southern Africa before the coming of the Europeans produced any pronounced or lasting effect upon the natural or the technical environment. Even the most devastating wars appear to have done little except bring about some population reduction. Settlements were destroyed and crops and granaries devastated, but these could as easily have been the results of natural catastrophes. What was laid waste was capable of regeneration.

A more serious sequence is that which follows the utilization and exhaustion of resources which cannot regenerate, or of which the period of regeneration is so long and unpredictable that its outcome cannot be anticipated. This is unlikely to happen when hunter-gatherers inhabit an area, since the minimum of utility of an area to them is reached, and they move on, long before the resources are seriously damaged. Only when other factors such as climatic changes are having an effect on the resources is it likely that the activities of hunter-gatherers will accelerate the process. The practices of pastoralists can be more serious. Unwise and excessive grazing of an area with a fragile or unstable flora may hasten desertification, and attempts to keep too large herds and flocks even on more favourable land can cause a rapid deterioration in its quality. The modes of grazing of the various types of domestic animal also have differing effects. Goats are hardy and will survive in most arid environments, but only because they are able to browse deeper and use more of a plant than other animals can. Plants so used are often unable to regenerate, die, and cease to function as soil retainers, leading to erosion and the incapacity of plant life to gain a foothold again in the same position. An excess of bovines in an area of regenerating grasses will destroy those grasses by nibbling the shoots too early and trampling down much potential fodder. Competition for grazing may easily result in pastures being used destructively early. Slash-and-burn agriculture, too, can be safely practised only when population density is low and rotation of sites is sufficiently long-term to allow for regenerative secondary growth.

Such processes are very evident in Southern Africa, and have been accentuated by reckless clearing of land and felling of trees and by political considerations which in South Africa have penned cattle-keeping peoples into inadequate or impoverished reserves and allo-cated better land to numerically much smaller groups. Though this has probably resulted in greater overall productivity, it is evident that the burden so placed on the poorer lands and peoples must in time cause the benefits to diminish, and that by the time that has happened irreparable damage will have been done to the environment (Downing 1978). The deterioration of the land is not a new process. Acocks (1953) points out that studies of modern vegetational patterns indicate major changes during the past half-millennium, and suggests that most of the plateau was originally covered with either forest or closed scrub. There are few trees on the highland now, many fewer even than there were a hundred years ago, and parkland vegetation is found mainly north of the latitude of Pretoria, where there has been relatively less Caucasoid settlement.

Another effect of human occupation is one which has tended to be overlooked or disregarded until relatively recently. Following land use, and its abandonment upon exhaustion of the soil from the standpoint of crop-raising, the secondary growth is often not agri-culturally regenerative of the soil, but it may favour the introduction, multiplication and/or spread of faunal types which are yet of value to the local human inhabitants (Feely 1980). This may be accom-panied by an extension also of relatively exotic flora, and can have a profound influence on the hydrology of an area. In the long term it may even produce an effect on later agricultural exploitation, whether by making it relatively easier, or more difficult, to clear, or by causing a change in soil types favouring one type of crop over another.

Up till the Second World War the region was heavily dependent on imported foodstuffs, but during the past few decades technical improvements in tillage agriculture have resulted in a production substantially above local demand. Maize accounts for about 70 per cent of the total crop area and 45 per cent of the area in the hands of Caucasoids. In the past much of the maize was grown as a single crop, without rotation and with inadequate fallowing, leading to depletion of the soil. Where there is pressure on the available arable land such practices must of necessity persist. Sorghum, less destructive of the land and of higher nutritional value, has lower yields and is thus less acceptable to subsistence farmers unable to expand their acreage. Even where land is in short supply, a certain proportion of it is used for cash cropping, mainly of tobacco. Much of the surplus maize grown by Caucasoids is intended as a fodder crop for cattle. There has been a recent tendency to move towards regional specialization in crop production, with increased yields of wheat, oats and barley, and towards market gardening in irrigated areas.

In a region chronically short of water irrigation becomes very important. The water needs of Namibia already exceed the extent to which the underground water, which is pumped for irrigation, can be replenished, and similar processes are happening on a smaller scale in other parts of the region. Lake Ngami in Botswana, when first described by Andersson (1856), supported a large variety of aquatic fauna; today it is a miserable film of moisture barely accessible across dry mud flats, having been pumped away to supply water to a mining company. Removal of the natural vegetation over large areas in the hope of securing quick returns from cash crops will inevitably lessen rainfall and reduce the stability of the soil. In the lifetimes of the present writers unlovely sugar cane has replaced the acacias and homesteads of the coastal hills of Kwa Zulu, and water problems have led to the construction of a dam which cannot be filled without flooding part of Swaziland. Prudence and conservation, though they show signs of coming, may well come too late.

The reduction in wild life throughout the area has been drastic and led to the virtually complete loss of what might, wisely deployed, have been a valuable food reserve for an increasing population. The mystique of hunting is strong among the Caucasoids and hardly less so among the Basters of Namibia and some Negro peoples, and the possession of firearms has resulted in indiscriminate slaughter. Land which is now used to graze exotic bovines and caprines once supported a far greater mass of indigenous ungulates. The economical ostrich is farmed in only one sector of the area, and there, ironically enough, on fairly good arable land which could be better used, and for its feathers and only secondarily its flesh. The native eland contributes to the diet of Soviet Central Asia but in Southern Africa has not been domesticated to any noticeable extent.

But probably the most considerable man-made changes in the environment of Southern Africa have resulted from concentrated industrialization and urbanization. Such changes have been in large part the results of the Industrial Revolution in Europe, and the technologies which have diffused under European influence progressively closer to the regions of primary production of raw materials. Southern African industry has been until recently almost confined to the Republic of South Africa, though factories of various sorts have existed for some time in southern Moçambique and Namibia and are being set up in Swaziland. It has been less the accessibility of raw materials which has promoted the accelerated industrialization of the region, however, than the availability of capital derived from the mining of diamonds and gold.

Mining has been responsible in this indirect way for a massive redistribution of the population. It is also directly responsible for producing the largest single concentration of population in the region, on the Witwatersrand, as well as the very rapid growth of a number of other urban agglomerations. The primary effect of mining on the human biology of Southern Africa has been somewhat tempered, however, by the efforts of the authorities to restrain migrant workers from bringing their families to the neighbourhoods of the mines. An elaborate system of laws regulates in South Africa the whereabouts of everyone not acknowledged to be of unmixed Caucasoid descent, and to these laws the nationals of adjacent states seeking work in South Africa must perforce submit. Those who return to their areas of origin often take with them new skills and some wealth; but many who have become accustomed to an urban life do their best, often successfully, to remain or return illegally.

As a result, much of the urban population not only in South Africa but in the adjoining countries, where the towns either in themselves or as transport centres represent the hope of economic benefit, is underemployed, underfed and embittered. Useful or gainful occupations have not grown in number as quickly as the inhabitants have. This is particularly marked in South Africa, where legislation reserving jobs for certain ethnic groups goes some way towards curbing productivity. South Africa's economic hegemony over almost the entire region consequently retards the pace of development even beyond its borders. By holding back the potential of its own inhabitants it detracts from its own proper position as a centre from which skills should radiate naturally into nearby states.

The changing population densities also affect the patterns of land use. The drift to the towns has led to labour shortages in the country-side and an increasing reliance on mechanized farming. Sources of energy are not abundant in Southern Africa, and the rapid increase in the energy requirements of the region has meant a dangerous need for imported fuels. For as long as the mineral wealth is able to pay for these there seems little effective inclination to take drastic measures to reduce the dependence. But the position is a perilous one, and should the precarious economic stability of the region give way a large and growing population will be faced with an environ-ment in which the veld has been degraded towards desertification, the forests reduced beyond effective regeneration, the game, except in a few vulnerable areas, has been hunted almost to extinction, the springs and underground waters drained, arable land eroded, and the very seashores so denuded of shellfish by crass holidaymakers that even the few fortunate people able to revert to gleaning a living from the edge of the sea will hardly prosper as much as did the ancient inhabitants of Matjies River Cave.

# 2
# EARLY MAN IN SOUTHERN AFRICA

It is believed that man and the apes descend by two great divergent lines from the *Dryopithecinae*, an arboreal primate stock whose fossil record dates to the Miocene epoch about 19 million years ago. This hypothesis is currently accepted by most students of the subject. With *Dryopithecus* the process of hominization, the path which leads through various hominids (or forms ancestral to man and the great apes) to the eventual appearance of modern man, may be held to have begun. Dryopithecine fossils are to be found in rocks in many sites of Eurasia as well as in East Africa.

The radiation which gave rise to the great apes seems to have developed soon after the dryopithecine phase. The fossil record is both complex and incomplete, but it does give clear evidence of the differentiation of the apes within this radiation. Today true apes are found only in the tropics of South-East Asia and Central Africa. In view of the relatively close affinity which man bears to the African apes (Le Gros Clark 1967, 1971) the dryopithecine ancestors of the hominids would in all probability have had an African provenance. Indeed, *Ramapithecus*, the fossil which may give the earliest indication of a dryopithecine divergence leading to the hominids, has representatives, previously known as 'Kenyapithecus', in East Africa as well as in India. There is still controversy on the exact affinities between these geographically separated remains. The hominid status of *Ramapithecus* also remains disputed, not surprisingly in view of the sparse and fragmentary nature of the remains. In some respects the mandibular and dental morphology suggests a primate closer to true hominids than any of the dryopithecine ancestors of the great apes. Nevertheless, recent molecular data (Andrews and Cronin 1982) seem to indicate that *Ramapithecus* and the fairly similar *Sivapithecus* belong more properly to a radiation which gave rise to the orang-utan, while among extant hominoids or man-like creatures man belongs to the same radiation as the gorilla and the chimpanzee.

## The evolution of man

Until relatively recently it was possible to contend very plausibly that man must have had his ultimate origins in Southern Africa. It was there, about two million years ago, that a form of animal intermediate between apes and man had appeared, and it seemed probable that in the same region descendants of this creature had progressed until they became undoubted human beings. If this had been indeed the case, the very process of hominization could have been linked to the challenges and opportunities afforded by the Southern African environment, and this description of modern man in that setting would have been of especial interest as an account of a region continuously inhabited for millions of years by his imme- diate ancestors. The first remains of australopithecines, the earliest representatives to be discovered of a genus now generally regarded as ancestral to the genus *Homo*, were discovered in Southern Africa. But it was recognized eventually that neither of the two species found there was likely to have evolved directly into humanity (Tobias 1965), and the search went on either for some congeneric species which did, or for an even earlier australopithecine which could have given rise to a radiation including both the Southern African species and a line definitely ancestral to man. It is still a matter of dispute whether hominid remains from Ethiopia represent the species in question (Tobias 1980; White *et al.* 1981).

Following a long gap in the geological record, remains of australo- pithecines begin to appear in a number of Pliocene sites in East Africa, at Lothagam, Koobi Fora, Kanapoi, Omo and elsewhere, about 3-4 million years ago. In consequence, an even firmer claim can be made for an exclusively African origin of the ancestral hominid stock. Spectacular evidence of the further evolutionary development of the australopithecine phase in the succeeding three million years has come from both East Africa and Southern Africa. This evolutionary differentiation led progressively to the emergence of the later and culturally more advanced genus *Homo*.

The two major taxa, or apparent divisions, of *Australopithecus* first identified, *A. africanus* and (Paranthropus) *A. robustus*, both closely related to those of East Africa, are well represented at five known Southern African sites, Makapansgat, Sterkfontein, Swart- krans, Kromdraai, all in the Transvaal, and at Taung in the Cape Province. All of these sites are to be found along the eastern extension

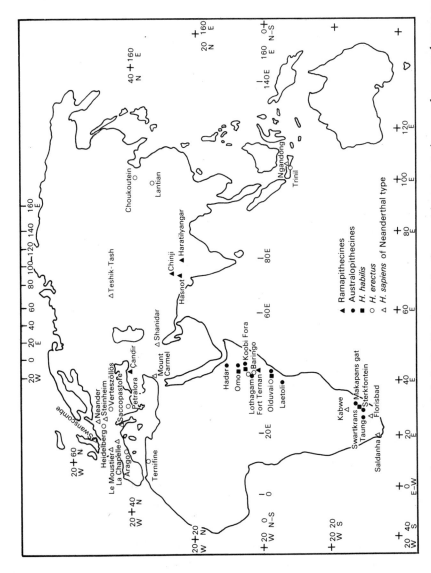

Fig. 2.1. The precursors of modern man and their distribution. Only the most important sites are shown, and some ascriptions are in doubt: the remains shown at Ngandong as neanderthaloid and at Swartkrans as habiline may be *H. erectus*, and Petralona man may be neanderthaloid rather than *H. erectus*.

of the Kalahari basin. *A. boisei*, (Zinjanthropus), which evolved later, seems however to have been confined to East Africa.

Although their dating is less certain than those of the fossils from Lake Rudolf or Omo, there are arguments in favour of an antiquity comparable with theirs for the South African fossils, particularly those at Sterkfontein and Makapansgat, which may well be as much as three million years old (Tobias 1978). There are nevertheless grounds for believing that *A. afarensis*, the oldest of all these hominid species and so far found no further south than Laetoli in Tanzania, may be ancestral to the other species and to man (Johanson and Edey 1981).

*Homo erectus* is the name given to a widespread species of man now recognized as having directly anteceded *Homo sapiens*. What is called *Homo habilis* is undoubtedly an older form than the fully developed *H. erectus*, but it may not represent a separate species or necessarily a variety ancestral to *erectus*. There is evidence that some of the australopithecines were tool-makers and thus habilines, though their brain volume was hardly greater than that of the great apes. Brain volume seems not to be the only, or even the most important, distinguishing trait among the higher primates, but it is a highly relevant one. There is no direct correlation between brain volume and intellectual capacity, especially when it is difficult or impossible to determine the relative volumes of the constituent anatomical parts of the brain.

Forms apparently transitional between *Australopithecus* spp. and *Homo sapiens* have been identified at Sterkfontein and Swartkrans in the Transvaal, as they have been at Olduvai and Koobi Fora in East Africa. Whether we should consider these remains as *Homo habilis* or as an early form of *Homo erectus* is still undecided. Also unresolved is whether *A. africanus* or an early *Homo* was responsible for the tools found at Sterkfontein. These implements appear to carry affinities with those found in Bed 11 at Olduvai, where there are also remains classified as those of *Homo habilis*.

The transition to *H. erectus* happened during the early Pleistocene. This process is well attested in East Africa, and may possibly have occurred in Southern Africa as well. Like earlier advances, it was marked by progressive enlargement of the brain. This hominid subsequently spread into many regions of the Old World. In view of the presence of these African forebears and of later archaic forms of *H. sapiens* also in Africa, it is perhaps surprising that more clearly identifiable remains of *H. erectus* have not come to light in Southern Africa.

The immediate precursors of *H. erectus* manufactured stone tools for chopping, paring, cutting and scraping by a technique sufficiently self-consistent for identification as the widespread Olduwan culture, called after the Olduvai Gorge in Tanzania. With the spread of *H. erectus* came the more elaborate Acheulian culture, in Africa and Europe, though in Asia *H. erectus* continued to employ Olduwan chopper techniques. Both these cultural traditions persisted, with much local specialization, through many thousands of years.

The appearance of the *sapiens* varieties of *Homo* was accompanied not only by a further enlargement of the brain, but by large developments in tool-making and tool-using. The process of 'sapientization' is heralded in such forms, which retain affinities with the *erectus* stage, as those from Swanscombe, Steinheim, Petralona, Arago and Vertésszöllös in Europe, from Choukoutien in China, and from Lake Ndutu in Tanzania. The cranium at the last-mentioned site was found in association with Acheulian tools and remains of butchered animals.

### The pace of human evolution

A major puzzle of the hominization process concerns the progressive increase in the size of the brain. Enlargement of the human brain happened very rapidly indeed. Evolution must have been unusually fast to produce trebling of the skull size in a mere ten million years, and most of this took place during the final million or million and a half. Various explanations have been suggested for this. Among the most credible are those most lately put forward.

It has recently been suggested, most particularly by Gould and Eldredge (1977), that the evolution of man may represent a specific example of a fairly general evolutionary model. This model, termed 'punctuated equilibrium', is characterized by long periods of absent or very gradual change, or relative stasis, intervening between bursts of rapid morphological evolution during which changes in bodily shape and proportions, and in the internal organs, happened in a comparatively short time. Contrasted with it is the more conservative model of relatively slow, even-paced evolution, now termed 'phyletic gradualism'. As Gould and Eldredge appreciated, and indeed stressed, their hypothesis would mean that gaps in the fossil record need not be considered as absence of evidence: such gaps could themselves represent stasis, and hence function as data. Intermediate forms between taxa might thus be regarded as having lasted for too short a time to leave fossil remains, since evolution from one form to another would have occurred very fast. This assumption would presuppose that the fossil record of early man and the earlier hominids is already virtually complete.

Cronin *et al.* (1981) have pointed out that it is improbable that all the relevant fossils have already been discovered, and treated this as grounds for dismissing punctuated equilibrium as the paramount principle of evolution. They nevertheless do not regard it as improbable that specific instances may be identified of both explosive and gradual evolution, at both the molecular and the organismal level. They reject punctuation or saltation as a complete explanation of the relatively rapid hominization process, and consider that apparent cases of punctuated equilibria in the fossil records may be due to errors in dating or morphological misinterpretations resulting in misleading taxonomies. They prefer to suggest (p. 122) that 'gradualism is, itself, an average of periods of horotely (fast change) and bradytely (slow change). Evolution of the Hominidae is still most reasonably interpreted by a model of phyletic gradualism with varying rates'.

Vrba (1980), applying the concepts to a number of other classes, and indeed phyla, besides mammals and man, points out that central to phyletic gradualism is the requirement that evolution should produce different sympatric species, while punctuated equilibria are more likely to involve allopatric speciation in small isolated populations. She suggests that direction in evolution is mostly towards increased specialization in species-specific characters, and hence that (p. 61) 'the sole motor ultimately driving evolutionary change seems to be the environment'. This would appear to lead to conclusions similar to those of Cronin *et al.* (1981). Nevertheless, her emphasis on environmental effects does open up an intriguing possibility suggested by the geographical location and the dimensions of the region in which transition from australopithecines to man would have occurred. An especially large spread of environmental factors is revealed when we consider the fluctuations in climate which have occurred during the Upper Pleistocene and even the present era, the Holocene. These are likely to have been somewhat more pronounced in the temperate latitudes of Southern Africa than they were closer to the equator.

## Climatic changes during the Upper Pleistocene and
### Early Holocene

Approximately one hundred thousand years ago a world-wide lowering of temperature, the preliminary phase of the last Ice Age, set in, and the polar ice-caps began to spread towards lower latitudes. In Southern Africa the extension northwards of the Antarctic ice and the consequent cooler and cloudier weather brought about increased rainfall and a greater availability of surface water, so that the vegetation of the western Cape was able to spread into the Kalahari and Namib Deserts. Decreased evaporation and lower temperatures permitted the expansion of the evergreen montane forests and their associated animal species to lower country and into parts of the interior plateau where today relict forest patches may be found in which lowland and upland species mingle. Such changes made the desert and arid steppe country more attractive for human occupation or re-occupation. The drying effects of the sub-tropical anticyclones and of the winds associated with the cold Benguela current, which are responsible for the modern aridity of the west coast, were pushed further north to affect the equatorial regions and cause a pronounced withdrawal of the lowland tropical forest, which was replaced over a great compass by open woodland and grassland more easily traversed by human migrants. Such movements were further facilitated by the locking up of vast quantities of oceanic water in the enlarged ice-caps, leading to a lowering of the sea-level and exposure of much of the continental shelf.

Processes of desertification in hitherto well-watered areas of east and west Africa, where long sequences of lithic technological development, Early Stone Age or Acheulian, had already occurred, could have persuaded the inhabitants of such areas to migrate. At the same time *Homo sapiens* began to replace *Homo erectus* in Africa. While *Homo erectus* was largely confined to an environment of grassland and open savannah, *Homo sapiens* was able to occupy parts of the equatorial forest and some of the drier, semi-desert areas as well. The period between sixty and thirty-five thousand years ago is marked by the beginning of regional specialization in stone tools and an increasing human occupation all over the continent, with a greater ability to exploit more varied habitats. The extent to which these processes contributed to selective adaptation, or could have resulted from prior adaptation, to environments from which various strains of migrant *Homo sapiens* had come, will be dealt with more fully later; but it seems very likely that the reduction in the overall harshness of the continent had as one of its results a number of fairly localized relaxations of selection, permitting more rapid growth in population size, perhaps to a degree which in places could have led to competition for resources and hastened technological innovation. The contrast of this industrial revolution with the very gradual development over the preceeding several hundred thousand years of the Acheulian tool tradition suggests, though not conclusively, that human intellectual abilities may have been improving quite fast at the same time.

By fifty thousand years ago temperatures in the tropical and sub-tropical regions had fallen to 5°–6° lower than they are today, and the period of lowered temperature lasted, with interstadial inter-missions, until some ten or twelve thousand years ago. The marine regression had reached its maximum, with the exposure of land as much as ninety metres below the present sea level, about eight thousand years earlier. Changes in the vegetation zones must obviously have led to a redistribution not only of plant foods but of associated animal species, and much of the human movement during the period may have reflected less a deliberate effort to accommodate to new climates than a search for familiar conditions and familiar prey. A warmer and more humid climate set in with the melting of the accumulation on the Antarctic ice-cap, and lasted for five thousand years or so. During this period game apparently became even more abundant and there were further consequent increases in the human population. At the end of the warm spell began the exsiccation of Southern Africa which has continued until the present day. The changes will have been more pronounced in the more arid areas, where even minor climatic fluctuations can have profound effects on habitability, and where a culture of nomadism or cyclical migra-tion is consequently enforced at the lower technological levels. The selective adaptation of the hunter–gatherers may well have been complete before the coming of later immigrants introduced new agents of infection and other hazards of rather greater complexity.

### Could the Southern African environment have had a bearing on the hominization process?

Adaptive selection will operate most vigorously, and, it appears, probably more rapidly as well, either on static populations exposed to a fluctuant environment or on mobile populations wandering in search of familiar resources or an ecological niche in which they will feel at home. Such populations, as Vrba (1980) points out, will be small; but this does not mean that a number of small populations, exposed consecutively to similar selective factors, would not react similarly. This would admittedly involve the conservation of com-patible mutations in different populations, but given sufficient time, and sufficiently vigorous selection, this is far from inconceivable. Mate exchange among such small populations would perpetuate and accelerate the rate of conservation of traits advantageous in ecosystems to which their members might recurrently be exposed, leading to the relatively rapid spread of such traits. The greater the variety of ecosystems to which exposure takes place, the greater the number of traits likely to be affected. Biological change can consequently be expected to be more rapid and more extensive in environments which are themselves subject to change.

Hominization involved not only point mutations (or single gene changes), epistasis (or gene interactions) and linkage disequilibrium (or the persistence together of sets of genes). There were also changes involving the chromosome, the structure in the nucleus of the cell on which the genes are carried. At least nine pericentric inversions, rearrangements of genetic material around the central point of the chromosome, as well as discrepancies in chromosome number, separate man from the modern pongid apes which he most closely resembles. Such divergencies can only be fixed in a breeding line by the most intense inbreeding (White 1978). This is most likely to occur where the range of available mates is restricted and sibling or parent-offspring mating is unavoidable or even favoured. We have no certain information about the exact size of early hominid populations, but the pace of human evolution requires that they should have been small and subject to intense selection. If the adaptive requirements were sufficiently severe, as they are more likely to have been in environments with strong seasonal contrasts, they would themselves help to keep population numbers low.

There can be little doubt that early man did show a tendency to migrate. The wide spread of remains of similar periods throughout East Africa cannot be accounted for by human over-exploitation of subsistence resources, which the same sites show to have been abundant. Each hominid wave consequently rippled out from the centre around which it had evolved. Klein (1977) points out that in Southern Africa, at some time between the early Pliocene and the mid-Pleistocene, the diversity of carnivore species was substantially reduced. At the same time, there appears to have been an increase in potential prey species. He suggests that the decrease in carnivore species was due at least in part to the arrival in the region and subsequent evolutionary success of meat-eating hominids.

The claims of A. afarensis to primacy in the ancestry of man seem strong at present, but it must not be forgotten that relatively few sites in Southern Africa have already over a relatively narrow area supplied a relatively large quantity and variety of australopithecine remains. There is every chance that when more intensive archaeological searches can be carried out in Botswana, Namibia, Zimbabwe and Moçambique more will be added to our store of knowledge about the process of hominization. A new dating of some remains from Border Cave, in Swaziland, has already led Beaumont et al. (1978) to postulate a pre-sapiens scheme of human evolution in which the savannah lands of sub-Saharan Africa are seen as 'the area where anatomically modern *Homo sapiens* originated at some as yet uncertain time' which they nevertheless claim to have been at least 110 000 years B.P. A similar suggestion has been made independently by Klein (1977).

### True man in Southern Africa

The transition from *erectus* to *sapiens* in Europe is attested by about 100 000 years B.P. in the various forms of Neanderthal man, accompanied by a modification of the tool-making tradition into a new style—the Mousterian. The Neanderthal stage of *H. sapiens* in Europe included varieties with progressively more developed modern features, leading eventually to the supersession of Neanderthal man by populations indistinguishable from modern man.

In Southern Africa, as in Indonesia, there is evidence of an analogous differentiation. Human remains from Broken Hill and Saldanha Bay are not fully *sapiens* in their morphology and indeed resemble Neanderthal man in a number of features. This stock has been termed 'Rhodesian' man, *Homo rhodesiensis* or, on the obvious analogy of Neanderthal man, *H. sapiens rhodesiensis*. In Africa, however, there is no evidence of the Mousterian tradition. The Saldanha skull bones were found in association with tools of the local variant of the Early Stone Age industry (also called Final Acheulian). These included cleavers, hand-axes of assorted sizes, bolas-like stones and pebble choppers. They were made from a variety of raw materials—quartzite, silicate, feldspar, or soft sandstone. The original *rhodesiensis* skull, found at Broken Hill in what is now Zambia, was associated, however, with implements belonging to the Flake cultures of the early Middle Stone Age.

From about 40 000 years B.P. onwards in the Middle Stone Age period, remains of *Homo sapiens* comparable to modern forms are widespread, though not numerous. They have been found in Zambia, the Transvaal, the Free State, Natal and the Cape Province. They provide evidence of the continuity of human occupation from the preceding *rhodesiensis* stage into the next era of the Later Stone Age.

Some adjustment backwards of Middle Stone Age horizons in Southern Africa appears to have been made necessary by the recent work on the remains from Border Cave on the boundary between Swaziland and Kwa Zulu (de Villiers 1973; Beaumont *et al.* 1978). The earliest human remains found there date from at least 49 000 years B.P. They are of a big-brained stock, undoubtedly *H. sapiens*

and without neanderthaloid traits, and showing enough morphological similarities to modern indigenous populations to suggest that
the fossil series may represent a relatively undifferentiated population possibly ancestral to both the Khoisan and the Negro peoples.
They are associated with well-developed Middle Stone Age artefacts.
It is not yet clear how this stock can relate to the relatively more
'primitive' hominids which also appear to have pursued a Middle
Stone Age technology during the early Upper Pleistocene era. More
recent studies (de Villiers and Fatti 1982) give a strong indication
that these remains may be even older than has been supposed.

During the Later Stone Age, from about 11 000 years B.P. until
the inauguration of the Iron Age about 2000 years B.P. (100 B.C.),
*sapiens* groups continued to extend through Southern Africa and
regions to the immediate north. Zambia, Malaŵi, Zimbabwe, the
Free State and the Cape Province have yielded skeletal evidence of
this occupation. It is possible to make quite early identification of
these groups as related to contemporary or near-contemporary
inhabitants of the region. Remains identified as Khoisan and dating
from the period from the beginning of the Iron Age up to the coming
of the Europeans in the seventeenth century have been recovered,
for instance, from many caves, middens and other occupation sites
in Zimbabwe, Zambia and the Cape Province.

### The continuity of human occupation

The fossil finds of sapient man in Southern Africa extending through
the Stone Age periods to modern times raise two major issues. In the
first place there is the question of the evolutionary relationship of
the 'Archaic' or neanderthaloid *rhodesiensis* man to the closely
succeeding following *sapiens* populations. Can *rhodesiensis* man
be seen as a stage immediately ancestral to both major sub-Saharan
African populations, Negro and Khoisan? This relationship would
be analogous to that claimed for Solo man in relation to the Wadjak
stock and Australian aboriginals in South East Asia, or that of the
Neanderthal 'spectrum' in Europe in relation to European *sapiens*
from Mount Carmel and Combe Capelle onwards. In the second place
arises the problem of the degree, nature and antiquity of the divergence between the presumed descendent peoples, and the extent to
which it was brought about by differing environmental adaptations.

A 'Rhodesian' ancestry could not proceed in simple rapid direct
transition to purely South African descendants and particularly to
a Khoisan stock, as promulgated by Tobias (1962*a*) partly on the
grounds of the short time gap between Broken Hill and the earliest

Middle Stone Age representatives. A further consideration is the fact that divergent as they are in a number of characters, the Negro and Khoisan peoples share sufficient similarities for them either to derive from a common stock originally or to have met and mingled so considerably in the past as to give that appearance. Brothwell (1963) suggests that both may derive indeed from 'Rhodesian' stock, and that the diversity of African fossils from the latter half of the Upper Pleistocene may indicate the selective adaptation to varying environments that has gone on since. The skeletal evidence interpreted from the time of Shrubsall is in general keeping with such relationships.

The morphological findings of Shrubsall (1907) and Stern and Singer (1967) are in keeping with genetic evidence which indicates strongly that San and Khoi must have shared a relatively recent common progeniture. This may be termed (as Brothwell has done) the proto-Khoisan (KS) stock or stage, and may have been in evidence in Late Upper Pleistocene times, perhaps contemporarily with, or very soon after, the stock represented by the Border Cave remains. To equate a proto-Khoisan stage with the existence of the 'Boskop' race would be misleading on various grounds (Singer 1958).

The evidence does invite the further inference of a still earlier stage capable of radiating into both the proto-Khoisan and into the Negro line which gave rise to the Bantu-speakers of Southern Africa. The scanty fossil material does give some support to such a Negroid–Khoisan 'matrix', but there are morphological objections to seeing in the Broken Hill and Saldanha rhodesioid remains the ancestral common stock suggested by Tobias (1962a) and Brothwell (1963). The strongly marked frontal and occipital skull ridging provides one obvious contrast to the earliest known Khoisan material. The closeness in time of the known *rhodesiensis* specimen to the earliest of the Middle Stone Age specimens is another objection, though not, of course, an absolutely insuperable one. We cannot suppose that the sparse remains we have of 'Rhodesian' man represent specimens that had only just evolved; it is just as likely that they might be coeval with the evolutionary stage which followed them. Nevertheless, it is possible to formulate a hypothesis which would constitute a modification of Tobias' proposition and analogous to the 'spectrum' or reticulate theory of voluntary modification in Europe or in South East Asia (Weiner and Campbell 1964). The fact that *rhodesiensis* carries *sapiens* features allows of the possibility that at the *rhodesiensis* stage differentiation in a *sapiens* direction was actively proceeding, just as Neanderthal man shows a graduation from 'classical' through intermediates to *sapiens* predominating over

Neanderthal. Of the human remains from the seven Middle Stone Age sites, Brothwell believes that two at least carried 'rhodesioid' affinities. The Florisbad skull can also be seen as a link between 'large-headed Khoisan' and 'Rhodesian' man.

This theoretical phase would necessarily have occupied a long period preceding the dates of Broken Hill or Saldanha and it is possible to generalize it as the ancestral matrix which later gave rise to the whole Negroid stock from which various regional strains including the Khoisanoid ultimately emerged. Where this differentiation took place is unknown. The relatively more numerous Negro populations in Central, East and West Africa suggest at least that it was not on the southern promontory. *Homo erectus* was clearly in evidence in East Africa; a particularly striking example is the nearly complete skull found at Koobi Fora in East Turkana in 1975 (Leakey 1976). Transitional forms leading to *rhodesiensis* are nevertheless not yet clearly evident in the East African material, except possibly in the skull found at Bodo in Ethiopia (Conroy *et al.* 1978). Thus the immediate forebear of *sapiens*, at a stage comparable to that of Swanscombe, or Steinheim in Europe, may remain to be discovered.

### Early evidence of modern man in Southern Africa: additional considerations

The human remains from all the Middle Stone Age sites in Southern Africa, according to Brothwell's review, fall into what he calls the 'large' Khoisan category. The original authors on the whole subscribe to this attribution. This points to the morphological differentiation of Khoisan stock within Southern Africa itself. The factors of isolation and the introduction of pastoralism may account to a considerable extent for the emergence of the recognisably divergent sero-genetic and morphological variants of the present day; these matters are discussed in later chapters.

There is an unbroken continuity and ubiquity of Khoisan occupation over the same wide area in the Late Stone Age. The presence of a Negroid component at a few sites (Hora, Cape Flats) has been claimed and at one, Border Cave, fairly well established, but the omnipresence of Khoisan peoples from Zimbabwe and Zambia southwards remains a persistent feature well into the Iron Age. Of the skeletal remains from Iron Age times until the recent past very few carry certain dates. Most of the material gathered by Shrubsall appears to be late post-settlement. But the remains from Layer A at Matjes River (Cape Province), Mumbwa (Zambia),

Bambandyanalo (Transvaal) (c.1055 A.D.) and Mapungubwe (Transvaal) (c.1400 A.D.) are nearly all attributed to Khoisan and are important in providing further evidence of Khoisan peoples inhabiting the same wide area without interruption, well on into the centuries when other people dominated and controlled the region.

On the distribution map provided by Tobias in 1957, no San peoples are shown in the whole of South Africa with the exception of the tiny Lake Chrissie remnant (cf. Potgieter 1955). This is deceptive, since there are isolated small groups and individuals still to be found in the northern Cape Province, many of them pretending, like the Khoi, to be 'Coloureds', even though their 'Coloured' neighbours discriminate socially against them (see chapters 4, 5 and 9). As late as the 1820's a number of young male Bushmen were captured and killed in the area of Graaff–Reinet and their skulls exhibited at the circuit court (Wells 1957). At least one of these was working on a farm in the neighbourhood. Wells concludes that the available skeletal remains provide 'satisfactory evidence of the actual physical type of the Bushman on the northern frontier of the Cape Colony during the early part of the nineteenth century'.

Shrubsall's writings (1898, 1907, 1911) make clear that many Bushman and Hottentot skeletons were obtained from the Cape during the nineteenth century. His material may also include Kalahari and West African representatives. At the time of European occupation of the Cape the distribution of the Khoisan populations was very different from that of the present day (cf. chapters 4 and 5). At that time most of the South African cul-de-sac, certainly the whole of what later became the Cape Colony, Bechuanaland (Botswana) and South West Africa (Namibia), was inhabited by Khoisan. Skeletal remains, rock paintings and contemporary records all testify to this state of affairs. The morphological evidence points strongly to the conclusion that throughout the period from about 30 000 years B.P. into the present day, the indigenous populations inhabiting the great triangle forming the southern portion of the continent included a great many of Khoisan stock, and that for many millennia this stock predominated.

Thus, from the Middle Stone Age and perhaps as far back as 35 000 B.P., as testified by the Florisbad remains (Tobias 1962a), Khoisan or Khoisanoid populations were in evidence over roughly the same large area extending from Zambia and Zimbabwe southwards through the modern Republic. No early remains have been forthcoming from Namaqualand and the south-western areas which for the Khoisan are today major areas of occupation, but this is hardly surprising considering the lack of obvious sites requiring excavation and the relatively small amount of archaeological research which has so far been feasible in these areas.

The paucity during this long period and over this large territory of any groups diagnosed with certainty as Negro is phyletically highly significant. Evidence of genetic affinity between present day South African Negro and Khoisan is set out in some detail in a later chapter. As described below, the genetic affinities attested by sero-genetic characteristics are reflected also in morphological features. Along with these common characteristics there has unquestionably been a high degree of differentiation. The skeletal evidence thus invites the conclusion that if Khoisan and Negro descend from a single common stock, the divergence was already present at least 20 000–30 000 years ago. The process of differentiation would have been such that the Khoisan groups became for a time dominant in Southern Africa. A divergence which took place originally outside the southern cul-de-sac of the continent led on to further local differentiation in Southern Africa of the Khoisan stock into the San and Khoi sub-divisions. The area of earliest Khoisan development may have been situated well to the north of the boundaries we have set for Southern Africa. There are skeletal remains from East Africa which closely resemble the 'proto-Bushman' remains found much further south (Wells 1952a; Tobias 1960).

### The pre-industrial environment and the scope for early primary industry

The needs of true man are variable, and the development of a technology will enable many more resources to be utilized than is possible in a toolless state. It is not clear whether any of the technologies employed by early man in Southern Africa had their origins there. On the whole, it appears unlikely. Certainly large cutting tools characteristic of the Lower Acheulian industrial complex have been found in association with pre-human remains at Sterkfontein and Swartkrans, and Upper Acheulian sites abound in the region, but earlier dates are probable for similar discoveries elsewhere in Africa. Clark (1970) has suggested that the use of deliberately flaked tools was directly related to meat-eating, and that the broadened range of manoeuvres made possible by these cutting and chopping instruments were 'both the outcome of and the agency for accelerated biological evolution' (p. 77). The very necessity of making a choice between stones to be worked, and the capacity to reason out the relative utility of alternative materials, would seem to signalize an important and possibly comparatively abrupt evolutionary step. The pre-Acheulian or Developed Oldowan instruments of north and east Africa are often of limestone or chert, while the Lower Acheulian tools in Southern Africa are mostly fashioned from quartzite.

Surface quartzites are abundant in Southern Africa, and the open woodland, occasional forest and grass savannah, which at the first coming of true man into the region characterized the plateau system and its fringes, supported into quite recent times an astonishingly rich ungulate fauna, a ready prey for hunters equipped with even primitive weapons such as stone-tipped arrow and spears. Part of the hunting technology of the surviving hunter–gatherers, the San, involves the use of poisoned arrows. The management and manufacture of the bow, from pliant wood or cane and fibre made from the intestines or skins of animals or from bark, is common to a great variety of human societies, and knowledge of the bow may have spread by diffusion or been arrived at independently by a number of primitive peoples; but the ancillary refinement of poison is more sporadic and dependent on a precise knowledge of the properties of sometimes very localized plants and insect species. The preparation of poison, too, is a fairly complicated procedure, and one demanding the taking of careful precautions. Selective processes could often have operated in the elimination of careless people, who made technological mistakes. Similar processes would lead to some reduction in numbers of the ignorant, the imprudent and the incompetent, both through the misuse or imperfection of tools and probably less commonly through encounters with the large predators, lions and leopards and cheetahs, which also followed the ungulate herds and flocks of ostrich. Venomous snakes and the ubiquitous scorpions of Southern Africa could also have taken their toll. It may be acknowledged in passing that the victims of these processes, like those who tested unwholesome or dangerous vegetable foods and as a result succumbed, could have been actuated by altruism centring on the groups of which they were members; but whether nobility or dull-wittedness or both were responsible, the outcome would have been an accretion of knowledge and skill to the survivors, and a sensible extension of technological competence.

No evidence of the use of fire has been found in any except the very latest Acheulian sites of Africa, probably owing to the accelerated breakdown of charcoals by soil fauna in tropical areas. The modern San use fire not only for cooking and heating, but also in hunting small game when the grass has dried sufficiently to be set alight. It seems improbable that their forerunners were unacquainted with its potentials. Fire-hardening of sharpened sticks may also have played a part in their weapon manufacture. There was certainly a sufficient supply of wood for fuel for a small scattered population of hunters, and even for later larger settled communities who practised the smelting and forging of metal. It is probably much later comers who must take much of the blame for a good deal of the deforestation of the highveld. Fire has been responsible for the degradation of a large proportion of the Southern African vegetation, especially where slash-and-burn agriculture recently has been or is being practised, as along the Kavango Valley, in northern and eastern Botswana, the eastern Transvaal and parts of Natal.

Along the coasts of the Cape Province a further and relatively localized resource was exploited. At the Matjies River site over thirty feet of shell and occupation midden represent eight thousand years of partial or total reliance on shellfish for food (Louw 1960). There are signs of a gradual shift from hunting to the collection of molluscs and a rather sudden introduction of fishing about five thousand years ago. Rock paintings indicate that this new technology did not include line fishing but relied on the trapping of fish in tidal weirs and spearing from rocks or rafts near the shore. It is possible, too, that wicker fish-traps for use in tidal pools and estuaries may have developed or been introduced at about the same time; but there is no clear evidence for this. Similar techniques may have been used along the banks of rivers. That an irregular but doubtless valuable source of food was found in the occasional stranded whale or seal is attested by Dutch and Portuguese chroniclers who encountered coastal hunter–gatherers.

The Acheulian large cutting tool tradition appears to have persisted latest in those parts of the continent, including Southern Africa, where the ecological conditions to which it was adapted went on longest. In Southern Africa it evolved into the Fauresmith culture, with a fairly specialized exploitation of the montane and plateau grasslands and, subsequently, into the Wilton and Smithfield cultures. Through the second and later evolutions the armamentarium came to include more efficient instruments of smaller size. An apparent reduction in the complexity of their manufacture and functions is less likely to be evidence of degeneration in the intelligence or skill of their makers than to represent the use of a greater variety of materials, many of them less likely than stone to have survived. The clever technological adaptations of unfamiliar materials made by the modern San may well be typical of the kind of ingenuity the hunter–gatherer has to have, perhaps not to survive, but certainly to flourish.

Survival would depend most on the availability of means of subsistence which communal knowledge and experience showed to be

suitable. Even where game is abundant, the difficulties of its procurement and storage preclude its use as the sole form of food. The arid
and semi-arid interior of the region is rich in water-retaining roots
and tubers, which are now and must always have been valuable
sources of moisture for hunter–gatherers. The bored stones, digging-
stick weights, so frequently found, show that the use of underground
foods, probably including insects and small burrowing animals and
reptiles as well as plants, dates from at least late Acheulian times.
These implements could also have been employed in the digging of
shallow seepage wells in the dry beds of rivers. An important source
of moisture for both man and animals, including possibly the large
cattle of the earliest pastoralists, is the hardy and ubiquitous *tsamma*,
a cucurbit the bitterness of whose flesh detracts hardly at all from its
value. Knowledge of the seasonable appearance of wild fruits and
nuts has probably dictated the cyclical migration path of earlier
peoples much as it does those of the San today. Some of the available roots and particularly the nuts of both the ground-hugging vine
and the small tree *Bauhinia* types, are of high nutritional value
(Truswell and Hansen 1968*b*). Eggs of reptiles, and especially those
of birds which build their nests on the ground, constituted other
valuable food sources, as did slow-moving grubs and tortoises and
some insect pupae. The San have assimilated the ostrich egg rather
strikingly into their technology, and it was open to their early precursors to do the same. The shell is used for carrying and storing
water, food and medicines, and for the fashioning of a great variety
of ornaments.

The first incursions into Southern Africa of peoples other than
hunter–gatherers appear to have been made by cattle-keepers. The
material culture of the early pastoralists does not seem to have
differed greatly in its use of resources from that of the hunter–
gatherers, though of course there would have been a different array
of implements including those used in the management of cattle.
Bark cloth was made by the San, who also use the skins of game for
clothing and decoration. Pastoralists would have had in addition
the hides of cattle which died or were killed. An elaborate use of
*riem*, dried strips of hide, characterized the culture of the Khoi and
to some extent that of the later pastoralist Caucasoids with whom
they came in contact and influenced. Unlike hunter–gatherers, however, the pastoralists possessed mobile wealth, and so were able to
trade, either among themselves or with metal-working peoples on
their periphery. There is evidence from Namibia (Sydow 1967;
Jacobson and Vogel 1979; Boulle *et al.* 1979) that the Khoi possessed
an individual style of pottery. The presence of pottery in Africa

often signalizes either the presence of metal-working or contact with metal workers, though pottery is associated with Wilton and Smithfield remains, and further north in Africa, with the Nachikufan Industrial Complex (Clark 1970). The early pastoralists occupied defined and on the whole well-watered areas, along the Vaal and Orange rivers and the southern and south-western seaboard, and except when migrating across intervening stretches of drier land were less dependent than hunter–gatherers on indirect sources of moisture. They seem not to have made any extensive use of stone, and it is probable that hardened wood, clay and ostrich shells, and possibly bone, supplied them with all they needed for implements other than the metal ones they could acquire through trade.

There is some confusion about the identity of the peoples who first exploited the accessible sources of metal in Southern Africa. The craft of working metal was almost certainly introduced into the region from outside, but by whom, when, and to whom is obscure. Traditional history, especially in Namibia, is here at variance with archaeology: the Bantu-speaking immigrants into the north-western parts of the region possess strong traditions of having used San and Dama smiths and claim that the rich and easily extracted copper of Tsumeb was controlled and exploited by the Dama before their own arrival. Copper beads have been found at a late Stone Age site in the Brandberg which has been dated to about 900 years B.P., well before the arrival of the Ambo and the Herero (McCalman and Grobbelaar 1965), but they could have been traded over a considerable distance since there is no other evidence to support their local origin. Much futher east, in Swaziland, iron artefacts found in association with stone implements have been carbon-dated to the fourth or fifth century A.D. (Fagan 1967b) suggesting the simultaneous occupation of the area by two peoples with different technologies. The first contacts of literate peoples with the San and the Dama have not, however, produced records of their having processed metal in early modern times, and there are no sites of unequivocally San or Dama origin which have yielded signs of metal working among them.

There are large deposits of easily accessible iron ores, of grades permitting their exploitation by sufficiently skilled peoples before European contact, in a number of parts of Southern Africa. With the exception of the single possible site in Swaziland, there is no evidence of Early Iron Age movement into the region. Early Iron Age communities certainly had their origins a long way from Southern Africa, and this sole example almost as certainly represents the belated spread of a technology rather than the specific incursion of a people (Tobias 1958). Iron Age occupation distinct from the

Early Iron Age tradition was established on the Witwatersrand at about the same time as the copper beads arrived at their Brandberg site and coincided with the establishment of an active Iron Age industrial complex just south of the Limpopo (Mason 1962; Gardner 1963; Fagan 1965, 1967a), where there were copious sources of both iron and copper. Even before this extensive mining and smelting of copper had been initiated at Phalaborwa in the eastern Transvaal, and this area continued to be worked in the same tradition until well into the nineteenth century.

It is customary in Southern Africa to associate together iron-working, the making of pottery, the introduction of agriculture and the spread of the Bantu languages. This question will be discussed at greater length in later chapters; it is sufficient here to reiterate that pottery may be associated with assemblages that greatly antedate those left by iron-workers, and to mention that linguistic evidence presently available suggests that the Bantu languages may have arrived later than the other significant introductions. To this set of supposed simultaneities, however, we are tempted to add the construction of dwelling-places intended for more than temporary occupation. The hunter–gatherers today construct crude windbreaks of branches and thatch very similar to one whose remains were found in Zambia and dated to the third millenium B.C. (van Noten 1965). The greater sophistication and closer economy of the Khoi pastoralists developed from this a structure in principle not unlike the black tents of the Turkic nomads of central Asia. A framework of sticks which curved over to form a dome-like cage, and which could be disassembled and transported, was fixed in the ground, and onto the outside of this skins or woven mats were fastened or thrown to make a snug and relatively impermeable portable home. There has been a significant retention of this pattern of dwelling by some of the less fortunate descendants of the Khoi; we have seen similar structures, but covered with sacking or flattened paraffin tins, in the depressed 'Coloured' villages around Kimberley and in parts of southern Namibia.

The buildings of the earliest Iron Age peoples included some of more permanent construction than any built by later Iron Age peoples in Southern Africa. De Vaal (1943), describing a Northern Transvaal site for the first time, referred to it as 'a Zoutpansberg Zimbabwe', and the geographical and cultural closeness of some of these stone ruins to the great achievements of the civilization that rose and fell in Mashonaland to the north indicates that they probably represented outposts of the Rozwi empire. Subsequent dry-course stonework has been confined largely to the lower part of the walls of cattle-enclosures, the upper part being filled by branches of thorn-bushes, or of dwelling-houses, forming a base for a framework of sticks over which mud or clay was plastered and allowed to harden. As the iron-working peoples moved southwards into somewhat drier parts of the country much of the stonework was dispensed with, particularly as suitable stones can often only be assembled with difficulty, and in the more alluvial parts near the coast there was some reversion to a more primitive type of dwelling resembling, in mud and wattle, a more permanent version of the Khoi tent. This was associated, however, with a somewhat tighter pattern of village organization and layout.

All of the developments referred to in this section involved the application or refinement of techniques, most of them almost certainly introduced from outside either by cultural diffusion or the immigration of peoples practising them, to resources already occurring naturally in the region. The presence of such resources either lured immigrants who were accustomed to exploiting them or enabled peoples already in the region to deploy their energies more efficiently. Nevertheless, no native Southern African animal was ever domesticated, no wild South African plant ever tamed and cultivated. This is hardly surprising, since of the fauna indigenous to the region only the eland was proved capable of being tamed, and that only under very special conditions, and the potentially most valuable native plants, such as the Bauhinia nuts, flourish only in ecosystems which are very circumscribed. The revolutionary introductions of animal husbandry and agriculture brought from outside the region demanded highly specialized tools and technologies. They were made possible and necessary by changes in the climate and vegetation which occurred some time after the first humans had already established themselves in the region.

## Cultural divergence and the differentiation of the populations

In Central and Southern Africa the industries associated with human remains are clearly related to the ecosystems, with the truly microlithic Wilton culture (cf. Inskeep 1967) predominating in open parkland and scrublands, the non-microlithic Smithfield (cf. Sampson 1967) extending into dry river valleys, and the Nachikufan (cf. Clark 1950) tradition being virtually confined to what were then forest areas.

To judge by their modern close resemblances, the divergence between the Khoi and the San must have happened relatively late. The separation of the Khoisan into two main stocks must have begun only when one segment accepted and started to exploit the bovine cattle, of either European or Asian provenance, the sheep, chickens and goats which certainly originated in Asia, and the domestic dogs which could have originated anywhere in the Old World. It is fairly easy for us now to decide which particular strain of proto-Khoisan would have accepted pastoralism. The close physical resemblances of the Khoi to the San suggest that the practitioners of a single cultural tradition were predominantly ancestral to both. The kind of ecosystem in which pastoralism would be practised most recently was that in which the Wilton culture existed. It is possible that the practitioners of the Wilton culture, who extended from the far south of Africa, to the Horn, may, with some contribution from the Smith-field folk, have followed various geographically discrete micro-evolutionary paths to develop into the Khoisan peoples while those who followed Nachikufan culture practices may, perhaps with admixture from the Tshitolians from the north and west (cf. Clark 1963), have contributed to the formation of the Negroes. Some Wilton materials are known, in fact, to have been produced and used by San during historical times (Mason 1962).

Language is an inconstant indicator of biological ancestry, but this hypothesis does at least help in part to account for the two peoples who speak click languages in East Africa. It is not hard to suppose that these highly individual phonemes evolved in the speech of those proto-Khoisan who practised the Wilton culture. Hadza and Sandawe, despite their phonetic similarities, show few contemporary indica-tions of being related, but if we accept either Swadesh's (1948) formula for glottochronological divergence, or some modification of it, the length of time which it is necessary to postulate between the period when there was uniformity among the forebears of the Khoisan and the East African, and the present day, is more than long enough to produce a considerable degree of dissimilarity, both linguistic and biological.

The Hadza and the Sandawe do not, for instance, show those morphological characters, the localized fat deposits and the hair distribution, which indicate most strongly that the major selective adaptation among the Khoisan has been to hot environments in which water is scarce. Several authors (Schultze 1928; Tobias 1960) have been at pains to point out that we have a large number of written records which show that San inhabited relatively well-watered, temperate areas during recent historical times, and that we conse-quently ought not to regard them as the natural human inhabitants of the areas in which they are now principally found. The implica-tion is that it is only on account of their having been displaced, and in some cases massacred, by later comers, the Caucasoids and Negroes, often with the enthusiastic co-operation of the Khoi, that the San have been forced to take refuge in the most 'inhospitable' parts of the sub-continent. Nevertheless, not only is the San physique peculiarly well adapted to the deserts and semi-deserts in which they now have their home; the technology which gives them additional support in such ecosystems is also one which has plainly evolved in the conditions under which it is now practised, and their languages abound in terms which would have little utility in more fertile surroundings.

## Probable interactions between the earliest Khoisan and the earliest Negroes

The wide scatter of the proto-Khoisan and Khoisan during the late Pleistocene and early Holocene may therefore have represented the outcome of the population explosion of a people which had already acquired its characteristic genetic profile during and in part as a result of intense selection in a hot, dry environment. This could have happened further to the north than the present South African deserts, and could have gone on in areas previously and subsequently wooded, or even in an area exposed by the locking up of water in the polar ice-caps during the Ice Age, and later covered by the sea. With the partial melting of the Antarctic ice-caps ten or twelve thousand years ago, a warmer and more humid climate set in. Game apparently became very abundant, and circumstances favoured increases in a human population freed of many of the rigours which had confined earlier generations. The spread of the stock which later gave rise to the Khoisan, from East and Southern Africa, partly in pursuit of game, partly away from the extending forests for which it was neither biologically nor technologically equipped, was probably inevitable.

Biological adaptation to desert and semi-desert ecosystems suits people to the savannah as well. The negative prior adaptation which would operate to prevent effective exploitation of a forest biotope would not hinder survival in open parklands and scrub; though steatomeria would conduce to clumsiness in thick undergrowth and certainly inhibit tree-climbing, it would confer no disadvantage in open country. Many of the fruits, and all the game, would be similar; and technologically there might be some selection relaxation, as food would be more abundant and there would be less need to utilize the extreme expedients of procurement, such as the preparation of bitter melons or animal rumen as fluid sources, or the search for and digging up of uncommon tubers.

It could only have been very early in this phase, if it had not been earlier, that the proto-Khoisan element in the proto-Negro separated from the main strain. The absence of Khoisan morphological traits from the Negroes of much of Africa implies, in fact, that the separation must have occurred before these had developed, and hence probably rather earlier than the widest distribution of the Khoisan. Whatever may have been the case, it would have been inevitable that at some time the Khoisan expansion would meet and clash with the expansion of the Negroes which seems to have happened not much later.

As has been pointed out in Chapter 1, there is often very little competition for resources between the agriculturalist and the hunter-gatherer. Only when the agriculturalist is dependent to some extent on wild foods, or when adverse conditions press the hunter–gatherer into preying on crops and herds, does this possibility even arise. But when two strains of hunter-gatherer are both moving into the same lands it can happen that the absence of sufficient resources to support both can lead to strife and the displacement or eradication of one or the other. The evolution of the Negro is unlikely to have resulted immediately in the establishment of agriculture. For many millennia, into the present day, there have been Negro hunter-gatherers. The earlier ones included the stocky strain which many Bantu-speaking peoples describe by names containing the radical *twa*. As such names suggest, this stock may have had some ancestral connection with the Pygmies (cf. Maquet 1961), while, as indicated in Chapter 6, the contemporary ones could in at least a few cases represent the survivors of unsuccessful migrations. In a number of instances the two strains may well have mingled. In modern times there is little or no evidence of competition between San and Negro hunter–gatherers, but it is surely significant that their ranges of habitation do not overlap. Well before the arrival of the European Caucasoids, the appearance of the Khoi as pastoralists or the great agricultural settlements of the Negro, there must have been some withdrawal of the San in the face of the first Negro expansion, the outward spread of hunter–gatherers.

When we read in modern records of San inhabitants of the savannah, or listen to accounts of them in the recent traditions of the Negroes, we find them referred to as tiny and threatened populations, few and elusive, in bitter enmity with the newcomers. The widespread legends of 'little people' could, it is true, refer as easily to small-structured Twa as to San, and the 'Wakwak' whom al-Idrisi encountered on the east coast of Africa (Freeman-Grenville 1962) may, as Nurse (1967) has shown, have been not San but Makua; but a relatively recent

San biological strain certainly survives in the Negroes of East and Central Africa (Jenkins *et al.* 1970; Nurse 1974). The accounts of the extinction of the 'little people' in Lesotho (Ellenberger 1972) and elsewhere in Southern Africa, and many Central African oral tales, plainly refer to San.

The first competition for the savannah, therefore, would most probably have been between desert-adapted San and forest-adapted Negroes, and is likely to have been only rather local until the Negroes began to settle. There seems nonetheless to have been some fall in the San populations of the savannah *pari passu* with the rise in the number of Negro inhabitants even before the need to alienate some lands from hunter-gatherer exploitation and devote them to agriculture made the agriculturalists in a sense the active opponents of the hunter-gatherers. From such opposition, though, the most usual hunter-gatherer strategy seems to have been a peaceful withdrawal; and it was not necessarily only San who withdrew in the face of the spread of agriculture. As the Iron Age moved southward through the continent, the main concentrations of San, followed by scatterings of Negro hunter-gatherers, would have retired before it. But while what the San would ultimately attain would be a haven, an eco-system to which above all they had undergone prior adaptation, the Negro hunter-gatherers could only maintain that precarious and marginal existence in which the descendants of those of them who were not absorbed by the advancing agriculturalists persist until the present day.

# 3
# MIGRATION AND ADAPTATION

In the foregoing two chapters there has been frequent mention of population movements within Africa, but no very explicit discussion of any underlying consistency of pattern in them. We have discussed the three main forms of subsistence practised by the inhabitants of Southern Africa previous to, and to a great extent overlapping with, the present phase of industrial development, and treated the question of cultural competition at some length. We have touched on the climatic changes in the region and adjoining ones, and on the effects these might be expected to have had on population differentiation within the species, and even on speciation within the genera comprising the hominids. In the course of this we have referred on several occasions to the relative abundance of resources disclosed by archaeological investigations. It must be remembered nonetheless that archaeological records, though reliable on a broad-time scale, can give little or no indication of precisely how available any particular resource was at any specific time.

Before proceeding to detailed consideration of the contemporary peoples of Southern Africa and their affinities, it is worthwhile giving some consideration to the general incentives which attract peoples to the region or induce them to remain there. In later chapters this question will be treated in more detail for each particular group. Here our object is to make a preliminary assessment of the ways in which the different modes of population movement are likely to have been dictated by, or have resulted in, contrasting patterns of cultural or biological adaptation to the environment.

## Trans-Saharan migration and gene flow

Record-keeping was introduced into Africa from the north, from the historically-minded peoples living around the Mediterranean. We have mentioned in the first chapter all the earliest accounts known to us of penetration of Africa by people more advanced culturally than the then natives of the continent. None proceeded very far to the south.

The barriers opposing further exploration of the continent by all these visitors were, of course, the same barriers which must to some extent have regulated the most recent diffusions of sub-Saharan peoples. Once the climate and vegetation had again stabilized themselves after the Ice Ages, any change which came about in them would be small in amplitude and could have resulted from human manipulations of the environment. Early in the Holocene era it is likely that the direct ancestors of all the present-day peoples of sub-Saharan Africa, with the exception of those living along the southern and eastern fringes of the desert and those later introduced by sea, were already there. There were certainly sapient human inhabitants of Southern Africa long before the Phoenician circumnavigation, and if the Phoenician sailors did indeed, as Cary and Warmington (1929) so convincingly suggest, sow and reap a crop at St Helena Bay near Saldanha, they very probably did make at least visual contact with them. San relics and San genes are found on this far southwestern coast and all along the plateau system which stretches from Southern into East Africa. The Ethiopians, from all accounts, extended from the Horn to the west coast in classical times; but there is no good reason to suppose that these Ethiopians had anything more in common than a black complexion. Desert and forest, cold high mountains and wide unfordable rivers, until modern times confined populations to particular areas and gave peculiarly explosive expression to some recorded migrations. There was abundant opportunity, in what had been for all of the Holocene an under-populated continent, for local evolution to take place.

The Sahara plays a unique and significant part in the history of the peopling of Africa. Like some other deserts, but probably to a greater extent than any of them, it fulfils the ethnological functions of a sea. It is open to traversing only by peoples living on its edges and at the island-like oases within it. Only they are likely to have acquired the technology and expertise which enable them to wrest a living from it, and to pass from one part of it to another. A significant aspect of their technology is or was the ability to make use of grazing animals, whether wild or domesticated. Where these were wild, the effect of the desert-dwellers on the productivity of their environment would be limited by their hunting ability. Few species would be hunted to extinction, since long before extinction was approached their human predators would have had to move on in search of alternative stocks. On the other hand, few wild grazing species would over-graze the grass and shrub cover, since being unconstrained in their movements they, too, would move on with any deterioration of pasture, and leave enough in place to regenerate in time.

Domesticated grazers, by contrast, have been responsible for the progressive expansion of the Sahara during historical times. Goats and sheep of the types kept by man are limited in their foraging excursions by the requirements of man; they are fenced in and constrained to graze some areas to exhaustion. Added to this, a marked climatic change about 3–5 millennia ago added to the desiccation of the Sahara, greatly diminishing the habitability of those areas which had previously been cultivated or even supported lakes and rivers and fishermen (Mauny 1961). As a result, by the time of the classical historians the Sahara was already a formidable barrier to population movement, which could only take place with the assistance of people possessing specialized knowledge of routes and researches. The introduction of the camel, probably by the Romans at around the beginning of the present era, must have revolutionized transport across the desert (Bovill 1968); but the relative advantage it conferred on its possessors would have been counteracted fairly soon by progressive desertification and the lengthening of the routes to be followed.

There were Negroes north of the Sahara during classical times, but it is impossible to estimate when they first arrived there. It could not have been as a result of the slave-trade, since there were abundant sources in Europe and Asia of slaves who could be obtained much less expensively than by protracted passage across the desert. They most probably represented relics of populations which had moved north before the period of major exsiccation, though possibly they could have been actual desert-dwellers who settled on the northern side rather than on their ancestral southern fringes. They appear to have been concentrated mostly in the Fezzan, in southern Libya, the country in fact, of the Garamantes, and the northern terminus of what appears to have been the shortest direct route across the desert.

Whether in the twittering speech of the troglodytes described by Herodotus as living on the desert fringes we can catch a hint of a Khoisan presence is a matter of debate. The wild young Nasamonians who crossed the desert were captured on the southern side by people of short stature, who, since they were black, we may suppose to have been Pygmies. The colour of the troglodytes is nowhere mentioned. They were hunted by the Garamantes, which may be the earliest record we have of the exasperation engendered in their settled neighbours by the pranks and predations of the San.

The Berbers, including the Tuareg, make up the most prominent modern group actually inhabiting the Sahara. Linguistically they are difficult to place, but their physical characters indicate them to be Caucasoid. They are a possible channel for the passage among others of the gene for Rhesus negativity, the allele $r$ or $cde$, into the Negroes, in whom it reaches unexpectedly high frequencies. There are Negroid elements associated with the Tuareg, popularly and credibly supposed to descend from domestic slaves taken by them during southward swings in their transhumant migrations, or traded among them. It must not be forgotten, too, that once supplies of slaves from Europe and Asia for the North African and Middle Eastern markets had begun (surprisingly late) to fall off, the consequent demand for Negroes at last rendered a trans-Saharan trade in them lucrative. Some of these certainly proved unsaleable, and were retained.

Movement in the opposite direction across the desert was rather less in extent though rather more explicitly recorded. Apart from the early jaunts of the Nasamonians and the later Roman campaigns against the Garamantes described by Tacitus, there is reasonable circumstantial evidence that the Carthaginians must have trafficked with the interior. Those useful and unscrupulous men served so often as explorers and trail-blazers for other powers that it would not be unreasonable to suppose that they conducted other more closely self-interested expeditions in addition to those which they permitted other people to know about. Yet there is no indication now in the morphology or the genetic profile of the sub-Saharan peoples to suggest that any biological contribution was made by any of these travellers to the Negroes. Not even the extensive voyaging of Jewish goldsmiths and merchants through the Maghrib, reinforced as it was by the flight of the Cyrenaican Jews from the Romans following their revolt in 115 A.D., was sufficient to produce any indelible token of gene flow southwards. The suggestion has been made, that the founding and ruling dynasty of the mediaeval empire of Ghana in West Africa may have been Jewish. There is no evidence, however, to support this, except the legend that the dynasty was 'white'. What southward gene flow there was seems to have happened mainly or exclusively through the intermediary Tuareg, until the Arab incursions began during the eleventh century A.D.

The bedouin Arabs had one great advantage over the previous conquerors of the Mediterranean littoral of Africa. Biologically and technologically they were already well adapted to desert conditions, and the Sahara was much less of a barrier to them than it had been to the Greeks, Romans and Carthaginians. Once the urbane missionaries of Islam had subdued Spain and tamed the Berbers, the first wave of invaders was spent; but they had opened the way to the tough and intractable bedouin tribes of the Bani Sulaim and Bani Hilal. These marauding nomads had been expelled from Arabia; they spread across Egypt, from which they were urged westwards into the Maghrib. The Bani Sulaim were content to remain in Cyrenaica, but the Bani Hilal penetrated far into the interior. They devastated many cultivated areas, thus permitting the further encroachment of the desert. They grazed their herds with abandon, and destroyed forests to give them new pastures. The careful technology which had permitted cities to survive in areas of tiny rainfall was uprooted with the destruction of dams, cisterns and aqueducts. The Bani Hilal drove the Berbers into the high country, and took command of the desert trade routes. Paradoxically, they opened the way for the efflorescence of Arabic scholarship which adorned the Maghrib during the succeeding centuries.

Even though this irruption was soon followed by a recrudescence and consolidation of Berber power under the Almoravids, the Arab influence and Arab techniques had been firmly impressed on the area. The great indigenous empire of Ghana had arisen on the other side of the Sahara, and though it was never fully converted to Islam it provided a centre for Islamic teaching. The faith spread through West Africa; with the obligation of *Hajj* or pilgrimage to the Holy Cities of al-Madinah al-Munawara and Makkah al-Mukarama in Arabia, the stage was set for regular gene exchange between Negro Africa and the rest of the Muslim world, particularly the Arabic-speaking and Arab-dominated parts.

The flow of genes appears nonetheless to have been mainly from rather than into the African populations. A noteworthy contrast between West and East African responses to Islamic influence can be seen in the positions adopted by Arabs in the two regions. The Arabs in West Africa, who had crossed the desert, established no regular constellation of Arab cities, possibly because they were themselves of desert-adapted stock but equally possibly because of the high culture already in existence there and the ancient cities which the Negroes had earlier founded. In East Africa the Arabs came by sea and established trading posts of their own; except among the Shona-speakers well to the south, and in persistently Christian and isolationist Abyssinia, there was no tradition of urban culture, and even in these cities lay relatively inaccessibly inland. Gene flow from the

Muslim Arabs would nevertheless not be likely to differ in degree between the spheres of Arab influence. In the west, however, there seem to have been relatively more, and more far-ranging, Caucasoids, mainly Berbers and Tuareg, who acquired a religion and an accretion of technology from the Arabs but little or no biological contribution. Both Tuaregs and Arabs, however, contributed some genes to their immediate Negro neighbours.

When we consider the populations of Southern Africa we are faced, of course, not only with the results of patches of micro-evolution, but also with an ongoing flux which blended the interim outcomes of these with the products of a variety of far-reaching social and political processes which may or may not have had biological causes but which generally had biological effects. There is little incontrovertible evidence that gene flow across the Sahara, or indeed across the north-western Indian Ocean, has percolated very far into any of the more anciently established peoples of Africa. Indeed, it appears to have been confined largely to the nomadic peoples on the fringe of the desert, and the urban peoples of the East African coast, especially those who have adopted Islam. Despite that presence of $\beta$-thalassaemia and Rhesus negativity in the Negro, the absence of a number of other Caucasoid gene markers strongly suggests that such gene flow has been minimal. Some genes, such as that for the Mediterranean type of glucose-6-phosphate dehydrogenase deficiency, might even have proved advantageous, and been selected for, in a tropical African environment. It seems likely, therefore, that what contribution there has been from the flow of genes originating to the north of the Sahara, would have occurred as a result of migrations across that desert well before the times of which we have written records and hence probably before the commencement of major exsiccation. Any such flow appears to have affected the Negroes of Southern equally with those of West Africa, as might be expected with a relatively recent expansion of the Negro peoples. The gene marker profile of the Njinga of Angola (Nurse et al. 1979), who live just beyond the limits of the Khoisan genetic influence to which the Southern African Negroes have been exposed, closely resembles that of the Yoruba of Nigeria (Ojikutu et al. 1977). This would argue a broad Negro genetic homogeneity extending southwards from West Africa into at least the South-Western Negro peoples. The Southern African Negroes differ from both mainly in the possession of Khoisan gene markers. There seem to have been strong non-biological forces limiting the spread southward of trans-Saharan genetic material during the past two millennia. What these forces were, with recorded history stressing apparently opposing trends, we are unable to determine.

## Sub-Saharan migrations and their motives

It is often held, by vernacular historians of Africa among others, that the motives of migration are fundamentally economic. There are certainly economic constraints on the directions they take; nobody could successfully traverse a desert without at least the preparations made by the Nasamonians, and even the crossing of a river or a range of mountains may prove too risky in loss of livestock or personnel for it to be assayed, or, if assayed, to succeed. In such contexts the fusion of the biological and the economic is undeniable, but it is not easy to say whether the motives extend beyond those of immediate individual well-being into a consideration of the consequences to the community if it is disturbed. It may be of some benefit to consider at this point the factors which may prompt populations of *H. sapiens sapiens* to leave one place and settle somewhere else.

There is only one factor which can be regarded as purely economic and without any biological implications at all, and that is the pursuit of wealth for its own sake. For this to operate as a motive a cultural background in which pure wealth is esteemed needs to be maintained and this is hardly likely to be possible during the mass migrations of peoples. Such an objective seems to have regulated the earliest trading across the Sahara, by means of which the Carthaginians obtained their famous abundance of gold, presumably from Guinea. No large numbers of persons would have been engaged in this. Small parties of prospectors or traders or other adventurers may, and quite often have, established states, but these have been states with well demarcated hierarchic levels, ruling invaders and subjected autochthones. None of the original polities of Southern Africa was strictly of this kind. Where there were hierarchies, as among the Sotho/Tswana peoples, the autochthones were materially too undeveloped to make their possession worth coming a long way to seize. This is not to claim that after the migrations had taken place there was no conscious amassing of property at the expense of the original inhabitants. Labour, and the particular skills of those longer attuned to the land, were and are readily exploited by later comers to an area; and people who see land as wealth have been quick to despoil others who have not thought to vest ownership in it.

We may distinguish, therefore, between the migrations of comparatively well-informed peoples, such as the literate Dutch and French settlers of the seventeenth century, who had been told of conditions at their goal and had a good idea where it was, and the migrations of Africans. Many of the migrants who came by sea, the

Cape settlers and the Arabs of the East African coast, did not at first see their new settlements as establishing a new homeland. They came, as explicitly as the Greeks ever did to Libya, as the Spaniards and Portuguese did to New Spain and Brazil, as colonists. A lifetime or several generations might be spent at the Cape, but home, at first, was in Europe. The extent to which their objective was actual exploitation and the gathering of wealth, or simply survival in equivalent or more advantageous conditions than those at home, would have varied from case to case. They would certainly not have emigrated without the anticipation of advantage and some idea of how it was to be procured.

In the migrations of the African peoples, on the other hand, we can discern two types of population movement, both depending on a range of chance, unforeseen and often unforeseeable factors. In one of these types the participants would be people who had never attained a culture of settlement, whose nature it was to wander, and who would not cease to migrate once the ultimate boundaries to their onward movement had been met, but continue to circulate through such lands as were available for them. In the other type the migrants would have been people whose culture was one of settlement and who would be looking for somewhere else to settle because the place in which they had hitherto lived had come to be intolerable to them.

Under conditions where communications were hardly long-ranging or explicit, the majority of migrants could have had only the vaguest idea of what they were travelling towards. Few, if any would have had any particular tract in mind as suitable for their own settlement. Any urge to move away from where they were to some place as yet unknown would have originated not from any desire to acquire reputed treasures but rather in the deterioration of previously agreeable conditions at home. The familiar has to become very uncomfortable before it is abandoned for the wholly unfamiliar.

Most of the changes likely to render an area unattractive would have had a fundamentally biological component. Hunger, whether as a result of population pressure or poor harvests, is probably the most familiar of such factors, and its importance is attested in both traditional and recorded history. Since a migration prompted by shortage of foodstuffs could hardly set out well provisioned, it is likely to have led to the readoption of hunting and gathering by an agricultural people. Pastoralists moving away from exhausted pastures in search of new ones could likewise be expected to wish to conserve their surviving stock and support themselves with what they could gather by the way. The optimum stable size of a hunting and gathering community has been estimated by Wilmsen (1973) as unlikely to be much more than fifty. We should consequently expect such migrations to occur less as the mass movements of peoples than as the sporadic progressions of a set of segmentary interrelated parties, perhaps exchanging mates or perhaps logistically and ritually constrained to endogamy. This picture is a familiar one in the traditional history of the Bantu-speaking peoples: we meet with it not only among the accounts, both oral and written, of migrations such as those of the Nguni (Bryant 1929; Omer-Cooper 1966, 1969; Barnes 1954a, *inter alia*) and the Sotho/Tswana (Ashton 1952; Schapera 1953; Legassick 1969; Livingstone 1856) which occurred within the purview of European onlookers, but also in the purely oral histories of such peoples as the Venda (Stayt 1931; Mudau 1940) and, further afield, the Maravi (Ntara 1965; Chafulumira 1948; Mwale 1962). It can also probably be made out in the not dissimilarly motivated extension of Caucasoids into the African interior known as the Great Trek (Walker 1934). Parties would disperse to hunt and forage and search for grazing, and reunite either temporarily for offensive or defensive purposes or mate exchange, or more or less permanently in cases where one or more parties had been reduced in number below a level where they could effectively survive on their own.

When we examine the probable goals of the two types of aleatoric migration, we must presume them in essence not to be dissimilar. The migrants who had never known a culture other than that of hunting and gathering, like those who had already attained a culture of settlement, would aim at either the maintenance or the recovery of an ideal familiar. Venturing into the unknown would not happen for the sake of the unknown itself; it would be directed towards gaining possession of, or firm rights in, an area as much as possible like that which had been abandoned, or as much as possible like what that had been prior to circumstances leading up to its abandonment. Even where the object might have been expressed as the wish for an improvement, for something better than the best which had been known up till then, the principal criterion could not without a considerable extension of the collective imagination have been other than the country ancestrally inhabited.

Hunters and gatherers are naturally more mobile than agriculturalists and even than pastoralists, but their requirements for a stable life are no less stringent. It is necessary that they can identify and know to be good. Their range is restricted by their experience; to change their food types needs a further process of adaptation, either social or biological or both. Agriculturalists bearing seeds, or pastoralists driving stock, however directionlessly they might wander, would be on the lookout for lands where they knew from experience of similar ones that their crops would flourish or for pastures which they could conclude by retrospective reference to be right for their beasts. And the lands which were right for their sources of subsistence would be most likely to resemble most closely those in which they themselves had undergone the most recent, the most specifically adaptive, portion of their own evolution.

### Pre-adaptation and prior adaptation

In the previous chapters we have occasionally used the term 'prior adaptation' without, however, defining it. In the context of migration the concept needs particularly careful clarification, since we believe that it may play an important part in the motivation and the success of many of the migrations into, and some of those away from, Southern Africa. It is not to be confused with the familiar genetic concept of pre-adaptation.

Pre-adaptation, as is well known, is the phenomenon of the chance occurrence of a trait in an environment in which it is of no adaptive significance, and its subsequent function as an adaptive asset following migration into another environment to which it preferentially fits its possessors. Prior adaptation, on the other hand, is the term we use here to denote the whole range of selective factors which equips a particular population to occupy a particular environment, and will consequently impel that population, when it moves away from the ecosystem in which its own special constellation of traits has evolved, to seek either consciously or sub-consciously for an environment resembling as closely as possible that which it has left, in which to settle. This may involve the rejection of ostensibly advantageous areas, in favour of those which in the eyes of people from other cultures and with differing biological histories would appear far less attractive. A notable example from Southern Africa could be seen in the relative underpopulation of the fertile areas of the western Cape at the time of the earliest European contact. The Khoi and San appear to have been relatively unimpressed by lands which the Dutch and French thought desirable, and to have withdrawn from them after only desultory resistance. As the Europeans advanced inland, however, they were more vigorously opposed by these people (Palmer 1966).

We do not wish to detract here from the pluripotential biological nature of man, the capacity to occupy a wider variety of ecological niches than those filled by other animal species, but we do want to re-emphasize the truism that the occupation of these niches almost certainly involves fairly intensive bouts of selective adaptation at a level well below that leading to species differentiation and akin to those which have as their consequence raciation. Minor races, biological strains, are engendered within each of the major races; and even though, as Cavalli-Sforza (1974) and numerous others have pointed out, we have insufficient evidence on the biochemical or monogenic level to distinguish conclusively even among the major races, characters of multifactorial origin permit, and at a cruder level even obtrude, distinctions which can be particular and local. Among such characters would be counted those which would signal the attainment of the goal of a migration. Monogenic characters would almost certainly play an additional part and could well accelerate differentiation between peoples; but with the exceptions of the established roles of Haemoglobin S, glucose-6-phosphate dehydrogenase and the Duffy blood group, all concerned in the response to malaria, and the possible factors involving persistent intestinal lactase activity, liver acetyltransferase and the major histocompatibility complex, conclusive information about the selective value of the monogenic polymorphisms remains sparse.

It is consequently possible to suggest that when peoples migrate they migrate towards ecosystems for which they have undergone prior adaptation, and that in the absence of other constraints migrations do not cease and settlements begin until such ecosystems have been reached. In the case of the hunter–gatherers we may presume that their even more directionless spread has been affected to some extent by their contacts with settled peoples, but that in general they would follow routes which provided them with that to which they were used, until at the southern tip of Africa they could spread no further. Nomad pastoralists, less confined to fixed areas than agriculturalists, would have their movements regulated almost as much by the needs of their cattle as by their own, and would establish themselves when they had attained a grazing range which experience or tradition told them was optimum for both. The first Caucasoid settlers in Southern Africa chose to settle in an area most similar to familiar Europe, and only moved out into, and proceeded to adapt to, less favourable conditions when it appeared that the capacity of the Cape comfortably to cope with further Caucasoid population expansion had been exceeded.

The concept of prior adaptation was probably first introduced by Fischer (1913). He sought to account for what he considered the superior adaptation of the Rehoboth Basters to their environment by suggesting that their Khoi ancestors were well adapted to it through evolution *in situ* while their Caucasoid ancestors represented the outcome of natural selection on the descendants of Caucasoid settlers who had migrated from the Cape to the Karroo. This would mean that the hybrids were impelled by biological requirements inherited through both ancestral stocks to occupy the lands which they eventually chose. Fischer also suggested that biological superiority of the Basters to both parent races might have resulted from the combination in them of different arrays of advantageous genes inherited from either parent race, since the ancestral genome on which selection operated would have been different in different races.

Prior adaptation, however, would not necessarily always be advantageous. Where migrations failed to reach an acceptable goal, or pressures on the migrant parties proved too great, and they disintegrated, the dispersed or disorganized members would be faced with ecosystems to which they were neither socially nor biologically adapted; and though it is to be expected that some individuals would possess the capacity and the genes to make the best of such changed circumstances, it is probable that large proportions of such migrants succumbed. Those who survived would have to adopt non-biological means of doing so. As Davis (1974) points out, sociocultural adaptation is often more rapid and efficient than biological.

There is evidence that this did occur during the peopling of Southern Africa. The so-called 'Black Bushmen' of northern Botswana are biologically Negro, cherish traditions of having once possessed cattle, but have been forced by not very clearly defined adversity to take up hunting and gathering, at which they are rather less successful than their San neighbours. The earliest Caucasoid penetrators of the Karroo relinquished agriculture, adopted nomadic pastoralism, and hybridized extensively with the Khoi; it seems likely that many of those lost to the scrupulous and ethnocentric genealogies of the Cape but whose names do not crop up again in the traditions of the Griqua or the Basters may likewise have died without reproducing.

## Migration, the spread of cultures, and cultural selection

Associated with migration, but not peculiar to it, are the spread of culture and the effects this can have on the promotion or inhibition of gene flow and gene exchange. Cavalli-Sforza and Feldman (1981) have pointed out that in order to explore these adequately it is desirable to use diachronic data, which are only available for a restricted number of human societies. Any synchronic study seeking to compare the extant situations in preliterate societies, or societies in which records are historically inadequate or misleading, is likely to founder on our ignorance of the cultural stages which led up to them. Furthermore, without such data it is next to impossible to formulate any generally acceptable rules for rates of cultural diffusion or other change. A relatively limited number of answers are available, however, to the most common problems with which human ingenuity is faced. There can be no doubt that similar situations in widely separated parts of the world, often at widely separated times, have given rise to similar solutions. The armamentary of man at war or hunting is an example. Nobody can say for certain how often the bow, or the spear, or the deadfall trap have been invented. It is at a much less rudimentary level of complexity that evidence of cultural diffusion must be sought. At that level there enters in addition the element of choice. A society may be faced with a set of cultural alternatives which are of varying robustness. Upon what choice it makes may depend both its own survival or prosperity, if the alternatives convey a sufficiently strong differential biological component, and the survival of a particular cultural trait. The persistence of the possibility of making further choices may be similarly contingent.

Cavalli-Sforza and Feldman emphasize that cultural selection in either of these senses may oppose natural selection, though on the whole it can be expected that they will operate in harmony. An example of such harmony would be the influence which prior adaptation will have on the goals of migrations. In many instances, though not invariably, cultural selection functions in a Darwinian fashion, towards the maximization of benefit to the society and/or the individual. Social and individual choice may nevertheless occasionally lead to sharp decrements of fitness. This occurs most frequently where there is conflict between fitness and utility, as for instance in cases of drug addiction, the social use of tobacco or the consumption of delicious but detrimental foods. These can operate quite sharply to shorten the reproductive span, depress the fertility and hence curb the fitness. 'A maladaptive trait which has become common can be eliminated by cultural selection only if it is perceived as maladaptive' (p. 344). There is, of course, also a considerable element of cultural selection in the acceptance or rejection of technological innovations.

Like many biological traits, both monogenic and multifactorial, a cultural trait may be selectively completely neutral. Unlike most biological traits, however, the transmissibility of cultural traits is not necessarily random. The element of choice may fluctuate without its fluctuations necessarily being related to alterations in the environment. Cavalli-Sforza and Feldman believe that this relative imprecision in the degree and nature of transmissibility is likely to result in relatively faster cultural than biological evolution. It is impossible to be completely certain of this, since we lack the data to assess the rate and quantity of the extinction or decline of societies which at some crucial point have made adverse choices. The most we can claim is that culture, and in particular technology, is the base on which the relative fitness of many contemporary societies rests. Its perdurability is yet to be demonstrated. It is quite conceivable that the current dependence on non-renewable resources, brought about by technology, may ultimately prove to be a Darwinian disaster.

Migration is often, but not inevitably, accompanied by cultural change. To what extent this may be seen as cultural evolution is not certain. In fact, despite the evidence they bring forward for its exist-ence, the nature of cultural evolution is not very distinctly defined by Cavalli-Sforza and Feldman. It is much easier for a soceity to retrogress culturally than biologically. Each step in natural selection is substantially irreversible; societies can experiment with beliefs or practices less expensively than nature can with mutations. Natural selection is necessarily unilineal immediately after each point of change; a cultural trait, by contrast, can easily be tried and rejected. The cultural practices assumed for the purposes of migration endure only as long as the migration does, and must then be succeeded either by a reversion to an earlier pattern, or by yet another pattern dictated by the new environment. The Khoisan-speaking Negroes who hunt and gather afford an example of both types of outcome. Nevertheless, it appears both from oral history and from the archaeo-logical record that in Africa the major cultural traditions have spread most commonly through migration rather than by diffusion of ideas, though that must have occurred along the fringe of the migration paths and at their goals. During these processes there has been a certain amount of cultural selection. Whether we would be justified in calling it cultural evolution is another matter.

### The size of migrant parties

The limitations imposed by the logistics of migration on the sizes of migrant parties would also have had an effect on the biology of the eventual settled community. Inbreeding and outbreeding would depend, of course, on the social rules and potential for contact of the parties. Traditional history on the whole indicates that where segmentation of an original migration occurred it usually persisted through at least the first few generations of settlement, and generally much longer. In some cases, as among the Maravi, the persistence of segmentation was not purely economic. It did not have as its object only the effective distribution of available resources, but was consciously aimed at promoting exogamy. In others, of whom the Ambo are a good example, segmentation was more political than social and produced a set of geographically discrete endogamous communities. In both instances what resulted was in fact a geographical restriction on the scope of mating, since subsistence agriculturalists are not likely to search very far afield for their spouses. The difference lay in the dispersion of exogamous Maravi clans through the lands settled by the Maravi, so that clan exogamy was everywhere possible within a political unit which was originally large and only segmented comparatively late, while the Ambo parties secured clan dispersion and clan exogamy by the inclusion in most parties of representatives of most of the clans but were politically segmented from the beginning, and endogamy within the somewhat dispersed political units was the result.

If we assume that each migrant party was approximately of the size Wilmsen (1973) suggests as optimal, this would mean that communities settled on the Ambo model would tend to be more inbred than those like the Maravi for whom reaching their eventual goal meant a breakdown of the endogamy of the migrant parties. The Maravi, migrating as clans, could not have observed clan exogamy during the migration: the Ambo, on the other hand, were well able to do so. In both cases, however, the outcome at the time of settlement would have been the same owing to the limitations of size of the parties. The degree of inbreeding then would have been the same in both; but the Ambo parties retained isolation, while the Maravi isolates broke down and were redistributed before reforming as new isolates.

It is consequently impossible to generalize about the effects of migration itself on present-day population structure; each particular case has to be dealt with on its own in the light of such information as we possess about the people concerned. Nonetheless, it is generally accepted that the overall expansion of the Bantu-speaking Negro peoples happened historically comparatively recently (Oliver 1966), and it can therefore be assumed that the majority of the genes present in the Negro populations of Southern Africa are inherited from a comparatively small nexus of common ancestors. Only a proportion of these genes would have been carried in each migrant party; natural selection during the exigencies of migration, if this was at all of long duration, would tend towards the elimination of some which were deleterious and might more readily have been retained under the easier conditions of settlement, but others of less or heterozygously imperceptible effect would be retained to be expressed at increased frequency once the population began once more to expand. This could account for any irregular patterns of inherited disease which might be seen in the various Negro peoples of Southern Africa. It could also account for a somewhat restricted range of polymorphic variation which nevertheless manifests itself significantly in differences in gene frequencies among the peoples.

Much the same applies to the San and Khoi, though they, being found nowhere else than in Southern Africa, offer less scope for close comparison and less field for speculation about the genetic polymorphisms present in their ancestors. Until modern times neither had ever adopted an unalloyed culture of settlement, though there are indications that, left to themselves, some sections of the Khoi might have done so. Any fluctuations that have happened in the sizes of their populations have either occurred too long ago to have been remembered, or have been occulted by the cultural impetus to claim membership of other population groups than their own.

We know that a smallpox epidemic early in the eighteenth century had a profound effect on the distribution and movements of the Khoi, who fled from the diseased area and thereby helped to spread the disease. We know that land hunger led to their displacement by Caucasoids in the northern Cape; less clearly, that on the frontier with the southeastern Nguni they did not migrate away but were assimilated. We have detected relative monomorphism for monogenic traits in the San, and a surprisingly wide range of alleles in the Khoi, while those of their traits which are polygenic appear fairly uniform between the two peoples. In later chapters we will consider the bearing of these factors on the size of the parties in which they used to, or still do, migrate.

Quite different limitations would apply to the size and structure of migrant parties coming by sea; and during the period of maximum seaborne immigration into Southern Africa these limitations changed enormously. The movement of persons by sea involves dependence on capital investment, either by the migrants themselves or by those who expect to profit from their migration, or both. The capital is expended primarily in the country of the origin of the migrants, and reinforces the colonial motives of the migration. Benefit is expected to accrue to the home country, either immediately in the departure of surplus consumers, or in the ultimate remitted product of their colonial labours. Whether the migrants left as slaves, indentured labour or free settlers, it would be most advantageous to transport as many of them as possible with the minimum of

capital investment. The sizes of such parties would be limited, therefore, by the numbers which could be carried in ships. As the sizes of
ships increased, and the cost of running them fell, the potential for
migration grew, though it was limited at the receiving end by the
occupations profitably open to immigrants. These considerations,
too, will be discussed in a later chapter.

### The *Mfecane* or *Difaqane*

One phenomenon involving the mass migrations of people in Southern
Africa needs particular mention. Early in the nineteenth century
there took place a series of political disturbances, most notably but
not exclusively among the Negro peoples of the region, which erupted
as military expeditions and resulted in the fragmentation of what
until then had been fairly large uniform patterns of occupation by
peoples with single sociocultural traditions and the establishment
in their place of several quasi-imperial Negro state systems. This
period is known in the Nguni languages as *Mfecane*, a Zulu word
denoting the upheavals which take place in time of war. The Sotho/
Tswana term *Difaqane, Lifaqane,* or *Dihakani* represents the assumption of this word into Sotho/Tswana languages. What the origin of
the disturbances was, is unknown; the rise of the Zulu state under
Shaka, the most notable symptom, has been blamed as a cause, but
this is unlikely. That it might have been due to the coincident rising
to prominence of a number of politically able and ambitious persons
among diverse Negro tribes is a possibility; but it seems more likely
that Caucasoid expansion, with its displacement of indigenous
groups, may have been ultimately responsible.

If so, it rebounded and set off the most notable of all the migrations of Caucasoids into the interior, the Great Trek, which, seen in
its proper perspective, is only one facet of the *Mfecane*. The sequence
of largely military Negro migrations proceeded towards the foundations of several large Negro states by the consolidation of a number
of small autonomous groups. If Negroes blocked Caucasoid advance
eastward, so too did Caucasoids prevent the onward spread of the
Negroes to the west. The Cape Nguni, the Negro peoples with whom
earliest European contact was made, occupied rolling grassland
which, though attractive to Caucasoids, was already thickly populated. They did not constitute a single state, but among them were
several states sufficiently large and well organized to put up resistance. Beyond them, the Natal Nguni, potentially more likely to
succumb to invasion, were ruthlessly welded by Shaka of the Zulu
tribe, a military genius, into an empire which, unable to expand

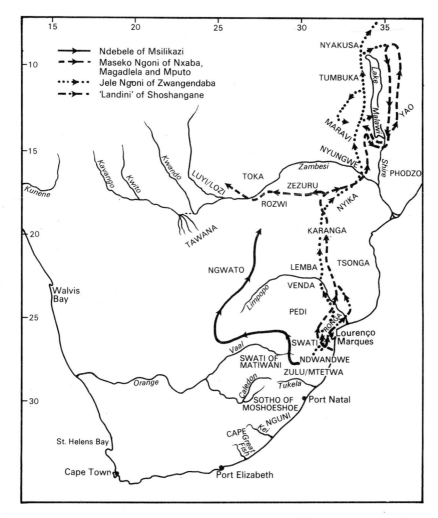

Fig. 3.1. The *Mfecane*, showing the principal routes of Nguni migrations following the rise of the Zulu/Mtetwa militarist state system, the consolidation of the
Swati and the destruction of the Ndwandwe military power under Zwide, early
in the nineteenth century. (Note that intersection of migration paths does not
necessarily indicate meeting between parties, though the Maseko Ngoni fought
with Soshangane's people, with the Jele Ngoni and with the Rozwi and Kololo/
Lozi.)

south-westwards owing to the relative strength and immobility of the
Nguni of the Cape, and checked to the north by Ngwane's Swati con-
federacy, was compelled to direct its attentions to the area along and
beyond the Drakensberg.

If the Cape Nguni had not been hemmed in to the west by the

Caucasoids, they might have given way to Shaka, and moved on. As it was the Zulu expansion set in motion further waves of activity among the Sotho/Tswana peoples to the west and north, resulting in the campaigns of the female general Mantatisi and her son and

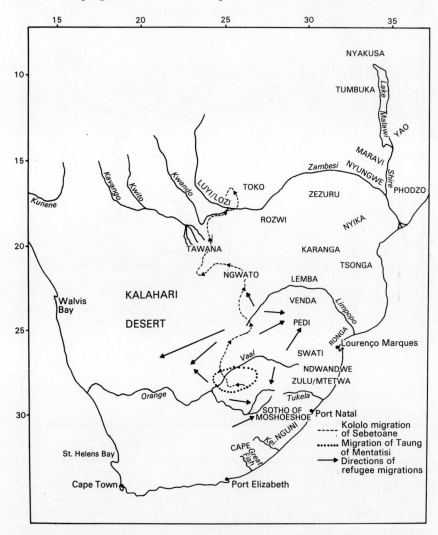

Fig. 3.2. The *Difaqane*, showing the principal routes of diffusion of Sotho/ Tswana peoples in response to military incursions consequent on the rise of the Zulu/Mtetwa state system. Note that the principal precipitating force was that of the Ndebele under Msilikazi, resulting in flight (—), consolidation (Sotho, Pedi, and Ngwato) or reciprocal military action, (Mantatisi, etc. . . . . . .) among these peoples.

successor Sekonyela, which led ultimately to the foundation of the Kololo kingdom among the Lozi in Zambia; the alliance among the Koena and Kgatla under the leadership of Moshoeshoe I which established the southern Sotho state of Lesotho; the temporary hegemony of the Zulu rebel leader Msilikaze in the northern Cape and southern Transvaal, until he retreated from the Trekkers to new conqeusts and the foundation of Matabeleland; and the displacement of the !Ora Khoi from the Vaal and upper Orange into a life of raiding along the lower Orange and into the northern Cape.

It was into this confused *Völkerwanderung* that the Trekkers poured their own unsuspecting but even more fundamentally disruptive contribution. The less well armed peoples who opposed them were overcome with relative ease, though not without much being made by the newcomers of their victories. The exploitation which followed may have been partly inadvertent. Land had not been treated as a commodity by its previous occupiers or users. They had not thought it possible that land could become the exclusive property of private individuals. Nevertheless, through misunderstandings, as much as cupidity, that happened now, and newly arrived Caucasoids assumed title to areas in perpetuity. For one or two generations these Caucasoids of the interior maintained unstable defensive peasant republics; and then the discoveries of gold and diamonds initiated that series of rapid environmental changes which have transformed Southern Africa into something very different from what it was even two centuries ago.

As far as the Negro peoples are concerned, the *Mfecane* has ended up in an apparent uniformity not dissimilar to that which preceded it, the main consequence having been a loss of land and a consolidation of sociopolitical entities largely along linguistic lines. In the course of the disturbances, however, the various Negro isolates and semi-isolates were broken down and exchange of genes among them took place, with the result that the modern relative genetic uniformity of the South African Negro peoples may reflect a considerable smoothing down of differences which could have existed two centuries ago. Overall, there was a net loss of genetic material, with the Nguni migrations into Central Africa and the Sotho/Tswana movement into the Lozi country. Similarly, the Khoi who remained south of the Orange River and along its banks made contact and intermarried with many more of their kind than might otherwise have occurred; and the San, as they were further displaced, were forced to form new entities from the relics of those which had been devastated.

The main effect of the *Mfecane*, whatever its causes, may well have been the extent to which it enabled the advancing Europeans to take advantage of the interim instability of most of the indigenous polities, and to establish their own technologically more developed states on the plateaux of the interior.

## Migration and natural selection

Some Southern African ecosystems differ appreciably from other ecosystems in Africa or those of Europe, and the extent to which natural selection continues to operate on the modern descendants of immigrants is naturally an important one. Prior adaptation, where this has occurred, would reduce the degree of natural selection, since each immigrant population would tend to choose for preference the ecosystem to which it had previously been adapted most closely; but population pressure has excluded certain immigrants or their descendants from such areas. The Ambo proclaim their satisfaction with their present lands as closely resembling those of their ancestors, but there are no other such cold arid mountains elsewhere in Africa as could have fitted the forebears of the Southern Sotho for those they now inhabit. Southern Sotho traditional history speaks of unusually high mortality in the early days of settlement (Ellenberger 1912). In addition, the original lands settled or migrated over by the various peoples have themselves undergone changes. Trees have been cut down, game hunted to extinction, overgrazed pastures have been eroded, and industry and urbanization have covered considerable tracts of arable country. People have drifted to towns and then been resettled in areas possibly ancestrally theirs but often to some extent despoliated, become less fitting for their occupation. Land has been opened up for new crops, sometimes thereby being made unusable for subsistence farming. A cash economy has disturbed the social and dietary habits not only of those peoples more anciently settled but also the peasant lifestyles of later comers. Medicine has changed old patterns of birth and death, and so indirectly introduced new types of competition for resources. It is not easy to identify in the midst of these new forces the ways in which natural selection may be acting.

This description of the peoples of Southern Africa is intended nevertheless in part as an attempt to elucidate some of the resultant probabilities. The affinities of the peoples, once determined, indicate the patterns of prior adaptation, and give some information about their potentials. At this particular time Southern Africa is populated by peoples descending from at least two, and probably three, of the world's major races, sharing several fairly contrasting environments. The ways in which they have variously reacted to these is of obvious importance to human biology, and may on further analysis reveal rather more about human adaptation than will at first be apparent from our mere description.

# 4
# THE KHOI AS MIGRANTS
# AND NOMADS

Of all the peoples of Africa the Khoi, considered as an entity, are perhaps the most elusive. Though they lived in and on the margins of what was for a time one of the more urbane of the remoter European settlements, they were rarely considered as interesting in themselves, apart from in their dealings with Europeans, until European contact had already produced considerable changes in them. Such changes may have begun with the very first contact of all: Bartolomeu Diaz, on his way down the South West African coast in 1487, put ashore several negresses, whose further history is unknown but may be conjectured (Theal 1910*a*). His later encounter with the Khoi at Mossel Bay was less productive, but in 1497 there was an exchange of goods for cattle on the landing of Vasco de Gama at St Helena Bay (Velho 1838). The Khoi were eager for trade, with the concept of which they appeared familiar, and particularly for payment in metal. It appears that the Portuguese saw no great difference between them and Negroes (Welch 1946), but then to the Portuguese the Negroes of the East Coast, the Amhara of Ethiopia, and the blond Berbers were indiscriminately 'Moors'.

Over nearly two centuries such casual contacts multiplied, first with the Portuguese and subsequently with the English, Dutch, French and Danes (Raven-Hart 1967). It is hardly conceivable that even before the establishment of the Dutch settlement at the Cape in 1652 there should not already have been appreciable gene flow from Caucasoids into at least the coastal Khoi. Contact was made not only in Table Bay, but in any of the few possible harbours of the coast. Mossel Bay, an anchorage affording some poor protection against winds and storms, Algoa Bay where Port Elizabeth now stands, and the sandy estuaries further east, were all either watering-places or the sites of wrecks. Relationships between the Khoi and the sailors were marred by misunderstandings and some loss of life on both sides, but more often accommodations were made, for trade or for water; Europeans accustomed to regarding running water as free to all men often failed to understand how jealous of their springs the Khoi could be. Nonetheless, when the first Dutch settlement was established the Khoi did not resist it, and when it began to expand fought only two half-hearted wars against it. Active resistance had ceased by 1690.

### The origins and migrations of the Khoi

A good case can be made out for supposing the Khoi and the San to be simply economically differentiated segments of the same people. When the Europeans arrived at the Cape they found some inhabitants who were pastoralists and others who were hunter–gatherers. As discussed in the next section, the distinction was in no way absolute and there is evidence that Khoi who lost their cattle reverted to a hunting way of life, while some San who became pastoralists would merge into the Khoi population. The view, for which the corroborative evidence is not entirely convincing, has been advanced (Jeffreys 1968) that the Khoi evolved in East Africa and only later migrated into Southern Africa. The fact that there are strong linguistic affinities between the languages of the Khoi and some extant hunter–gatherers of Central Botswana (the Naron, G/wi and G//ana) would suggest rather that the Khoi evolved in that region of present-day Botswana in proximity with these San, and only subsequently migrated to the south.

Elphick (1977) gives a comprehensive review of the evidence for the theory that the Khoi represent the descendants of hunters who acquired cattle in northern Botswana, and speculates on the circumstances surrounding this pastoral revolution. His map 1, (p. 16), however, shows as the probable area of evolution of the Khoi one which is today occupied not by the San but by Khoisan- (Tshu-Khwe-) speaking Negroes (see Chapter 6). Expansion and migration to the south-western tip of the Continent could well have been dictated by the needs of the Khoi pastoral economy. The bush grasslands of the south-western cape provided Khoi with more than most pastoralist societies require, and they evidently prospered there. The hunter-gatherers already inhabiting the area could not have felt too threatened because they survived in a symbiotic state with the Khoi. Without guns the Khoi, presumably, did not deplete the area of game, so that both types of economy could co-exist; and the area was rich enough to support both wild and domesticated ungulates.

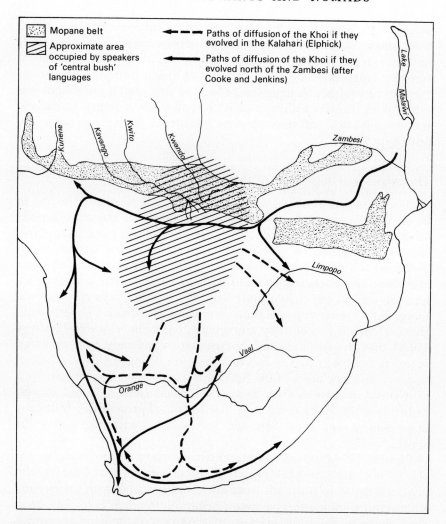

Fig. 4.1. Alternative origins for the Khoi, modified from Cooke (1965), Jenkins (1972), and Elphick (1977). The two former incline to an east African origin; the last-named suggests that they evolved from and among the forebears of the present 'Central Bush' speakers

The relationships of the Khoi languages were with Tshu-Khwe tongues of the Botswana San; even their difficulties in communication may have held them at a distance from the Cape San, speakers of Southern Bush languages (Bleek 1927), so that as long as the populations remained small there was no consciousness of any clash of interests.

It is possible (cf. Jenkins 1972) that Khoi migration to the western Cape may have post-dated by some centuries their occupation first of northern Namibia and later of Namaqualand. The grazing north of the Auas mountains is good, though it seems to have been relinquished by the Khoi some time before the coming of the Herero, and not to have been claimed again until the nineteenth-century northward migrations of the Orlams had put severe pressure on the capacity of the country to the south to support the Khoi herds. The alacrity with which proprietorship of the northern lands was then asserted in the face of at least several decades of occupation by the Herero, and the bitterness of the consequent wars, seem to indicate that the Khoi believed that they had ancestral rights to the disputed areas. Evidence for this original Khoi migration path has also been provided by rock paintings (Cooke 1965).

Elphick considers the possible interaction which might have taken place with Bantu-speakers if the Khoi had evolved in an area in or near northern Botswana. He concludes that some Khoi must have arrived in the south-western Cape without any evidence of their culture being influenced by the Bantu-speakers, since they made no attempt to cultivate the soil in an area which would have been eminently suitable; certainly more suitable than northern Namibia. This contention is not entirely convincing, since simple observation of the agricultural life would hardly furnish immediately convincing evidence that it was superior to pastoralism. It does nevertheless help to support the possibility that the emergence of the Khoi as pastoralists antedated the arrival of the Bantu-speakers in the area to the south of the Zambesi. If the cattle were brought south by Negroes they might have represented only a small advance party who soon became assimilated by the Central San, who evolved into the Khoi. This could account in part for the possession by the Khoi of slightly more 'Negro' physical features than the San.

### The modern distribution of the Khoi

Naturally enough, the earliest European contacts delimited only the coastal distribution of the Khoi, and not even that with precision. To the Portuguese, those in the best position to judge, there were few or no differences among African peoples, so that today we cannot be certain whether the peoples of the Namibian coast in the late fifteenth century were Khoi, San, Dama or even Ambo or Herero, and along the south-eastern seaboard the extent of Negro penetration and Khoi retreat remains obscure until the seventeenth century. In 1622 some Portuguese sailors whose ship, the *São João Baptista*, was wrecked near the mouth of the Keiskama River met people who, from the description of their physical and cultural characteristics, were Khoi. Later, near the mouth of the Bashee, they came across Negro agriculturalists (de Brito 1735). When the Dutch made their settlement at the Cape, the only peoples they encountered were the Khoi and the San, though they learnt from the Khoi of the Birikwa, the Negroes of the interior, and other, vaguer peoples to the north, wearers of robes. Such stories either represented direct contacts of the Cape Khoi, traditions of uncertain date, or evidence of extensive trade or cultural transmissions through middlemen. As the Dutch themselves pressed into the interior during the first century and a half, they met only Khoi and San.

Khoi conflict with the Birikwa or Sotho/Tswana occurred relatively late in their relationship. It was not until enough had been learnt about European weapons and methods of waging war that they felt themselves able to challenge the more numerous Negroes of the Highveld. Long before that a more intense and inevitable confrontation had taken place with the Cape Nguni. The more easterly of the coastal Khoi shared many of their values and customs with the Xhosa. Both prized their bovine cattle, which were integral to rituals aimed at the placation of ancestor-spirits. Both were hunters, though the Xhosa used spears rather than bows. But whereas to the Khoi their flocks and herds, and the watering-places that served them, were of paramount importance, the Xhosa were agriculturalists as well. There were close analogies among the Xhosa to the social and political institutions of the Khoi. They, too, had patriclans and patrilineages, councils of which membership was hereditary among the heads of patriclans, and a central focus at the court of the chief, where legislative, judicial and executive decisions were decided. They, too, expanded by the fission of patriclans and chiefdoms; but their ideology was one of tillage, of land, and as they spread they slashed and burned. Accommodations between adjoining chiefdoms of Khoi and Xhosa were easily made, perhaps too easily; the similarities tended to obscure the seeds of conflict. As time went on, however, the Khoi adopted a more settled way of life, and learned agriculture; but the relative weakness of their attachment to the land placed them at a disadvantage. There were exchanges of population, often reciprocal marriages, sometimes the incorporation of Xhosa refugees into the growing and relatively settled Khoi polity of the Xonakwa, sometimes the assimilation of outlying Khoi populations by the Xhosa. Intermediate between the Xhosa and the less changed Khoi further to the west, the Xonakwa steadily became more mixed; Khoi gene flow occurred deep into the Cape Nguni, but Negro gene flow into the Khoi does not seem to have progressed beyond the Xonakwa themselves.

The Xhosa, however, were growing in numbers faster than the Khoi were, and their superior military organization enabled them to extend their lands by conquest. Some Khoi doubtless retreated before their advance, and carried incompletely assimilated Negro techniques of agriculture westwards; but many smaller tribes remained at first as clients of the Xhosa, and then were absorbed. Concurrent with the resulting gene flow was the modification of the Nguni languages of the Cape by the incorporation of a number of Khoisan elements, most notably the click sounds. Whether the click sounds of Zulu and Swati entered those languages by a similar but earlier process, or whether they filtered back from the earlier to the later Nguni immigrant waves by some purely cultural means, is uncertain. It seems probable that the rare clicks which occur in Southern Sotho but not in Northern Sotho or the Tswana languages may have been borrowed from the Natal Nguni, though the more surprising occurrence of a click in all the Kavango languages except Mbukushu may be attributable to contact with San rather than Khoi.

Today there are no avowed Khoi left anywhere along the coast east of Cape Agulhas, though their apparent disappearance probably represents political expediency rather than biological extinction. To be a 'Cape Coloured' has always, since Caucasoid-controlled government was extended in these parts, been at least marginally more advantageous than to be a 'Native', whether 'Hottentot' or 'Bantu'. Yet the processes which have formed the 'Coloureds' have not been to any appreciable extent a feature of the history of these shores. By the time they were opened to Caucasoid settlement the importation of slaves had ceased, and in the households and fields of the farmers the slaves were greatly outnumbered by the Khoi. The households of the farmers, too, were mostly run by Caucasoid wives; the shortage of Caucasoid women did not outlast the term of the Dutch East India Company, and had, indeed, been substantially relieved by natural increase and relatively diminished male immigration even before the first British occupation in 1795. Miscegenation could hardly have occurred on a sufficiently large scale to bring a whole new population into existence overnight, even if we credit the few sailors briefly ashore in Port Elizabeth or Mossel Bay with phenomenal potency and fertility. Their shore leaves would have been longer in the protected harbour at Port Natal; yet the hybrid Negro/Caucasoid population of Durban is minute compared with the numbers of 'Coloureds' in and around Port Elizabeth. The Khoi showed early on that they were receptive to European customs and beliefs. Helped on by the missionaries and the easy sacrifice of their own language in exchange for Afrikaans or, on occasion, English, they have not simply blended with, but to a large extent substituted for, the hybrids they are now popularly supposed to be. The 'lost tribes of the Cape' are not lost; they are hardly even metamorphosed; rather are they metempsychosed, the willing victims of the self-inflicted brain-washing of their shrewd forebears.

Events seem to have followed a similar course elsewhere as well. Smallpox struck severely at the Khoi tribes of the western Cape during and after 1713 (Mentzel 1921), extinguishing some, reducing the numbers of others below the level of viability, forcing reassemblies and regroupings in which there were great breaks in tradition. Coming just as European influence was extending, and at a time when a labour shortage in the Dutch-controlled area offered opportunities to learn new techniques and acquire new consumer goods, the social disruption produced by the epidemic would have been a strong incentive to some of the survivors to abandon tribal life altogether and seek a closer association with the Caucasoids. There would have been greater opportunity for these to interbreed with slaves and bachelor Caucasoids, and there is sufficient evidence that they did. At least one Khoi tribe, the ≠Karixurikwa, incorporated non-Khoi elements, runaway slaves, deserters from ships and the army, as it reformed, and went on to develop into the Griqua kingdom whose history belongs among those of the hybrid peoples. The !Ora, on the other hand, moved inland without much attested admixture, and eventually took up a position along the valleys of the Orange and Vaal. The smaller tribes to the east and north-east remained in their fastnesses and ultimately assumed a 'Coloured' identity, as, too, did many of those who inhabited or moved to the Boland, Cold Bokke-veld and Cedarberg, while others went to make up the Nama of Little Namaqualand. In Great Namaqualand, north of the Orange in what is now Namibia, there have certainly been Khoi for many centuries, but the impact of the smallpox epidemics is difficult to assess. At the edges of the area of Dutch control missionary activity induced bodies of Khoi to settle and adopt agriculture. Some of these mission stations, such as Genadendal, Elim and Wupperthal, run by the Moravian Brethren, were virtually tiny independent states. When they emerged into the wider polity their congregations were no longer 'Hottentot', but 'Coloureds'; and as such they have remained.

Sometimes the change of identity seems to have happened quite without outside intervention or any particular foresight, simply by means of the adoption of a way of life and a language which made communication easier and relations more advantageous with Cauca-soid officials or neighbours. The Richtersveld community, for instance, (Carstens 1966), though historically known to have received some Caucasoid gene flow, retains a gene marker profile which justifies its inclusion among the Khoi. These people are still nomad pastoralists, but became Christians early, assumed European ways learned from the Caucasoid nomad pastoralists of the colonial fringe, took to hunting with firearms and migrating in wagons, and are now accepted as 'Coloureds'. In any case the Khoi are remarkably free of ethnocentricity. For as long as they have been known to Europeans they have been extremely receptive of recruits of any kind. The now resettled Riemvasmaak community, Khoi when visited by van Warmelo (personal communication) nearly fifty years ago, was, when we described it on the eve of its final dissolution, made up of an assortment of Khoi and hybrids of all sorts, with the politically dominant element Negro (Nurse and Jenkins 1978).

Today the great majority of those who will admit to being Khoi live in Namibia. With the relatively few and scattered Khoi to the south of the lower reaches of the Orange they constitute the Nama; a name probably not connected etymologically with that recently adopted for the country, and indeed used uncommonly by the Nama themselves for themselves. They prefer the appellation Khoi-khoin, real people; a semantic preference shared with any number of other populations. The Nama traditionally fall into three principal divisions inside Namibia, though foreign domination, settlement and the loss of many of their customs has caused the distinctions to become blurred. In the south, around Warmbad, lived the !Gami≠aun, a name translated literally as Bondelzwaarts, the black-bundle people. Around Walvis Bay and in the dry valley of the Kuiseb River as it crosses the Namib Desert, were the ≠Ounin, those who went ahead, known also from their main article of diet, the *!naras* melon, as the !Narenin. Their Dutch name combines an alternative meaning for ≠*oub*, summit, with this second name: they are more familiarly known as the Topnaars. The Bondelzwaarts and the Topnaars were small independent tribes; the rest of the Nama were included in a loose confederation, the leading tribe of the six included in it providing the overall name of //Khouben, Roode Natie or Red Nation. This is not indeed a translation of their own name, the meaning of which is obscure, but of the generic /Awakhoin by which the Nama as a whole sometimes distinguish themselves from both blacks and whites. The //Khouben keep a headquarters for themselves and the confederation at Hoachanas.

The principal associated tribe, the //Khou-gõan or children of the //Khou, is known in Dutch by the Dutch surname assumed by the ruling family, the Zwaartboois. They have been very mobile during historical times, and in a little over a century have wandered from

Hoachanas in the south-east, up to and across the central girdle of the Auas mountains, westwards down the Swakop Valley, across the Dama country through the Erongo Mountains and eventually north to the fringes of the Kaokoveld, where they have mixed with an offshoot of the Topnaars. The //Hawoben or Veldschoendragers, wearers of hide shoes rather than sandals, lived to the north of the Bondelzwaarts, and between them and the Roode Natie proper were the !Kara-gei-khoin, a name difficult to construe but possibly meaning Great Stabbers. The latter used to be known as the Franschmannsche Hottentotten, but are now more familiar as Simon Koper's People after their leader in the 1904 insurrection, following which many of them fled to British territory. Most live in Botswana today. The //O-gein, Groote Doden or Great Dead, given that name neither *post hoc* nor *propter hoc*, are now extinct. They used to live to the west of the !Kara-gei-khoin. The sixth tribe, from which the Nama of Keetmanshoop descend, appears to have had little cohesion and no strong identity, and to have been known either as the Tseiben or the !Kara-!oan.

The tribes which fled northwards out of the Cape were known collectively as the Orlams, the shrewd people; a name not Dutch, though now Afrikaans; derived impurely from the Malay. Afrikaans seems to have been as much their language as Nama was, and the Khoi of Namibia, correctly or incorrectly, imputed mixed descent to them. They consisted of the /Khowesin or Beggars, better known as the Witboois from their going into battle with white cloths around their hats, who settled near Gibeon to the north-west of Keetmanshoop; the Gei-/khauan or /Hei-/khauan, Great Wanderers or Grey Wanderers, called the Amraals after their leader Amraal Lambert, who went north to Gobabis; the Lesser Wanderers, the ≠Kari-/khauan, who, with the !Aman or Green Men, went to areas to the north and east of Gibeon; and the //Aixa-//ain or Angry Ones, better known as the Afrikanders, who occupied the fertile valleys around the hot springs of Windhoek.

Today not all the Khoi of Windhoek are //Aixa-//ain; the community there includes elements of all the tribes, Nama and Orlam. Most of the Khoi still live in the south, concentrated at Keetmanshoop, Berseba, Tses and Gibeon, but the Topnaars still inhabit the valleys which stretch up through the Namib and spread to the southern Kaokoveld, the remains of the Roode Natie have drifted into the Rehoboth Gebiet from the east, and there are still Bondelzwaarts at Karasberg and Warmbad. Though tribal differences seem to have disappeared and never to have been particularly important, there has apparently been little gene flow between the different communities even when they acknowledge joint membership of a common group. This is probably due less to the complex traditional kinship system unravelled by Barnard (1975) than to the abandonment of nomadism and the consolidation of quite small communities in the neighbourhood of Christian churches.

The only other Khoi peoples to have retained some measure of identity, if we exclude the heavily hybridized Griqua, are the !Ora of the Orange and Vaal valleys. The name is said to betoken smoothness, straightness and by implication righteousness; their reputation, however, is one of marauding and banditry as fierce as were any of the wars of the Nama against the Herero. There is a possibility that the name is an eponym, and that the present tribes represent to some extent the descendants of the Gorachouqua or !Gora-//xaukwa, men of the hamlet of the chief !Gora or !Ora, known to the Dutch as the Tobacco Thieves. They fled early from the Cape; oral tradition claims that those who stayed behind tried to lure them back by sending women after them to be their wives. During the eighteenth century travellers to the middle reaches of the Orange reported a Khoi people known as !Ura-mã-//eis, the high standing tribe, and today some !Ora and Griqua bear the surname Hoogstaander or use the name as a general term for the !Ora. In the interior the !Ora either underwent fission or, like the Roode Natie, assimilated a number of other tribes into a loose confederation. The latter were more probably the case, as the smallpox epidemics are remembered here, and some modern divisions of the !Ora claim also to have been originally San. They comprehend also elements of the /Hõana, or Cat tribe, the //Utena or Springboks and the Karoshebbers, ≠Nam-//'eina, owners of skin quilts, as well as the ≠Xami-//'eis, the Book or Paper tribe, the /Nũisin- or /Husi-//'eis, Scorpion or Spider Tribe, the Links, Left-hander, //'Aremã-//'eis tribe, and the so-called Gei-!Orana or Great !Ora, also known as Taaibosch. The last-named certainly split, into at least eight divisions which it would be tedious (and typographically infelicitous) to enumerate in detail. One of them, Seekoei or Hippopotamus, !Xau-//'eis, like the anomalous Bitterbos, Bitterbush, /Gumtena, may have included a substantial San element, and another, the ≠'Õxokwa, Smalwange or Narrow-cheeked Men, was possibly originally the Cochoqua of the Dutch records and hence existed at the Cape independently of the !Ora. To the north of the middle Orange lived the Othersiders, /Nu-//'eikwa, divided into the Regshande, Kx'am-//oana, Right-handers, and the !Geixa-//'eis, Towenaars or Sorcerers, rather more loosely part of the confederation (Engelbrecht 1936; Maingard 1931, 1964).

During the late eighteenth and most of the nineteenth centuries these peoples variously acted as intermediaries between the coast and the interior, harried the other Khoi, one another, and the Sotho/ Tswana, with whom they also intermarried, and intervened in the affairs of the Griqua and the Basters. Curiously, there seem to have been no clashes with the advancing Caucasoids, on whom they were probably dependent for a good deal of the knowledge and equipment which gave them a military edge over their less developed rivals. They were never numerous enough, moreover, nor sufficiently internally united, to evolve into anything like a powerful military state, though occasional adventurers tried, sometimes with Caucasoid connivance, to displace more legitimate chieftains. When diamonds were discovered in Griqua territory, and the Griqua were displaced by the British, !Ora went to swell what came to be known as the 'Coloured' community in Kimberley, and as Caucasoid farms spread up into the desert some of them took employment, again often under the guise of 'Coloureds'.

Today the number of self-styled !Ora is small and diminishing; many find it to their advantage to call themselves Griqua, by which means they can gain prestige and retain their language; the Griqua, an unusually receptive nation, do not resent, but rather welcome, this trend. The old divisions persist on a scattered, irregular front in the western Transvaal, northern Cape, eastern Free State and possibly southern Botswana, and can be elicited by careful collation of names and genealogies and by patient questioning. It is possible even to come across in the *macedoine* of the 'Coloureds' of the Witwatersrand the occasional individual with a characteristic Khoi appearance, who will confess to a grandparent or two who spoke the language, and who may even remember more of it himself than he will admit.

The !Ora language, like Xiri, the language of the Griqua, is in the process of replacement by Nama, from which it was once quite distinct. The question is again one of prestige; Nama is a language in which there are books, one which, in Namibia, is taught in the schools. Almost all the Khoi who can write a Khoisan language will write in Nama, even if their own language is Xiri or !Ora; many of of them will have been taught to write it by migrants from Namibia, who drift back and forth across the border in search of work and do not scruple to seek hospitality from their fellow Khoi.

The surviving Khoi can consequently be said to consist of three main divisions: the Nama of Namibia, whose numbers are beginning to increase again after the wars and famines of the nineteenth and early twentieth century; the !Ora, who maintain a precarious identity along the valleys of the Orange and Vaal, and whose numbers are uncertain owing to their frequent reluctance to admit to being who they are; and the submerged multitude which it is now impossible to separate from the 'Coloureds' except on careful genealogical appraisal. The account given here of the Khoi cannot be considered exhaustive; it is necessary to consider not only these sparse details, but also the descriptions of the 'Coloureds' in Chapter 8.

## Government and economy

The Khoi have often been described as having been divided into tribes, and in their context the term has an unusually classical connotation. The tribes were not distinguished from one another by speech or custom, but do appear to have possessed a real or classificatory genealogical cohesion. Their elementary corporate constituents were patrilineal extended families, often polygynous, all belonging to larger patrilineages. Patrilineages were grouped by descent from a common ancestor into exogamous patriclans, and the tribe was made up of a series of nomadic associated patriclans loosely linked by representation in a council headed by the senior member of the senior lineage of the principal one. The patriclans were graded in order of primogeniture, the principal one being that from which all the others had originally sprung. The chief, or head of the council, remained fairly static; it was at his house that the council periodically met and exercised its judicial and executive functions. The patriclan was nevertheless more potent than the tribe; the chief was no more than *primus inter pares*, and it was always open to any patriclan to proclaim its own independence as a new tribe. The functions of the chief were limited to the waging of war, the conclusion of peace, and powerless presidency over the council. Neither he nor the councillors were distinguished by any marks of rank, until the intrusive suzerainty of European powers imposed upon grateful recipients such tawdry insignia as brass badges of office and copper-headed staffs (Schapera 1930).

Even before that suzerainty had been exerted effectively its influence upon the government of the Khoi had become considerable. Khoi political traditions were hallowed by nothing better than custom or convenience, with no supernatural sanctions (Hahn 1881; Vedder 1928b, 1965). To the Khoi the might of the invaders would appear to stem, in part at least, from their forms of government; and benefits could arise from imitating them. The modifications, being unlettered, were necessarily crude, and coloured, as time went on, by the teachings of missionaries and a sometimes garbled familiarity with the Bible. Succession to the chieftaincy came to follow more closely the rules of primogeniture, and hence it was more often necessary to institute regencies or appoint deputies. The executive and judiciary were separated to some extent from the chieftaincy, with a magistrate and leaders in war and in domestic policing being chosen from among the council; later still, Christianity was entrenched with the *ex officio* additional function of the council members as elders of whichever church held sway in the area. The chief assumed or was given the Dutch title of 'Kaptein' (Schapera 1930).

The ultimate flowering of the Khoi style of government took place not among the Khoi themselves, but, as will be described in a later chapter, in the viable but static autonomous states founded by the hybrid Griqua and Basters. Nomadism, or perhaps more accurately, though pedantically, transhumance, permitted little effective central control over the parties which went to make up the tribe. Though there was a fixed gathering-point at the home of the chief, from which all the parties set out on their annual cycle of movement, and though the position at any given time of any particular party was known to all the rest, there was no way in which any central authority could be exercised except during one short period of the year. Decisions which the heads of clans or families could not take, and the settlement of disputes among them, had to be stored up for months until the council met. Such dispositions would hardly favour stability in the tribe. They would make fission easy; no party or individual expecting that the judgement in a dispute would be adverse, could be blamed for shifting elsewhere or asserting independence by actual revolt. That a multitude of Khoi tribes should share a relatively small number of languages, all quite closely related, is scarcely surprising (Maingard 1964; Nienaber 1963).

The nature of the country dictated the patterns of movement. In the coastal areas, where good pastures were not rare, the tribes were more compact; in the dry interior they spread over larger areas. The pastures of the patrilineages and patriclans, and their cycles of migration, were apportioned by the council and were all contained within the compass of what can be regarded as the tribal lands. Proprietorship in the land, whether of the tribe or of the individual, was not, however, asserted: the only titles were to land use, and their locations could shift from year to year. As they moved from one grazing-land to another, the Khoi carried with them their *omi*, semi-circular tents made of grass mats which could be tied over a portable framework of long light pieces of supple undressed wood, structures very similar to those erected by the nomads of Central Asia. These were the only dwelling-places they had. Their flocks and herds wandered unpenned, and only the trained draught-oxen were tethered. It was unlikely that the beasts would wander far from a watering-place. The possession of water-holes was more stressed than that of grazing-land, especially on the plateaux and in Namibia and the north-western Cape, and fighting occurred more readily over springs than from any other cause.

The Khoi depended for their clothing and much of their subsistence on their large and small cattle. Meat and skins they obtained principally from their sheep and goats; their large, long-horned straight-backed cattle were used mainly for milking and draught purposes, and were slaughtered only on ceremonial occasions such as initiation ceremonies or the installation of a new chief or councillor. Hunting played some part in their lives, and was generally conducted in much the same way as it was by the San, whose methods are described briefly in the next chapter. Sharing of the meat of the chase, however, did not follow precisely the egalitarian San pattern; certain pieces were the due of the chief or the head of the patriclan, though what was left, and indeed food in general, was distributed to the community at large, and the hunter or other producer could not appropriate the greater part.

Preparation of the skins of game and cattle was undertaken mainly by the men. They were not tanned, but softened by beating and rubbing. Cloths and *karosses*, or blankets, were sewn with sinews, usually the back-sinews of oxen, using a bone awl. Whole skins were used as coverings for the dung or mud floors on which the *omin* were erected. Hide was cut into strips and plaited into thongs for whips and a variety of other purposes. Animal tendons were sometimes used for bowstrings, but much of the string required was prepared from bark fibre. This string could also be netted into bags or used for sewing mats together. Pottery was made by some groups who had access to suitable clay, but it appears not to have been fired (Rudner 1968). Most of the domestic implements, such as milking pails and bowls and dishes, spoons and pestles and mortars, were made of wood, though horn and tortoise-shell receptables served for fat, medicines, cosmetics and snuff. There is no record that the Khoi ever worked in stone except for the manufacture of soapstone pipes for tobacco and hemp.

Trading went on among the Khoi groups and with adjacent peoples before the appearance of the Europeans. The Khoi were dependent for such metal as they used on the Negroes, principally the Sotho/ Tswana, with whom their relations seem to have been good. It is possible that the Khoi name for goat, *birib*, obviously derived from a Bantu language, and their name for the Tswana, *birikwa* or goat-men, indicates that their goats were first obtained by way of trade with those peoples. The Sotho/Tswana seem also to have introduced the use of tobacco among them, and to have provided it in exchange for aromatic herbs such as *buchu* and perhaps cannabis, as well as for cattle. During the nineteenth century the Nama Khoi of Namibia were bartering cattle with the Ambo and Herero for copper and iron, and those of the Cape were acting as trade intermediaries between the Caucasoids and the San.

The impact of the earlier Caucasoid settlements on the economy of the Khoi may be regarded from two, to some degree opposing, aspects. On the one hand the Caucasoids displaced them, sometimes gradually but often urgently, from their pastures and their springs, pushing them back into ever less clement lands, depriving them ultimately not only of any chance of expansion but even of the possibility of independent existence. On the other hand the situation of the Khoi between the Caucasoids and the peoples of the interior presented them with the opportunity to avail themselves of Caucasoid innovations before the San or the Negroes could. The Khoi were not slow to grasp their chance. Within a few generations they adopted the wagon, European domestic tools and clothing, and firearms, and became at least as determined as the Caucasoids that the San should be exterminated. While acting as middlemen in Caucasoid trade with the interior they also sought, and for a time with success, to impose a hegemony of their own on the Negroes of the interior. This attempt was most blatant and notorious in Namibia, but the smaller domination of the !Ora over the Orange and Vaal valleys survived in pockets almost until the end of the nineteenth century.

## Physical characters: morphology

Eighteenth-century scholars, avid for curiosities, found in the strange shapes of the Khoi, and particularly the women, much to satisfy their taste for the grotesque. It is surprising to find that they did not exaggerate. Steatopygia as enormous as that portrayed naked in the old engravings is still often concealed beneath the profusion of petticoats of Topnaar or Rooinasie women, producing the effect merely that, with the rest of their Victorian costume, they have retained the bustle; and the tablier, the elongation of the labia minora, is more often present than absent.

These two anatomical distinctions, unique to the Khoisan, attest most forcibly the similarities between the Khoi and the San. That the two peoples are biologically closely related has been recognized for a long time. Peringuey (1911) suggested that since the very beginning of Caucasoid settlement at the Cape the distinction between them had been made quite arbitrarily. 'Those who had herds of cattle, and were not too small, were Hottentots; the others, who lived mostly by the chase or kept no cattle or sheep, and were small, were Boschiesmen.' Few morphological differences between them have been noted, except in overall size, and Schultze (1928) believed that the physical characteristics they possess in common are too numerous for them to be regarded as separate races. Drennan (1929) found the dental differences to be negligible.

Like the San, the Khoi are small in stature, though generally not as short as the San are. Their steatopygia, the isolated deposition of fat over the buttocks, tends to be accentuated by their postural peculiarities, most notably a lumbar lordosis. Steatomeria, or isolated fat deposition along the thighs, is frequently present. Their yellowish-brown skin coloration is often as light as, or lighter than, that of the San, whom Weiner et al. (1964) showed by skin reflectance studies to be the lightest in colour of the unmixed indigenous peoples of Southern Africa. The 'peppercorn' appearance of their head hair is produced by tight spiralling of the tufts, and is reflected in the pubic and, where present, the chest hair of the men, though it is not so marked in the pubic hair of the women. Greying appears late, and baldness is unusual. They have slender bodies, with small hands and feet and upper extremities which are short in comparison with the lower. Both extremities are short relative to the trunk and are thin and not heavily muscled. In young people the skin is taut and smooth, but, as in the San, wrinkling occurs early, possibly because of a general deficit of subcutaneous fat except over buttocks and thighs, and is especially marked on the face, neck and abdomen.

The most marked difference from the San is in head shape. On the whole, the head is long, narrow and low, though the platycephaly is not often as pronounced as it sometimes is in the San. Schultze (1928) noted that with greater stature the head in the Khoi tends to become larger and higher. Though in both peoples the faces are small, flattish and orthognathic, contrasting strikingly with the large prognathic faces of Negroes, the faces of the Khoi tend to be longer and narrower than those of the San. The lips are thin and their tendency to project sometimes produces a false impression of prognathism. The low, narrow forehead, high, prominent cheekbones and relatively narrower mandible and pointed, retreating chin give something of a triangular appearance to the face; the pronounced frontal bossing so often seen in the San is less common in the Khoi. The short, broad nose is nevertheless usually higher and narrower than in the San, though in both peoples it tends to be flat, with wide nostrils

directed forward. The eyes are often slanting, with prominent epicanthic or other folds. The irides are almost uniformly dark brown
in colour. The ear is somewhat larger than in the San, but still
unusually small. Absence of the lobe is common, and overrolling of
the helix can be so pronounced that the upper margin of the pinna
forms a straight line (Tobias 1955a, 1966, 1972a, b; Singer and
Weiner 1963).

Steatopygia is somewhat commoner in the Khoi than in the San,
and so is the tablier, though the genital peculiarity of the males,
ithyphally or horizontal projection of the non-erect penis, is relatively rare in the Khoi. The breasts of pubescent girls are small and
conical, with areolae so swollen and prominent as almost to resemble
balls resting on the apices of the breasts. With suckling of the first
child this appearance diminishes, though the areola remains unusually
prominent until with repeated children sagging of the breasts sets in.
Eventually they become quite limp, and hang flat against the body
(Krut and Singer 1963).

There is now, once the effects of admixture are discounted, as
considerable a morphological as linguistic uniformity among the
Khoi groups. The measurements carried out by von Luschan (1907),
Schultze (1928), Tobias (1955a), Fischer (1913) and Grobbelaar
(1956), as well as the rather impressionistic accounts of their physique
furnished by Fritsch (1872), Schultze (1907), Passarge (1908) and
Dornan (1925), and the quasi-scientific appraisal by Galton (1853),
who 'measured' steatopygia from a distance with a sextant, as well
as numerous travel narrations from the earliest contacts onwards, all
convey much the same overall information. The physical conformation of the Khoisan is so distinctive that it almost certainly bears on
their physiology. Whether it has come about on account of the prolonged operation of genetic drift during their period of isolation
from all other races, or whether it has some adaptive significance,
is by no means clear yet.

## Physical characters: genetic markers

The most striking genetic marker differences between Khoi and San
lie in the gene frequencies in the ABO and Rhesus blood group
systems, as well as in the haptoglobins. The blood group B allele in
the ABO system assumes a very low frequency in the San (Central
Bush speakers included), never higher than 0.043, and often lower.
In the Khoi it has a higher frequency, sometimes five times as high,
as is the case for the Sesfontein Topnaars (0.228) and Tsumis Nama
(0.276); in the very large Nama sample of Zoutendyk et al. (1955)

and the Keetmanshoop Nama (Jenkins 1972) the frequencies are 0.167 and 0.159 respectively. Southern African Negroes show slightly lower frequencies of $B$ than those found in the Khoi but Zambian and Zimbabwean Negro populations have comparable frequencies (Elsdon-Dew 1939). In the Rhesus system $R^0$ ($cDe$) assumes a very high frequency in the San populations, 0.80 to 0.96 being the approximate range, whereas it is significantly lower among the Khoi, 0.60 to 0.81; $R^1$ ($CDe$) is generally below 0.02 in the San (the !Kung of Ngamiland are an exception with a frequency of around 0.15) whilst it is around 0.15 in the Khoi; $R^2$ ($cDE$) is virtually absent among the San (the G//ana of the Central Kalahari Reserve surprisingly have a 0.125 frequency) while it seems to have a consistent 0.06 frequency in the Khoi; the $r$ ($cde$) allele is of very low frequency among the San, except for the !Kung and Naron, but assumes frequencies as high as 0.172 in the !Kuboes Khoi and 0.163 in the Nama tested by Zoutendyk *et al* (1955). The Southern African Negroes have significant frequencies of $R^1$ (0.025–0.157), $R^2$ (0.071–0.172) and $r$ (generally above 0.100) so it is conceivable that the interaction between Negroes and the hunter–gatherers who became herders resulted in some Negro assimilation with an alteration in the frequencies of the genes in the direction which, under the influence of genetic drift, we see today. The haptoglobin $Hp^1$ gene frequency effectively separates almost all San populations from all Khoi. Almost all the San populations, and 14 have been tested to date, have frequencies within the range 0.185–0.304, whereas the Khoi range is 0.529–0.733, a range very similar to that for Southern African Negroes.

If the San who evolved into the Khoi had a similar genetic makeup to the other San, and if the gene frequencies of the latter have not altered substantially over the years, it is difficult to say how the Khoi should have diverged so radically. Gene flow from Bantu-speaking Negroes must have taken place on such a large scale that it is difficult to explain why the influence should have been restricted to such a small proportion of the Khoi genome. Morphological characters would surely have also been influenced to a discernible extent. In many systems the similarities are striking. It is, of course, possible that the genes under discussion might have conferred some strong selective advantage on pastoralists so that, once introduced (together with the cattle) into the emergent Khoi population they increased in frequency to the present high levels.

It is difficult to suggest any further candidates for the roles of genetic markers recognizable now or in the past subject to discriminating and particular selection which would help to demonstrate the

Khoi as a distinct people. Low frequencies of the two common African variants of glucose-6-phosphate dehydrogenase, the presence of adenylate kinase variation, high frequencies of the ABO allele miscalled $A^{bantu}$ and of the Duffy antigens, and relatively elevated levels of the acid phosphatase allele $P^r$, bespeak an affinity to the San much closer than to the Negroes. Nevertheless, these peculiarities are not constant; two of the samples from the Rehoboth Gebiet in Namibia have lower frequencies of $P^r$ than many Negro peoples. This can be accounted for by the gene flow which obviously occurs from the Negro Dama into the Nama in that area. Since the Dama are traditionally slaves of the Nama, and have themselves been the recipients of little or no gene flow from their masters (Nurse *et al* 1976), it would appear from this that where miscegenation occurs it is the custom of the Khoi to acknowledge the hybrids as Khoi; a tendency which accords very well with the 'receptivity' mentioned earlier and noted especially among the Khoi-oriented but no longer biologically Khoi Griqua (Nurse 1978, and see chapter 9). This directional gene flow is particularly noteworthy in the Sesfontein Nama, who in the 6-phosphogluconate dehydrogenase system lack the common variant $PGD^c$ but possess $PGD^R$, which reaches relatively high frequencies in the Dama. In the same population the frequency of $AK^2$ in the adenylate kinase system reaches an unusually high level, while the Dama are almost unique among Negroes in having a high frequency of this variant (Nurse *et al.* 1976). Such receptivity could also account for the relative paucity of signs of gene flow from the Khoi into the San, whereas flow in the opposite direction has been more considerable.

A variant which in Southern Africa might be exclusively Khoi is the $PGM^6_1$ allele of first locus phosphoglucomutase. This is found in three Khoi samples and at rather lower frequencies in Southern African Negroes. A similar variant is described in persons of Chinese descent, as well as sporadically in other peoples. The Negro population with the highest frequency is the Kgalakgadi of Botswana, part of whose evolution has taken place in an environment similar to that in which the Khoi have evolved. This does not necessarily mean, however, that the allele has any selective significance, since it could have attained its present frequencies by founder effect or drift. Considering the present-day spread of its distribution in the Khoi, the gene flow is more likely to have happened from them to the Kgalakgadi than vice versa.

Some of the least equivocal but nevertheless puzzling evidences of the affinities of the Khoi are provided by the Gm results reported for them by Steinberg *et al* (1975). The Khoi possess the characteristically San phenogroup $Gm^{1,13,17}$, but in lower frequencies than the San, and the usually Caucasoid $Gm^{1,2,21}$ in lower frequencies than the Caucasoids; they appear sometimes to lack $Gm^{1,21}$, which is common to San and to Caucasoids, and they possess $Gm^{1,5,14}$ at frequencies which suggest that it may be more of a Khoi than a Negro marker. This phenogroup is nevertheless missing in the Sesfontein and Kuiseb samples. Of the more characteristically Negro phenogroups, those including $Gm^6$, neither is present in the Kuiseb sample and the frequency of $Gm^{1,5,6}$ is low in all other samples tested, while $Gm^{1,5,6,14}$ is particularly high in Kuiseb and low in Keetmanshoop. $Gm^{1,5,13,14}$, the phenogroup with the highest frequencies in Negroes, reaches relatively high frequencies here too, usually exceeding those for $Gm^{1,13,17}$, which never happens among the San. In addition to $Gm^{1,2,21}$, the Khoi possess another Caucasoid phenogroup, $Gm^{3,5,13,14}$, though this has not been found in the Sesfontein Nama. It consequently appears that the Khoi are most probably an amalgam of a predominantly San stock with some Negro and Caucasoid elements. Whether the Caucasoid elements were acquired in East Africa prior to a southward migration, as the hypothesis of Jeffreys (1968) suggests, or, as seems more probable, represent gene flow just before and following the Dutch settlement at the Cape, cannot be determined with any certainty. The receptivity of the Khoi to fugitives, so finely shown in the history of the Griqua, may once more have been responsible. It is not inconceivable that both these hypotheses may be true.

### Genetic implications of Khoi pastoralism: lactose absorption

Another polymorphic trait which at first sight may appear to show descent from Caucasoids is the ability to metabolize milk sugar, an inherited ability found principally in Western Europeans and some African and Asian dairying peoples (Simoons 1978) but not confined to them (Flatz and Rotthauwe 1977) and possibly selectively not primarily concerned with lactose at all (Cook and Nurse 1980). The condition is dominantly inherited, and the responsible gene is known as $PHILA^+$, (Persistent Hereditary Intestinal Lactase Activity).

Elphick (1977) argues that the Khoi are descended from hunter–gatherers who acquired stock in or near northern Botswana and who then migrated southwards in search of pasturage for their cattle. Movement to the north and east was not possible because the areas were already occupied by the agricultural Bantu-speaking Negroes. The exact dating of such a change in economy is unknown but it is not likely to have taken place more than 2000 years ago. The expansion of the Khoi southwards could have been rapid. Pastoralists, unlike hunter–gatherers, could probably have tolerated a relatively rapid increase in population size; the availability of milk would contribute to this increase in spite of the fact that the individual members of the population would probably, in view of their descent from hunter–gatherers, have all been lactose intolerant (i.e. PHILA-negative) (cf. Simoons 1970).

If the stock acquired by the San who evolved into the Khoi had been introduced by significant numbers of immigrant pastoralists it is highly probable that the genes of these immigrants were readily incorporated into the genome of the hunter–gatherers. These genes may well have included those for the PHILA-positive trait. Whether this represents the source of the PHILA-positive character in those of the present-day Khoi who have been tested for it, the Nama of Keetmanshoop, is impossible to affirm with certainty; there is some evidence of both a genetic and linguistic nature to support such an hypothesis. The genetic evidence has been set out in the preceding section. It must be remembered that the extant pastoralists of East Africa and the Sudan/Sahel are among the peoples who do possess the $PHILA^+$ allele, and that it is historically possible for them to comprise very early proto-Caucasoid elements, as well as to have received trans-Erythraean and even possibly trans-Mediterranean gene flow. Ehret (1967) has suggested that a number of Hottentot words, including those for cow, goat, ram, milk and ewe, are related to Central Sudanic languages which, he believes, were spoken in central Africa before the emergence of Bantu-speakers.

Elphick (1977) lists a number of factors which he interprets as suggesting that the Khoi of the western Cape had not inhabited that area more than a few centuries before 1488 when they were seen there by Bartolomeu Diaz: (i) very low population density in the seventeenth century; (ii) apparent absence from the good pastures in the vicinity of the mountains of the central Cape Province (though it is possible, as pointed out in an earlier chapter, that prior adaptation to a more arid ecosystem may have made these less attractive to them than they were to prove to settlers from Europe); (iii) the continuing uniformity of the Hottentot language over such a wide area; (iv) the paucity of rock paintings of sheep and the absence of those of cattle in the western Cape and Namibia; and (v) the fact that remains of cattle have not yet been discovered in late Stone Age sites in South Africa.

It is to be hoped that archaeological sites containing Khoisan skeletal remains and domestic animal deposits will be found and, by modern techniques of age assessment, give more accurate dates for the pastoral revolution in Southern Africa. Meanwhile, we can simply reaffirm the striking biological similarities between the Khoi and the San. These biological similarities do not include comparable frequencies of $PHILA^+$, though the presence of the gene in several San populations suggests that it could have arisen by mutation or gene flow in the San prior to Khoi differentiation, and in the latter increased due to the selective advantage it conveyed once some measure of dairying was adopted (cf. Bayless 1971). This would thus be an example of pre-adaptation rather than prior adaptation.

Milk was a fairly significant component of the diet of the Khoi but was not always available. During the dry season there was sufficient only for the calves (Elphick 1977). The milk was not drunk fresh but was stored in cowhide sacks with the hair side inwards; men drank only cow's milk but women and children were allowed to drink the milk of ewes as well. A Nama child was assigned a cow for his or her own use and was expected to milk it her or himself. Such a practice is, or was until quite recently, found also among the Herero (Wellington 1967).

Only one Khoi population, that of Keetmanshoop, has been tested for the lactase polymorphism. In it, 9 of 18 adults but only 4 of 31 children were found to be able to metabolize lactose. The difference between children and adults was significant at the 5% level ($x^2_{[1]} = 4.18$), but only just so (Nurse and Jenkins 1974). This was hardly enough to justify postulating induction of the enzyme in adults, or to reveal any possible selection or selection relaxation.

### Disease among the Khoi

As far as we have been able to determine, only one review of disease patterns among the Khoi has ever been published. Though some of the material presented by Helman (1957) is necessarily anecdotal, and though he gives only approximate percentage rates when he gives any quantitative data at all, he undoubtedly possessed great experience of the Khoi, and his account is of considerable value. His observations were confined to the Khoi groups in and around Keetmanshoop in Namibia, and strongly coloured by the environment. For instance, he comments on the high prevalence of brucellosis, the rarity of pulmonary tuberculosis and intestinal helminthiasis and the complete absence of amoebiasis. It is almost certain that in all these instances the degree of exposure was a limiting factor. Malaria and schistosomiasis, similarly, do not occur in the area, and no assessment of Khoi susceptibility to them could be made. Syphilis and gonorrhoea were rife in the 1950s, and borreliosis or relapsing fever was common. Malnutrition is described by Helman as occurring in its severer forms; twenty years later, in our much more casual visits to the area, it was not apparent.

Helman reports a relatively high incidence of epithelioma of the glans penis, associated with intact prepuce, but does not mention carcinoma of the cervix uteri. Instead, he found the commonest malignant disease in women to be epithelioma of the oral mucous membrance, probably associated with the use of cheap metal mouth-pieces for their pipes. He also comments on the occasional occur-rence of *onyalai*, a peculiarly African manifestation of idiopathic allergic thrombocytopenic purpura characterized among other signs by haemorrhagic blisters in the mouth. He describes the frequent occurrence of an acquired type of epidermolysis bullosa of unknown aetiology. He records that during an outbreak of louse-borne typhus, presenting mostly among young adults and with a mortality of around 5 per cent, the vectors were found not only on the clothes and persons of the patients but also, most unusually, in the sackcloth walls of some of the huts or tents.

No consistent pattern of inherited disease has been described for the Khoi, and we are unaware of anything unusual among them except for the occasional cases of lipoid proteinosis, which may nevertheless reflect a heritage from one or more Caucasoid fore-bears (Gordon *et al.* 1969).

### Gene markers, selection and drift: smallpox and blood group B

Following the contention of Vogel *et al.* (1960) that blood group B individuals are better able to resist smallpox, Tobias (1966) has suggested that the smallpox epidemic of 1713 might have been responsible for the preferential survival of such people. The high B frequency in the Khoi of Great Namaqualand (present-day Namibia) would require that they, too, should have suffered to a similar extent from the ravages of the epidemic. The smallpox virus was probably introduced into the Cape on the clothes of the passengers who arrived in ships from India. It spread though the agency of slaves who washed the clothes of the passengers, and who were the first victims ashore. From them it spread to the Caucasoids, first at the port and later out into the countryside. Both groups could have passed it on to the Khoi.

The Khoi possessed less resistance to the disease than the Whites, and its ravages in them were much more marked; they died in their hundreds, and those who were strong enough fled to the interior, in this way spreading the infection to their kinsmen far afield. One report suggests that 'scarcely one in ten of their number had survived the catastrophe' (Elphick 1977). Theal records that 'Strangers who had visited the colony before 1713 and who saw it afterwards, noticed that the Hottentot population had about disappeared' (Theal 1909 (in 1907-10), p. 433). The San, on the other hand, were more fortunate, and, 'owing to the isolation of the people, they, the Bushmen, escaped the disaster which overtook the higher races'. Nevertheless, it would take very little contact to spread the disease, and the San were not all that isolated. If the *B* gene were in fact protective, one would expect that the San, who have it at relatively low frequencies, would have suffered a good deal more than either historical records or their present genetic profile suggests. In any event, the disease recurred several times during the ensuing century, and its effects on population numbers and differential survival must have been profound. The Khoi went through a demographic bottleneck, and it is quite possible that the resultant drift may be responsible for most, if not all, of their major differences from the San. But if the superior rate of survival of the latter was due to genetic factors, we need to look elsewhere than to the B antigen of the ABO system.

### Social change and gene flow

At the same time as the smallpox epidemics, though growing in magnitude during the eighteenth century while the epidemics diminished, another cause of the decline in numbers and social dissolution of the Khoi was operating. Long before it became politically advantageous for the Khoi to masquerade as 'Coloureds', they were already losing genes to the growing hybrid community, owing to the acquisition of Khoi wives by those Caucasoid men who had lost out in the competition for the relatively small numbers of Caucasoid women (see Chapter 8). Their loss these Caucasoids passed on to the Khoi men, their offspring to an intermediate society. The bottleneck was narrowed, and once again drift was brought to bear on the genetic profile of those who remained Khoi.

Such outward gene flow represents not only a net loss of genetic material; unless it is absolutely random it can lead to the especial leaching outwards of less common alleles. It consequently represents a transition towards monomorphism and, in theory at least, a loss of hybrid vigour. It may be argued whether in conditions where the selection is for adaptation to an extreme environment, too great a degree of heterozygosity may not represent the retention of some alleles which may be at least marginally disadvantageous. In fact, all, or almost all, of those populations which appear to have evolved as isolates under conditions of rigour, and which have been investigated for a variety of genetic polymorphisms, have tended towards the monomorphic. They have also shown reduced multifactorial variance, inasmuch as they incline towards uniformity in gross traits as well. These latter may admittedly be more exposed to adaptive selection than monogenic characters are; they may also represent phenocopies governed in different stocks by different arrays of genes.

This possibility is explored further in the chapter on hybrid communities. There is no indication from the Khoi populations thus far investigated for gene markers of any trend towards monomorphism. Rather is the contrary the case; we have abundant evidence to suggest that gene flow into the Khoi has been far more considerable than any from them. The loss of vigour which they may appear now to be undergoing is due to no drop in hybridity as yet. Its causes are far more social and economic than that, as will have been apparent from the outline of their history given at the beginning of this chapter.

# 5

# THE SAN YESTERDAY AND
# TODAY

The San, or Bushmen, are the best known and the most numerous of
the remaining hunter–gatherers of Africa. Like the Pygmies, with
whom, however, they appear to have little or no close biological
connection, they must have evolved for a considerable time in relative isolation and without much close contact with the Negroes. In
addition, it is virtually certain that they, like the Pygmies, do share a
common ancestry with the Negroes, though in the case of the San
and, for that matter, the much more closely related Khoi, it is by
no means sure that the remote ancestry of both races is sole and
identical. In any event, there has been ample opportunity during
recent times for gene flow to occur both into and from the San,
though perhaps not as copiously as has been the case with the Khoi.
Despite this, the physical characters of the San, to a greater extent
even than those of the Khoi, have remained distinctive enough for
a San, or even a proto-San, morphology to be postulated for a
number of fossil remains (see chapter 2).

Until the first reliable estimate was made by Tobias, it was widely
believed that the San were dwindling towards extinction. Tobias
(1955) estimated their numbers at about 53 000. Lee (1979) gives
estimates which would appear to call this into doubt, until it is
appreciated that his conclusions are based on very restricted data. In
the twenty-five years which have elapsed since 1955 there is likely
to have been both an increase in the numbers of people who are
biologically and socially San and a somewhat more gradual increase
in the numbers of those who nevertheless opt politically to belong
to other groups.

The reasons for the abandonment of their own identity by individual San are several. In the first place, those San who cease to hunt
and gather abdicate the protected status enjoyed by hunter–gatherers

in many countries where they still survive, and enter to some extent
into competition with other groups at a level where adherence to
their own group would be disadvantageous. For instance, the average
Caucasoid farmer in Southern Africa, with however much nostalgic
approval he may commend the traditional life and abilities of the
San, would nevertheless regard them as too flighty, too inconstant,
for employment, except perhaps temporarily or sporadically as
herdsmen. In Botswana no San may be engaged for service in the
South African gold mines, a major source of cash income. This
regulation, devised by the mining companies, is aimed at keeping the
'Bushmen' from being 'spoilt'. The San are ingenious enough to
evade it by protesting stoutly that they are Tswana. The cultural flux
through which the San are currently proceeding consequently makes
misleading any estimates, such as that of Lee (1976), of the proportion still supporting themselves by hunting and gathering. This
naturally varies from group to group and country to country. Lee's
claim that 5 per cent of the San still engage in hunting and gathering
would be wildly excessive for the San populations of South Africa,
while in the face of the situation in Angola, or indeed along the
Kavango valley in northern Namibia, it is ludicrously small. Indeed,
without rather more precise knowledge than any of us possess of the
precise whereabouts and identity of all San populations, attempts
to make such estimates have little point. It is not very long since
a 'new Bushmen tribe', discovered in the Central Kalahari and
having a puzzling dearth of young men, was later found to be the
source of the majority of the semi-skilled labour employed by a
particular Witwatersrand gold-mine (A. Traill, personal communication).

### The modern distribution of the San

Even with these eccentric divergences from genetic reality, however, it is possible to identify a fairly substantial total of avowed
San. Probably somewhat more than half live in Botswana, across
which they are widely distributed, being found in all areas of the
country except, somewhat curiously, the strip of land lying to the
east of the railway line. As one would expect of people who live
mainly in the desert, they are rather thinly and irregularly scattered,
and by now should amount to appreciably more than the 24 652
enumerated in the 1964 census. That figure was estimated by Silberbauer (1965) as probably about 20 per cent lower than the reality;
he pointed out that not only was census-type ascertainment of a
shifting population difficult, but that a number of Negroes readily

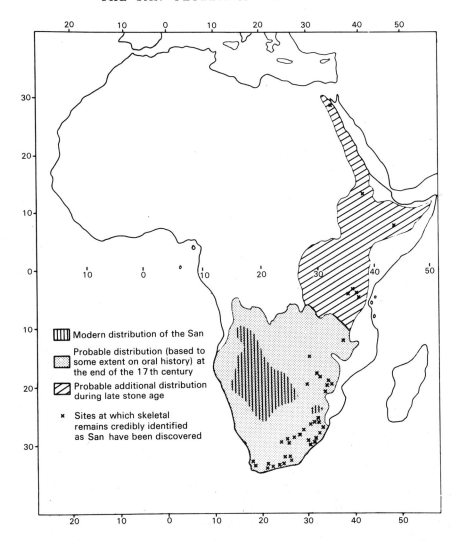

Fig. 5.1. Modern distribution of the San, and suggested past distribution. This map is greatly modified from that presented by Tobias (1964), for the following reasons: (a) neither history nor archaeology has furnished any unmistakable evidence of occupation of the eastern coastal plain, or of the Congo Basin, by San; (b) no San remains have been identified for certain west of the Nile below Khartoum or of the White Nile above it; (c) San were still living on the Bie Plateau, shown here as their furthest northward modern extension, in living memory, and probably still are.

accepted San as fellow-Negroes on cultural grounds, so that once San start to practise agriculture they become even more difficult to identify as San. The more rigid ethnic criteria politically supported in Namibia make them rather easier to identify there, but even then it is unlikely that the total of 21 909 given in the 1970 census includes all the San in the country. Even today it is not impossible, or even particularly difficult, for small unobtrusive hunter–gatherer bands in Namibia to evade contact with the authorities. Very little is known, for instance, about those small groups which are said to inhabit or to have inhabited several parts of the Namib desert. Very little more has been recorded about them than their names (Schapera 1930), and if they still exist they are singularly elusive.

The San populations known to be present in Namibia are the relatively large numbers of widely scattered !Kung or Zhu/oasi of the north-east of the country, the Hei//om of the Etosha area and G!oakx'ate of the Aminuis Reserve in the southern part of the country, as well as a number of Nharo along the southern half of the border with Botswana. The Strandlopers or Beachroamers were for a long time thought to be extinct, known only for their coastal shell-fish middens, but have been re-identified (Jenkins 1972) in a party which has moved inland and settled at Sesfontein in the Kaokoveld. Some of the G!oakx'ate have also settled, and become retainers on the farms either of Herero, inside the Reserve, or of Caucasoids around it. The rest of the San of Namibia, in spite of some fairly determined official attempts to make them settle, still subsist largely on what they can hunt or gather. This tends sometimes to be seasonal and, even when it is not, to be augmented quite often with food-stuffs purchased with money from the sale of artefacts or labour, especially in those areas which are nearest to road systems.

The !Kung are the principal speakers of what used to, and may yet again, be called the 'Northern Bush' languages. They not only make up the largest group in Namibia; they also comprise all, or very nearly all, the true San in Angola. Estermann (1956) estimated their numbers there as probably more than 5000, while Guerreiro (1968) thought that 6000 would constitute a gross underestimate and that the total might be close to 10 000. The Portuguese, however, do classify the Kwengo, who are Negro hunter–gatherers (Nurse and Jenkins 1977b), as San. The Bantu-speaking neighbours of the Angola !Kung divide them into moieties, the Kwankhala and the Sekele, but the division seems to have little real significance (de Almeida 1964). Both groups maintain, or until recently maintained, contacts with !Kung-speakers in Namibia and Botswana, whom they appear closely to resemble in manner of life.

The !Kung of Botswana similarly had close connections with the !Kung of Namibia, until a high fence was put up along the inter-national border. The fence does not stretch so far to the south, or

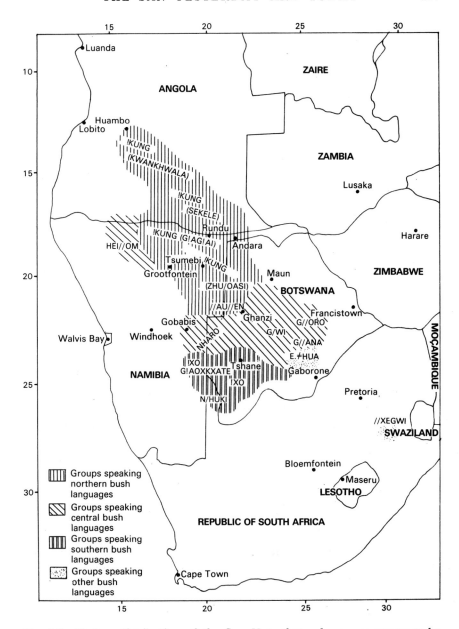

Fig. 5.2. Modern distribution of the San. Note that only groups proven to be biologically San are included here; such groups as the Kwengo and the 'Masarwa' of north-eastern Botswana, including the Denasena or 'Danisan', though they, like the Dama, speak Central Bush languages, and are, like the Dama, biologically Negro.

interfere so much with travel along the line of the road between Gobabis and Ghanzi, as to prevent contact across the border by the //Au//en, the 'Northern Bush' speakers lying to the south of the !Kung. These in their turn border on the Nharo, who are speakers of a 'Central Bush' or Tshu-Khwe language, one of the Hottentot (Khoi) language family; the Hei//om, whose rivalry with the !Kung keeps them well on the Namibian side of the border, speak a language of the same family. Tshu-Khwe languages are spoken by most of the hunter–gatherers of the central and north-eastern Kalahari; of these the G/wi and G//ana and associated groups are San, while the more northerly peoples lumped together by the Bantu-speakers as 'Masarwa' are Khoisan-speaking Negroes (Chasko *et al.* 1979; Nurse and Jenkins 1977*a*). The 'Southern Bush' languages are spoken by the G!aokx'ate in Namibia and by the more numerous !Xŏ in Botswana, as well as by the remnant San population in the Gemsbok National Park on the borders of Botswana, Namibia and South Africa, and probably by the barely extant San groups hiding themselves among the 'Coloureds' of the northern Cape. There are probably a few other Botswana San who also speak languages of this family though one of the most dramatic turns in Khoisan linguistics of recent years has been the conclusion that the eastern ≠Huā, also denizens of the south-central Kalahari, speak a language which cannot be fitted into any of the previously accepted families (Traill 1973). They, like the tiny remnant of //Xegwi living around Lake Chrissie in the eastern Transvaal (Potgieter 1955) add hardly at all to the total number of extant San. The //Xegwi population has now sunk below a viable size, and is drifting towards, or may even have reached, extinction. The overall total, however, would almost have doubled itself since Tobias's 1955 estimate, if it has enjoyed even the rather modest rate of increase, for the Third World, of a little under 3 per cent per annum. It is possible, however, that San population growth has been occurring at a rather slower rate than that.

During the relatively short period of time during which Nancy Howell (1979) observed the !Kung of Dobe the Crude Birth Rate (CBR) (live births per thousand persons living) varied between eighteen in 1966 and seventy in 1967, the latter being higher than any national rates; the average CBR over the six-year period 1963 to 1969 is forty-one, which is also the rate for Southern Africa as a whole. With an average rate of immigration into the area of twenty-two per thousand the addition to the population per year averages sixty-three per thousand. The Crude Death Rates (CDR) for the same period ranged from six to twenty-six per thousand with an overall rate of sixteen. Out-migration averaged eight per thousand per year.

The growth rate based on these birth and death rates is two and a half per cent per year which would cause the population to double in twenty-eight years. If migration is included, the growth rate is 3.9 per cent per year giving a doubling of the population every eighteen years. Tanaka (1980) in his study on the G/wi and G//ana of the ≠Kade area arrived at a growth rate of between 1 and 2 per cent per year for these San.

Howell (1979), by the judicious application of other demographic techniques more appropriate to the study of a small population like the !Kung of Dobe, concludes that a realistic growth rate is one half of one per cent per year. Although the current fertility rate among younger women is slightly lower than that among the older women (probably due to gonorrhoea) there has been, during the period of observation, a lowering of the mortality rate due perhaps to the effects of smallpox vaccination and to the increasing control of epidemic diseases in the area.

It would appear then that these two San populations are growing but not at a dangerous rate. In the absence of other data we may take them as paradigmatic of other San populations and confirmatory of earlier conclusions (Tobias 1955b; Guerreiro 1968; Silberbauer 1965; Nurse 1977b; Nurse and Jenkins 1977a) that the San are in no immediate danger of biological extinction. If the San disappear it is most likely to be because of assimilation and intermarriage with other groups.

## The languages of the San

As may be seen from the foregoing, the San are divided into what are ostensibly a number of separate peoples; and the principal distinction among these is in language. The San or Bush languages were originally classified by Bleek (1927) into three main groups, the Northern, Central and Southern. She commented on the similarity of Nharo or Naron, one of the Central group, to Hottentot, but did not extend the argument to include Hottentot in the same group, and it was not until the reclassification published by Westphal (1963) that the position was further clarified. He divided the many languages of the San into major moieties, the Tshu-Khwe group, or those related to Hottentot, and the Bush group, those whose relationship to Hottentot was not apparent. He went on to demonstrate a further disparity in the Bush group, which he showed to consist of four linguistic traditions, Bush A, B, C and D. Bush A comprehends what were formerly known as the Northern Bush languages, while B and C represent subdivisions of what used to be the Southern group, and

Bush D is the language of the small, possibly by now extinct, population known as the //Xegwi of Lake Chrissie in the Transvaal, and of the /Xam of the Cape Province, of whom none remain. Traill (1973) has challenged Westphal's view that these four linguistic traditions are discrete and only very remotely related to each other. He favours what amounts to a reinstatement of the classification of Bleek, with a tripartite division of the Khoisan languages.

Traill bases his argument to some extent on the singularity of the phonology of the Khoisan languages. The click consonants are unique to Africa, and apart from in Southern Africa and adjoining areas occur only in Hadza and Sandawe, two languages of Tanzania. The extent to which these latter two are related to Khoisan is unclear, but several Bantu languages of Southern Africa have incorporated click phonemes, and even if Hadza and Sandawe do turn out not to share their derivation either with Khoisan as a whole or with one or other of the Khoisan divisions, their sharing of these curious speech sounds must surely indicate significant contact at some time in the past. Indeed, if the four Khoisan linguistic traditions are as distinct from one another as Westphal (1963) claims, their common consonants do at least indicate some familiarity among the ancestral stocks from which their present speakers are descended.

It is less surprising that so many Khoisan languages have become extinct during historical times, than that so many have survived. On the whole they are limited languages whose semantic spread is rather specialized. Nama, which has already accommodated the transformation of its speakers into pastoralists, and which is among the richest of these languages, may point the way either to linguistic enlargement of other Tshu-Khwe tongues, or provide all the Tshu-Khwe speakers eventually with a *lingua franca*. The San adoption of settlement seems to result not so much in the modification of their languages as in their becoming bilingual: one can hardly expect Bush languages to include an agricultural vocabulary. This may go on either to rapid expansion of the lexis of some of the more widely spoken of these languages, perhaps with some simplification of syntax and even eventually creolization, or to their being abandoned by their speakers in favour of languages more adapted to the new way of life. In either case it is likely that some of the languages with the fewest speakers will quite shortly be lost altogether.

### The social and political economy of the San

The basic political and social unit among the San is not the 'tribe', which is essentially a linguistic entity, but the band. Every band is completely autonomous, and governed by a spontaneously emerging hierarchy based on age, experience and abilities. Major communal decisions are ostensibly made by the men, but attention, even if not always obviously, is almost invariably paid to the opinions of the women, who generally do not hesitate to make them known. In a few bands there are, or used to be, rather shadowy hereditary chieftains, whose functions seem to have been mainly military. On the whole, however, adroit advantage is taken of the varying talents and abilities of the members, and appropriate exploitation of these secures an effective and economical use of resources and helps to promote communal resistance to natural stresses.

Wilmsen (1973) has discussed the organization of hunter–gatherer bands as a special case of more general interaction and spacing processes. For the hunter–gatherers of Southern Africa, resources tend to be both concentrated in space and diffused with time, so that the most efficient form of group exploitation comes about by central localization of the population. Ideally, a migrant band should follow a cyclical, roughly circular, route, around a perimeter the radius of which is equal to the radius of possible exploitation. As Nurse and Jenkins (1977a) point out, however, this can only be the case where the resources are available at a constant rate of appearance and exhaustion, and amount at all periods to the same total utility. In Southern Africa such regularity does not occur, and the migration cycles of the San and the other hunter–gatherers are consequently irregular in scope and duration. Each band is a land-owning group, and is restricted to its own hunting territory; but especially in the context of gathering this territoriality is subject to quite extensive modification, and not only may hunting parties spill across boundaries in pursuit of particular beasts without provoking resentment, but where resources are concentrated in excess of the requirements of any one band territories may actually overlap. Where resources are sparser, however, the boundaries are wider and more jealously preserved.

Wilmsen (1973) states that within the band 'the most stable identifiable units are (1) the band itself, (2) the household–residence unit composed of one or more nuclear families plus extensions and (3) the task group composed of active adults drawn from these households'. The balance between the resources available and the requirements of procurement roles limits the size of the band. This balance is a generally favourable one. Not only are the families kept small by birth regulation (discussed in a later chapter) in the interests of mobility, but the vegetable resources provide a reliable, all the year round, source of food (Tanaka 1976). Often, gatherers are dispersed so that the greatest possible return can be derived from stable resources, while concentrating the hunters naturally enables them to work as a team and deal more effectively with mobile game. There is never any uniform abundance of any one resource, and consequently procurement roles sometimes have to be shuffled or recombined in response to changing circumstances.

Marshall (1961) has described how reciprocity conventions operating among !Kung kinships serve to maintain the stability of the band. Sharing of consumer goods is one of these. Stability of the band is ensured on a day to day basis by the 'rota' of food collecting among the women of the band (Lee and DeVore 1968b). These reciprocal services tighten the energy load and must contribute strongly to emphasizing the value of group solidarity. Many of the social judgements made within the group, and group decisions about outsiders, are based on the extent to which meat and tobacco are made freely available and on the readiness with which gifts are handed out with the deliberate purpose of averting jealousy and promoting goodwill. Restraint in consumption, and the careful avoidance of grounds for offence, are aspects of good manners. Criticism is often extremely frank, but for it to reach its most outspoken it is necessary that particular degrees of kinship, real or conventional, should exist between the critic and the person criticized. Resentment or loss of temper, indicating abandonment of these licensed release mechansims, constitutes extremely unsocial behaviour, and invites ostracism. Life is largely conducted in public, but the privacy of both subgroups and individuals is respected; derogatory speech except in the public interest is considered disruptive, and is permissible only among relatives and close friends. The stability of the group depends on observance of the precise permissible degree of familiarity between members and on the recognition and observance of the duties implicit in each particular band (Marshall 1959, 1961, 1962).

It is uncommon to find more than three generations alive in any one band at any one time, and recognition of kinship is circumscribed by generation length and band size. The existence of a common ancestor principally signalizes the acknowledgement of cognation, and, as Silberbauer (1965) points out, this means that during the lifespan of an individual the bounds of the kin group extend at first through the two ascending generations and gradually shift until it is the second descending generation which constitutes their outer limit. Both patrilateral and matrilateral kinship are recognized by most of the San, who usually observe the widespread principle of the avoidance of proximate and linking of alternate generations. This principle appears to develop wherever the members of a society necessarily come into close and frequent contact. There is quite rigid demarcation between cognatic and affinal kin, with a classification network defining the members of the kin group in relation to each individual. Birth order determines hierarchy among siblings; the relative seniority of the parents not shared determines it among half-siblings. Since these kinship systems are necessarily classificatory, the role invariably transcends the individual, so that the form of relationship recognized between individuals often bears little resemblance to their actual biological kinship.

A man may have one or more wives; where he has more than one, they do not, as among some settled groups, form separate households. Should any domestic disharmony, such as squabbles between co-wives, temperamental incompatibility between spouses, or an expressed preference for a different partner, arise to divide a household, this is treated as a threat to the concord of the band, and divorce is encouraged. Since women have the same rights as men, they are equally entitled to dissolve marriages. As a result housholds consisting of a married couple but with children belonging to only one parent are fairly common. Divorced and widowed persons re-marry fairly frequently and reconstitute households. Marriage is not a private matter, but the business of the whole band, and the whole community has to consent to a marriage before it can be permitted. Band endogamy is preferred to exogamy, but in practice the small size of bands often means that individuals reach marriageable age when no suitable partner is available, and one has to be found in another band. In such a case, the rule is that the husband should join the wife's band for a probationary period, during which the marriage can be dissolved simply by his departure. Once the first child is born the couple can, and often does, join the tribe of the husband (Marshall 1959).

Men usually hunt in pairs, and the hunting bond is a very strong relationship which is, oddly enough, of some biological significance. Very young men, immature and unpractised hunters, change partners frequently, seeking for someone whose behaviour in the chase adequately supports or complements theirs; but with maturity the bonds gain in strength and permanence, until each pair reaches stability. The classificatory level of the relationship in kinship terms is unclear, but the offspring of hunting partners are regarded as siblings. Marriage to a child of a father's hunting partner is considered incestuous, but the sense of kinship does not persist into subsequent generations. One result of this type of bonding is that when a man changes bands, he takes with him not only his own family but his hunting partner and his family as well.

The economic regulation of migration varies considerably among linguistic groups. It naturally depends to a considerable degree on the resource potential of the region inhabited by the group, the availability of food plants, the presence of water and the where-abouts of the main concentrations of game. Some social differences are apparent in variations in value emphases among different groups; the !Kung, for instance, migrate between waterholes (Lee 1965), while the G/wi (Silberbauer 1965) and G//ana (Tanaka 1969, 1976) depend a good deal more for water on the *tsamma* melon and other plants, and perhaps on the rumen of animals. It is these resource limitations, more than anything else, which restrict the size of the band, generally to about 50 individuals, and place constraints on the social organization. It is impossible to exchange membership in one band permanently for membership in another, but an individual or a family can belong to more than one band and follow an idio-syncratic migration pattern, changing bands at points of migration intersection. Membership of a San band confers rights but does not impose obligations (Marshall 1960), though failure to conform to what is regarded as decent behaviour can lead to ostracism and voluntary departure from the band of the individual or group ostra-cized. In order to join a new band it is necessary to establish heredi-tary membership rights, or to contract a marriage with one of its members, or simply to be accepted publicly by the members as a whole.

There must naturally be mechanisms to keep the band to its most effective size. When it shrinks too much below it, its members tend either jointly to fuse with another contracting band or individually to seek membership in other bands where they have claims. When it becomes too large, there is similarly either gradual erosion of its membership or fission of a segment. No decision is made, no policy followed in too much haste. There are preliminary explorations of the new territory, negotiations with other bands or smaller groups for the formation of new groups, until eventually the change takes place with as little trauma as possible, pressure is relieved, and the original band is back to a more wieldy size. Silberbauer (1965) has estimated that such fission generally occurs once in every two or three generations, though this does seem to argue a rate of popula-tion growth somewhat in excess of what is generally accepted for the San; it would mean that even before the current 'population explosion' the net annual increase in the population was of the order of 2.5 per cent.

## Physical characters: morphology

The San share with the Khoi the distinction of diverging more from the human mean in shape than any other race of man. This does not mean that they are in any way 'primitive' or 'subhuman' anatomically. The case is, in fact, rather the reverse. If what is most essentially human is plasticity, totipotentiality, the paedomorphic anatomy of the San, even less highly specialized than that of the Khoi, would represent human development at its most advanced. Of course, this is not so; the child-like San is not child-like in all features, but incorporates some which are extremely unusual and unlike those of any other races.

Since the San are almost certainly of relatively recent common origin with the Khoi, they naturally share a number of traits which have already been described as characterizing the Khoi. They are smaller in stature, but, as Tobias (1962b, 1972b) points out, they are possessed of similar postural peculiarities, most notably a lumbar lordosis, and have similar isolated deposits of fat over the thighs (steatomeria) and buttocks (steatopygia) (Krut and Singer 1963). These are less common in them than they are in the Khoi. In skin reflectance studies, Weiner et al. (1964) showed the San to be the lightest in colour of all the unmixed indigenous peoples of Southern Africa; their skin colour is usually a light yellowish-brown, though it often tans to a deeper shade and still more often is obscured by encrusted dirt. Occasional individuals among the Khoi may be lighter, but the colour of the San shows rather less variation than that of the Khoi. The tightly spiralled tufting of the hair over the vault of the head, with a consequently patchy exposure of the scalp, is highly characteristic; the head hair is looser and ostensibly thicker on the temple and occiput. Few men have chest hair, but where it occurs it, too, comes in 'peppercorns'; their pubic hair generally takes the same form, though that of the women is occasionally looser. In the rare men who can grow them, the beards are thin and straggling and never very long. It is unusual to see any greying of the hair, even in those individuals known to be of fairly advanced age.

In youth the skin is smooth-textured and elastic, but becomes wrinkled and loose quite early in life, probably due to the extreme localization of the subcutaneous fat. That this is so concentrated over the buttocks means that there is relatively little elsewhere in the body. The consequent deep and early wrinkling is particularly noticeable on the face and abdomen. It is often delayed among the settled San, especially those on the farms, whose high-calorie diet over a long period can lead to their becoming noticeably chubby.

The San have two genital peculiarities which they share with the Khoi and which are sometimes found in Southern African Negroes, most probably on account of admixture. A majority of the San men manifest ithyphally, or horizontal projection of the non-erect penis (Drury and Drennan 1926), while many, but not a majority of the San women display macronymphia, the tablier or 'Hottentot apron' described for the Khoi in Chapter 4 (de Villiers 1961). On the whole, the proportions of individuals showing these traits differ between San and Khoi; more San than Khoi men have ithyphally, while more Khoi than San women have macronymphia. As in the Khoi, the areolae of the breasts of pubescent girls are prominent and swollen, and though this appearance diminishes with child-bearing it can persist up until parturient sagging leads to limpness and flatness of the breast.

The skulls of the San tend to be short, pentagonal and relatively broad, with pronounced flattening of the vault and a tendency to bossing in the frontal region. The glabella is small, and more definite on the dry skull than in the living, while supraciliary ridges are either absent or only slightly present. The San are a small-faced people with pronounced orthognathism, and their faces convey an overall impression of flatness. Their noses are broad and their lips thin. Their dental differences from the Khoi are negligible (Drennan 1929): among their other paedomorphic features are a low frequency of Carabelli tubercles on the upper first molars, and of shovel-shaped incisors, a reduced prevalence of overbite and more common underbite, a relatively high incidence of spaced teeth and a rather low incidence of well-accommodated teeth in both the upper and lower jaws (Tobias 1972a). The cheek-bones are prominent and the chin fairly pointed and rounded. The eyes tend to be slanting, with marked medial epicanthic folds; they almost all have dark brown irises. The ears of the San are unusually small and commonly lobeless, with very frequent overrolling of the helix. The body is slender, with limbs which are thin and not heavily muscled. There is a tendency for the upper extremity to be rather short in comparison to the lower, while compared to the body both extremities are short. The hands and feet are small and well adapted for fleetness and dexterity.

The San have been quite extensively measured and examined and the degree of morphological uniformity among the various groups is obvious from the studies of the Nharo and //Au//en by Tobias (1955b, 1956), of a southern group by Dart (1937) and of northern groups by Lebzelter (1931) and Wells (1952*b*) and of the //Xegwi by Toerien (1958). Their dermatoglyphics have been studied by Cummins (1955) and Tobias (1961). Though great claims have been made for this inherited trait, and though it does appear to distinguish the various San linguistic groups from each other and from the Khoi, its genetic basis and adaptive significance are not clear enough for any definite conclusions to be drawn.

## Physical characters: genetic markers

The monogenic characters of the San demonstrate at once their distinctiveness, their remarkable similarity to the culturally different Khoi, and the extent to which either a common ancestry or extensive reciprocal gene flow with the Negroes can be deduced. Since the Negroes are so widely distributed in Africa, and the surviving San so relatively few in number and restricted in location, it is a simple matter to sort out several traits which depend on the extent of known historical contact between the people. There are a few genes electrophoretically or serologically definable which appear to be restricted to, or to have originated among and spread out from, the Khoisan peoples, and others which appear similarly to characterize the Negroes. There are also traits which are widespread in all divisions of both peoples and which have been among the most powerful arguments (Singer and Weiner 1963) for their common origin from a single stock.

Into the former category falls several traits which were first identified in Southern African Negroes and only later found to be very much commoner in Khoisan and absent or very rare in Negroes elsewhere in Africa. The most obvious example of this is the ABO allele miscalled $A^{bantu}$, which homozygously, or heterozygously with an $O$ gene, produces a very weak variant of the A blood group phenotype. It used to be thought (Jenkins and Corfield 1972) that the $P^r$ allele in the acid phosphatase system was equally typical, but the phenotype it produces has recently been identified in specimens from Nigeria (Ojikutu *et al.* 1977) and in the Babinga Pygmies (Santachiara-Benerecetti *et al.* 1977). Spedini *et al.* (1980) suggest that it is a selective marker negatively correlated with humidity rather than an indication of Khoisan admixture, but Jenkins and Dunn (1981) point out that, if all the known distribution of the allele is taken into consideration, their argument falls away. Sensabaugh and Golden (1978) have shown that different phenotypes of acid phosphatase are inhibited to different degrees by folic acid and various folates, but these properties have not yet been defined for the R phenotypes.

The case for a Khoisan origin of the $PGM_2^2$ allele at the second phosphoglucomutase locus remains fairly good (but see chapter 6). Though it has been detected in Negroes from parts of Africa remote from those inhabited today by the Khoisan, it reaches much higher frequencies in the Bantu-speaking, though oddly enough not the Khoisan-speaking, Negroes of Southern Africa. It may, however, provide a better argument for a common than for a disparate origin of the peoples. The Kell blood group system, using the common antisera against K and k, is unrevealing in both Negroes and San, since when K is found it almost invariably indicates Caucasoid admixture. The Sutter system, part of the Kell complex, may one day prove more informative, since occasional Js (a+) individuals have been found in some of the San, but none of the Negro, populations tested so far.

The Gm system has proved to have great power to discriminate among populations; the allied Inv or Km system appears to have little or none, and contributes nothing distinctive to the genetic profile of the San. As between the Khoisan and the Negroes Gm provides answers which can be either equivocal or emphatic. The commonest haplotype in the San is $Gm^{1,13,17}$, and the $Gm^{1,21}$ haplotype is also present at fairly high frequencies. Unlike the Khoi, the San do not possess $Gm^{1,2,21}$, but like the Khoi they do have low frequencies of the typically Negro alleles which produce Gm(6). $Gm^{1,5,6,14}$ is the commoner of these, and is present in all the San populations so far investigated, except the Tsumkwe !Kung. It seems that the other haplotype, $Gm^{1,5,6}$, may be a more reliable indicator of Negro admixture; it is not found at all in the northern !Kung or in two of the Botswana !Kung groups, and reaches only low levels in the other San populations, whereas its frequency in the Negroes tends to be high (Steinberg *et al.* 1975). Reciprocally, $Gm^{1,13,17}$, and rather more reliably $Gm^{1,21}$, can be used to measure San gene flow into the Negroes (cf. Jenkins 1972). Two haplotypes rarely found except in the San are $Gm^{1,5}$ and $Gm^{1,5,13,14,21}$; the former is found in the Nharo, //Au//en and northern and Ngamiland !Kung at rather low frequencies, while the latter occurs in three !Kung populations, reaching quite high frequency in the Dobe !Kung, as well as in the //Au//en. In contrast to these distinctions, the haplotype by far the commonest in Negroes, $Gm^{1,5,13,14}$, is also to be found at high frequencies, usually about half those in Negroes, in the San. This might be a more potent indicator of a common origin were it not that this same haplotype is also commonest in Melanesians, who are in most other systems quite unlike Negroes. This suggests that its frequencies represent rather an example of parallel selection, and that it is not to be taken too seriously as an indicator of genetic affinities.

The first and most striking monogenic similarity to be remarked between the Khoisan and the Negroes was the high frequency of $R^0$ in the Rhesus system. This is by far the most frequent Rhesus allele in both peoples, though it is not by any means unique to them, since it is found in Melanesians at low frequencies and at roughly similar frequencies in Caucasoids, being a little elevated above the Caucasoid average in the Ashkenazi Jews. The main interest in this system here, though, is the relatively high $r$ frequency in Negroes and its virtual absence from the San. This could have some bearing on the alternative hypotheses for Negro origins which will be discussed at greater length in the next chapter. It is possible nevertheless that the common possession in this system of the V antigen by both Negroes and Khoisan, and its complete absence from Caucasoids, may be more significant. Much the same might be claimed for the Henshaw antigen in the MNSs system, and for the transferrin allele $Tf^{D1}$. The latter is, once again, found not only in Melanesia, but throughout Oceania, and its presence in Southern Africa is difficult to account for except by postulating selection for it. On a lesser scale, this can be invoked to account for the Henshaw and V situations as well.

A marker which does not appear to be subject to obvious selection is the Kidd blood group allele $Jk^a$, of which both Khoisan and Negroes have high frequencies, while $Jk^b$ is more common in Caucasoids. The frequency of this allele tends nevertheless to be higher in the San, and in those Negro populations which have received the largest genetic contribution from the San, than in either the Khoi or the Negroes as a whole. Adenosine deaminase, a red cell enzyme whose close association with the operation of the immune system means that it is almost certainly subject to selection, has thrown up in both San and Negroes in Southern Africa variants having diminished activity (Jenkins et al. 1976). What is of especial interest in this connection is that the variants do not appear to be identical, which tends to suggest that reduced adenosine deaminase activity could carry some selective advantage in the environment of Namibia, where both variants were found. Though the Negro variant does not seem to reach polymorphic frequencies, the San one does among the !Kung of Tsumkwe (Jenkins et al. 1979). Adenylate kinase may be polymorphic for roughly similar reasons. The only Negro people in which the frequencies of $AK^2$ are anything like as high as in the San are the Dama, who are long-established hunter–gatherers (Nurse et al. 1976). It is probable that where the allele occurs in other Negro populations, as it does at very low frequencies, it represents gene flow from the San. The converse is probably the case with $PGM_1^2$ at the phosphoglucomutase first locus and the $PGD^c$ variant in the 6-phosphogluconate dehydrogenase system. Both of these are fairly common in Negroes and are sometimes not present in San populations, and are hence likely when they are found to represent Negro contributions to the San gene pool.

The most striking Negro monogenic characteristics seem invariably to be introductions into the San rather than obvious evidences of common ancestry. Among these are $PepA^2$ and $PepD^2$ at the peptidase A and D loci. These are found relatively rarely in the San, who nevertheless have in the latter system a characteristic allele of their own, $PepD^3$. That Haemoglobin S does not occur in the San is hardly surprising, since the Negro peopling of Southern Africa seems to have preceded the spread of the mutation in response to malaria. Very little sickling is found in Negroes living to the south of the Kunene, Kavango or Limpopo, and that almost solely along the northern fringes of the region. There is evidence that thalassaemia determinants are present among Southern African Negroes and it has recently been established that the !Kung of Tsumkwe have a significant frequency of the rightward deletion type of the $\alpha^+$-thalassaemia haplotype, known as $-\alpha^{3.7}$ (Ramsay and Jenkins, 1984). The San have an individual haemoglobin variant of their own, Haemoglobin D Bushman (Wade et al. 1967), which is present at polymorphic frequencies in the //Au//en and at very much lower frequencies in the Nharo and the !Kung of Botswana. It does not appear to possess any selective significance. The fact that the $Gd^A$ allele in the glucose-6-phosphate dehydrogenase system, which like Hb S and the thalassaemias is concerned in the hereditary defence system against malaria, is commoner in the San than the African deficient variant $Gd^{A-}$, seems to indicate that the former represents an earlier mutation which may have taken place in a common ancestral stock or may have entered the San through gene flow. Neither of these contributes very much to the question of common ancestry, but the third of the important variants shown to protect against malaria does.

It has been known for a very long time that Negroes were peculiarly resistant to *P. vivax* malaria, but it was only with the demonstration by Miller et al. (1976) that there were good grounds for supposing that the Duffy blood group antigens acted as receptors for *P. vivax* merozoites that the reason for this became clearer. One of the sharpest distinctions between Negroes and the rest of humanity is the relative rarity of Duffy blood group antigens among them and their commonness or universality in all other peoples. The San possess both $Fy^a$ and $Fy^b$, the frequency of the latter tending to be slightly higher than that of the former, whereas in the Southern African Negro populations the frequencies of both are so low that the contrasts in their frequencies can hardly be significant. It is probable that all the Duffy positivity found in the Southern African Negroes represents gene flow from the San.

Probably the most ecosensitive of all gene marker systems will turn out to be those whose loci cluster together on chromosome number 6 to form the major histocompatibility complex. Not much has been published as yet about these systems in the San. Botha *et al.* (1973) and Nurse *et al.* (1975) described the position of the antigens determined at the A and B loci in two populations of !Kung; there was no discussion of the C locus, and of course the D locus or loci were at that time barely recognized. Like the Negroes, the San appear to have a high frequency of HLA-Bw17, though since then this has been shown to comprehend several antigens. Other common traits in these systems are relatively elevated frequencies of HLA-Aw30, and low ones of HLA-A1 and HLA-A11. It is possible, though not certain as yet, that the San may differ from the Khoi in lacking HLA-Bw16 and HLA-B27. Interestingly, if this is the case, it would tend to make the Hei//om incline more to the Khoi than the San, a situation not borne out by their haptoglobin frequencies. There is one 'bridging' antigen between HLA-A10 and HLA-Aw29 which appears to be uniquely Khoisan (Nurse *et al.* 1975). It has been given the designation HLA-Aw43.

In the haptoglobin system in general the San tend, like the Caucasoids, to have higher frequencies of $Hp^2$ than $Hp^1$, while among the Khoi and Negroes the position is reversed. The main exception to this is the G!aokx'ate, of whom so small a sample was drawn that the confidence limits of the calculated frequencies are very large. This system, like Gm, is of most interest in illustrating the extent to which the monogenic markers are beginning to parallel the morphology in the divergence between the Khoi and the San.

## Patterns of disease among the San

Most of what is known about patterns of disease among the hunting and gathering San has been summarized by Nurse and Jenkins (1977a). The settled San are exposed to much the same environmental hazards as the settled Negroes, whether rural or urban, and may be expected to react to them in much the same way (see Chapter 7). When settled they tend to assume the identity of Negroes or 'Coloureds', like the

Khoi, and it is consequently unlikely that if any contrasting long-term trends do ensue it will be possible to identify them.

The unusual metabolic features of the San which have been recorded could all be environmentally induced and most could stand at one or other extreme of the Negro range. The San are relatively intolerant of glucose, having low fasting blood sugars but a prolonged rise following glucose ingestion (Joffe *et al.* 1971; Jenkins *et al.* 1974), and it is feared that clinical diabetes mellitus might result from increased consumption of sugar. (This is known to have happened in at least one case.) Similarly, their low serum lipids (Truswell and Hansen 1968*a*; Truswell and Mann 1972) are almost certainly diet-dependent; evidence is accumulating that San employed on farms and eating a farm-labourer's diet do have higher cholesterol levels. It must not be forgotten that the fat of the wild animals of the desert contains higher proportions of polyunsaturated fatty acids than that of domesticated meat animals. The systolic and diastolic blood pressures were found by Kaminer and Lutz (1960) to be lower than in Caucasoids, Negroes, Australian Aborigines and Eskimos. Once more, the values rose in the semi-settled San. Truswell and Hansen (1975) are of the opinion that low salt intake and absence of obesity, together with relative freedom from mental stress, might be responsible.

The small size of the San may be due less to any inherited factors then to an overall energy deficit during the growing period (Truswell and Hansen 1975). Despite this, overt signs of malnutrition are rare among them except in the settled state or in advanced age. The evidence provided by Tobias (1962*b*) for a secular increase in San stature is undoubtedly related to the changed diet consumed by the San working on farms, and this in its turn will lead to higher food needs for the larger body. The hunting and gathering San certainly appear to be in dietary equilibrium with their environment. Though spare in build, they are generally remarkably healthy.

The precise range of infectious diseases to which the San are exposed is not known. Pulmonary tuberculosis appears to be spreading from the settled San into the nomadic groups. Extravenereal syphilis is found even in the remoter G/wi and G//ana of the Central Kalahari (Nurse *et al.* 1973), and hepatitis B has certainly been present in them for some time. Influenza spreads regularly from the settlements, and though the effects are rapid and severe it appears not often to be fatal. Kokernot *et al.* (1965) found evidence of the presence of quite a wide range of arboviruses in the San. Malaria sweeps the Kalahari in epidemics dependent on the rains. Though no San case of trypanosomiasis has ever been reported, the disease has occurred in visitors to areas of San habitation, and along the north of the Kalahari is a wide strip of bush infested with the tsetse-fly vector. One of the more striking infectious diseases is a form of favus, shown by Murray *et al.* (1957) to be due primarily to infection with a fungus. Purulent conjunctivitis is frequent, and the occurrence of trachoma in the San was confirmed by Brontë-Stewart *et al.* (1960), who also recorded a probable case of urinary schistosomiasis in the G!ang!ai, who live near the Kavango River. The only other helminth recorded is the hookworm *Necator americanus.* Truswell and Hansen (1975) state that tonsillitis, rheumatic fever and mitral valve disease are relatively common.

Most other disease is probably the result of trauma, to which lightly-clad people in thorny surroundings are naturally very exposed. Puncture wounds and gashes of the feet are frequent. Fights with wild animals sometimes leave the survivors with bizarre deformities, but signs of old injuries due to warfare or within-band fighting are hardly ever seen. The only recorded malignant disease is of the female breast, and the prevalence is uncertain but unlikely to be high. Inherited disease of a severe nature is unlikely to be seen very often, as infanticide, the traditional way of disposing of obviously defective infants, is or has until recently been widely practised. A famous pituitary dwarf may be seen in one band in south-western Botswana. We have described a case of Treacher–Collins syndrome, and one of mental retardation associated with microcephaly and polydactyly, among the !Kung. Supernumerary nipples and rudimentary non-functional breasts are also sometimes observed (Nurse and Jenkins 1977*a*).

## San technology in a period of change

It has been doubted whether the physique of the San is specifically adapted to a desert environment. Their ingenuity certainly is. The weapons they have devised for the hunt, though they are in recognizably universal cultural traditions, are more suited to their use than similar weapons introduced from outside ever could have been. Their bows are simple, of cut and roughly shaped wood tapering almost to a point at the ends, strung with wood fibre worked into a cord, or with two thin sinews twisted together, attached by a simple slip-knot and wound around the bottom end of the bow, where it is kept in place by a small leather or sinew collar. It is to this end that the cord is first attached; the wooden stave is then bent and the cord similarly fastened to the other end, and the bow kept permanently strung. The arrows represent more specialized techniques, depending on the prey which they are intended to bring down. San arrows are generally rather fragile, and it is the way in which they are used rather than their sturdiness which makes them effective. Many of them are poison-tipped, the preparation of the poison being rather complicated and necessitating a fairly sophisticated appreciation of safety precautions. In addition to bows and arrows the San armamentarium comprises throwing-sticks, which consist of a carefully selected knob and shaft, with the shaft often sharpened to double as a digging-stick, and spears, possibly originally of sharpened hardened wood but in recent times almost invariably with an iron blade inserted into a long wooden shaft. The blade is obtained by barter. Spears are not poisoned.

Large game is hunted by pursuit. The poisoned animal does not succumb as soon as the arrow strikes it, but flees, becoming steadily weaker, and is eventually overtaken by the huntsman and the *coup de grâce* is administered. Smaller burrowing animals are taken in their holes by means of a long barbed stick, which is poked down the hole into the animal's flesh, either transfixing it there while someone else digs down to it, or if firmly placed being used to drag the animal to the surface. The San also employ several methods of trapping. Pitfalls are constructed in game paths in the approaches to waterholes. They are long and deep, and generally traversed by an earth wall so situated that a large animal will fall to be suspended on it, the limbs hanging ineffectually free. Sometimes pointed stakes are inserted into or replace the wall. Deadfall traps, where bait is exposed, and when it is disturbed or touched an appropriately placed block of wood or stone falls and stuns or kills the animal, are not uncommon. Snares are widespread among the more northerly San but seem never to have spread into the far south. They are constructed usually by bending down a supple twig or sapling by means of a cord attached to a bait and ending in a noose. When the cord is released, the animal taking the bait is caught by the neck or by a limb and jerked up into the air. Where they have the opportunity the San also fish, using funnel-shaped traps of woven reed, or spearing the fish from canoes, or driving them into artificial culs-de-sac made of stone, earth or reed fences. Nets and lines are not used in fishing. The only domestic animal traditionally kept by the San was the dog, which was used in the pursuit of game and was not eaten. Fire was also used in hunting: spread in a semi-circle, it forced the game against a line of beaters who would drive it to where the hunters lay in wait. Fire also helped to kill off snakes and scorpions.

Collecting is the principal pursuit of the women, but it is not uncommon for men also to collect. The only implement used, apart from the large net or skin bag receptacle, is the digging stick, which where the ground is hard may be weighted with a perforated stone. Gathered foods are mainly nuts, berries, melons, roots, or bulbs, and very small game, mainly insects but including tortoises and frogs and some other reptiles and amphibians, which are collected rather than hunted. Firewood is gathered at the same time. It is probable that the foods collected by the women make a substantially greater contribution to the dietary of the band as a whole than do the animals hunted by the men. If a kill takes place far from the camp, the men will eat as much as they can before transporting all of the rest that they can carry back to the camp. Vegetable food is much less liable to spoilage, and in some parts of the Kalahari the most important source of protein is the *Bauhinia* nut. The main energy sources are roots and bulbs; there are few suitable leaf vegetables (cf. Tanaka 1976).

The domestic utensils of the San are few and simple. They make, and sometimes decorate, wooden spits for roasting meat; food is sometimes cooked in wooden dishes (the coals being placed on top), and wood is also made into receptacles for fat and other fluids, eating vessels, spoons, pestles and mortars. Mortars are perhaps more usually of stone, or of wood with a stone bottom, and some pestles are calcrete. Reed mats similar to those made by the Khoi are sometimes used in the construction of huts and shelters. The uses of ostrich egg shells as receptacles for water show a fine under-standing of the capacity of their porosity to keep fluids cool: water-filled shells are sometimes buried and the site marked in anticipation of returning at a drier time. Broken into small bits, these shells are also used in the manufacture of beads. Small stone drills are used for making holes through them, and larger ones for scooping out the bowls of the stone pipes in which tobacco or cannabis may be smoked. Tortoise-shells of various sizes serve as spoons or scoops or receptacles for the *buchu* powder which both San and Khoi women use as a cosmetic. It is said that at one time some of the southern San made clay pots; none do so now, and the techniques of manu-facture of those stone instruments which demonstrated San con-tinuity with earlier cultures have been completely lost.

For all purposes for which cloth or leather serve, the skins of slaughtered animals are used. Karosses, the quilted fur blankets which are also made by the Khoi, clothing, bags and quivers for arrows are all made of these. The stomachs of large animals may be made into water-bags. Animal tendons serve as bowstrings and sewing-thread, and, like the wood fibre, can be used for making nets. Knives, some drills and the points of arrows or spears may be made from the hard bone of the limbs of animals or ostriches; horn can serve some of these purposes as well as being fashioned into much the same small implements as are made from tortoise-shell.

Of recent years there has been so swift a series of changes in the availability of materials that San may often be found putting their traditional skills to quite bizarre, though ultimately practical, uses. Metal, and above all iron, has become freely and copiously at their disposal in forms far more easily worked than were the lumps of un-forged iron which they used to buy from the Negroes and hammer into shape with stones. It takes less effort to barter either worked or unworked skins or other produce of the wilderness, for cheap cloths, metal crockery and cutlery and receptacles made of tin or plastic, than laboriously to follow the traditional methods of manu-facture. A market has grown up for San artefacts, and there seem to be too few San still capable of satisfying the demand and reducing the need to make domestic appliances. Nevertheless, where the appliances are relatively specialized, or where San tastes in design do not accord with what is commercially available, the new materials may be reworked in local style.

The San are leaving the world of barter, however, and being drawn into the cash economy. As hired workers they are faced with new ways of resource procurement through the medium of currency. In a number of ways the social practices of the settled, and even some of the migrant, San are changing to accommodate no longer to their own wild environment, but to the tame environment of new and standardized implements. In this environment they are not yet exactly regimented, but faced with novel pressures, less like their old exigencies but not yet entirely conveniences which reduce effort. Doubtless the remoter San could survive without the plastic buckets and briar pipes which they now possess, without Okapi knives and enamelled mugs, and even without the metal tips which they now seem to consider essential for their arrows. But it is possible that loss of the opportunity to settle and a narrowing once more of their economic horizon might make it necessary for them to readapt to what may have been their own formative ecosystem.

# 6
# THE NEGRO INCURSIONS

The Negroes could be the latest or the most ancient of the major races of man to have emerged. Until recently, the relics most widely accepted as undoubtedly Negro (Boule and Vallois 1932; Trevor *et al.* 1955; Szumowski 1956) were at once so recent in date and so widely dispersed as to admit of a single and simple explanation. The isolation in which the race had its apparent origin seemed to have been located where conditions did not permit the preservation of human remains. Throughout the Upper Pleistocene climatic changes, despite the fluctuation northwards of the south-western dry zone, regions of humid tropical forest persisted, and survived until modern times. With the return of warm conditions they spread to cover a far greater area. Where organic growth is rapid, so is decay. The ancestors of the modern Negroes, it was believed, having adapted to life in or on the edges of the forests, were in all likelihood confined to their vegetable fastness by the ease with which they could there procure food. The food resources even of tropical forests, however, are not inexhaustible, and an increasing population could have been responsible for the almost simultaneous appearance of Negroes in two widely separate drier parts of the continent, at Asselar in Mali (Boule and Vallois 1932) and Jebel Moya in the Sudan (Trevor *et al.* 1955), in about the fifth to sixth millennium B.C.

Several other fossils appeared to substantiate the hypothesis of a recent origin for the Negro. The 58 skeletons of apparent Negroes from Jebel Sahaba near Wadi Halfa in the Sudan were among the oldest: they dated from 10 000 to 12 000 B.P. (Wendorf 1968). The relics found near Khartoum and dated at 8 millennia ago (Deevy 1949) confirm the relatively early presence of Negroes in Nubia. Although the Negro affinities claimed for the Iwo Eleru remains from Nigeria (Brothwell and Shaw 1971) are hardly substantiated by comparison with a large range of Negro skeletal material (de Villiers and Fatti 1982), Chamla (1968) has brought together a number of Saharan cases which she claims to show features typical of the West African Negro. Eleven burials, dated as recently as 3300 years ago, from Kintampo in Ghana (Flight 1970) provide almost all the rest of the most ancient undoubtedly Negro remains from West Africa. East Africa has proved much more revealing, and supplied at least one set of remains which, if Gramly and Rightmire (1973) are correct, is very much older. The fragmentary cranium, which they have nevertheless pronounced to be Negro, from Lukenya in Kenya, is around 17 600 years old. The material from Ishango near Lake Albert in Zaïre, described by Twisselmann (1958) may be as much as 9000 years old, though some evidence suggests 4700–7500 as being more probable. The Elmenteita and Nakuru material from Kenya (Leakey 1935; Rightmire 1975) is now accepted as dating from at least 7000 years B.P., and is almost certainly Negro, as are the approximately 5000-year-old remains from Lake Eyasi in Tanzania (Brauer 1976). Further south, in Malaŵi, the Negro ascription of early remains is less certain, but in Zambia and Zimbabwe, at Kalemba and in Inyanga, Negro skeletal material has been found which may be as much as 45 000 years old (de Villiers and Fatti 1982).

With these tentative estimates of great antiquity we approach the ages lately claimed for some Southern African material, more especially that from Border Cave (Beaumont *et al.* 1978; de Villiers and Fatti 1982) which is now recognized as dating from possibly as much as 110 000 years B.P. Such an estimate in relative isolation will of course be controversial, and we have no intention of discussing here the grounds for the age assessment. What is of present reference is the bearing it has on the question of Negro origins. These are best considered after the weighing of other evidence as well.

## Physical characters: morphology

The Negroes, like the other major races of mankind, are distinguished by a number of traits all of which singly may be found in other races but which tend in them more commonly to associate together. In addition, they possess a few characters, most of them determined at the level of single gene loci, which are so frequent in them and so rare elsewhere that their occurrence in an individual may credibly be held to betoken Negro ancestry. The majority of such characters appear to have become established in them by the relatively recent operation of the forces of selective adaptation to African environments.

The skin colour of the Negroes is brown to brownish-black; some individuals are yellowish-brown, and some are very dark indeed. They are on average of medium height, with the sitting height of approximately half, and the span greater than, the stature. On the whole they tend to be laterally and muscularly built, though some of their subdivisions show pronounced linearity. Except in time of famine or in extreme old age they do not have much wrinkling of the skin. Moderate deposits of subcutaneous fat are usual. Generally they have ovoid, proportionately long skulls of low to medium height, with well-marked glabellae and supracillary ridges and little or no frontal bossing. Their faces are moderately broad to narrow, with fairly prominent zygomatic arches, and tend to be moderately high. Prognathism is pronounced, and there is a high prevalence of median diastema. Only rarely do they have pointed chins; these are more often rounded, and the bigonial diameter tends to be high. The broad to very broad nose has a low or slightly raised bridge, in contrast to the concavity often seen in the Khoisan, and is straight in profile. The Negro ear is of medium size with no marked over-rolling of the helix, and usually has a free and well-developed lobe. The lips are often thick and protuberant. Characteristically, the hair is black or dark brownish-black, lightly curled to woolly or frizzy, and of very varying texture. There is no obliquity of eye form, and vertical eye-folds are more common than oblique ones, which are probably due to Khoisan admixture. Eye-colour is brown to brownish-black, though some Negroes do have light eyes, often associated with pathological processes (Soussi-Tsafrir 1974). The occasional finding of tablier and steatopygia in the women (de Villiers 1968) is most common in those groups which have been the recipients of much Khoisan gene flow or which practise artificial elongation of the labia minora. The breasts tend to be rounded and regular in contour in the young women, but in older women come to sag into long flaps. The areolae, though large, are not distinctive.

## Physical characters: genetic markers

The monogenic characters of the Negroes include at their most typical a number which are strongly ecosensitive. The best-known of these is probably the sickle-cell trait, caused by the presence of Haemoglobin S in the erythrocyte; it is not, however, very frequently found in Southern African Negroes. The Negro variants of glucose-6-phosphate dehydrogenase, Gd(A) and Gd(A⁻), are; and so is the amorph in the Duffy blood group system (Fy(a-b-)). Colour blindness frequencies are moderate, and there is almost universal absence

of the gene for persistent adult intestinal lactase activity. It is mainly
in the remaining gene marker systems that we hope to find some
indication of the relative antiquity of the Negroes, by making com-
parisons with the distributions of the two groups, Caucasoid and
Khoisan, with which we know them principally to have had contact.

The ABO blood groups afford little evidence of the original
affinities of the Negro. Caucasoids, Khoi and Negroes have much the
same frequencies of $A^1$ and $A^2$; San frequencies are lower. The allele
miscalled $A^{bantu}$, so much more typical of Khoisan than Negro
(Jenkins 1974), is uncommon or missing in Negroes except in
Southern Africa and probably attests to later gene flow rather than
common descent. The frequency of $B$ is moderately raised in Khoi
and Negroes, somewhat lower in Caucasoids, and lowest in the San.
The Henshaw antigen of the MNSs complex is absent from Caucasoids but present in Khoisan and Negroes. Frequencies of S are much
higher in Caucasoids than in the other populations. In the Rhesus
system absence of the D antigen is found in all four peoples, though
it is responsible for only low frequencies of $r$ in most San populations. Several San populations are virtually monomorphic for the
Rhesus system, with $R^0$ the sole allele or at a very high frequency.
This allele more than any other has been used as evidence for a common origin for Khoisan and Negro. The possibility that its high frequencies result from very early drift, or peculiarly African and
hitherto unidentified selective factors, cannot be excluded. $R^1$, the
most characteristically Caucasoid allele, is present at varying frequencies in Negroes and Khoisan. In general the E antigen, which
when determined by the $R^2$ and $r''$ alleles is quite common in
Negroes, particularly those in the western portion of Southern Africa,
and of moderate frequency in Caucasoids, is found rather more
rarely in the Khoisan; its highest level is in the Khoi, and in some San
populations it is not present at all, or represents obvious Negro
admixture. It is possible to claim that in the Khoisan there is more C
than E, in Negroes more E than C; but the significance of this is
obscure. A much more informative component of the Rhesus system
in this context is the V antigen, discussed in the previous chapter.

Among the similarly informative blood group markers also discussed in Chapter 5 is Kidd. San frequencies of $Jk^b$ tend to be higher
than those in Khoi or Negroes, but not as high as in Caucasoids,
though the Pedi of the northern Transvaal and the Tsonga of southern
Moçambique have frequencies even higher than those in the San.
Variation in Kell is found hardly at all in Negroes or Khoisan. When
it is it may most readily be ascribed to gene flow from Caucasoids.
Much the same may be claimed for the $Lu^a$ allele of the Lutheran

system. Lewis, P and Secretor, which have been extensively investigated, show such variable frequencies among populations, even those which from other evidence appear to be closely related, that it is evident that they are probably subject to fairly intense local selection.

The haptoglobins give a somewhat confusing picture unless we grant that they, too, are subject to selection. The commoner allele in San and Caucasoid is $Hp^2$, but in Negroes and Khoi it is $Hp^1$. The frequencies of $Hp^2$ among Herero-speakers with little or no San admixture are high enough to distinguish them from other Negroes. The transferrin allele $Tf^{D1}$ is found in Negroes and Khoisan, reaching its highest frequencies in the Khoi. It is not found in Caucasoids. The group-specific component of serum, Gc, has not been tested in a sufficient number of populations to provide much data here. One immunoglobulin polymorphism, Inv, has given variable values suggesting that it, like Lewis and P, may be peculiarly ecosensitive.

Each of the major races possesses a characteristic Gm profile, and some include haplotypes which are unique to themselves. The antigen Gm(6) is found almost exclusively in Negroes and is present as the haplotypes $Gm^{1,5,6}$ and $Gm^{1,5,6,14}$, though the commonest Negro haplotype is $Gm^{1,5,13,14}$. This is also the commonest in Melanesians, whose outwardly Negro-like morphology suggests that this haplotype may be an example of parallel selection and consequently not to be taken too seriously in a search for origins. The serogentic profile of the Melanesians is otherwise quite unlike that of the Negroes, and their skeletal morphology is very different, though the fact that they too possess $Tf^{D1}$ may be indicative of the ecosensitivity of yet another monogenic marker. $Gm^{1,5,13,14}$ is also very common in the Khoisan, and is absent from Caucasoids. The Caucasoid markers $Gm^{1,2}$ and $Gm^{3,5,13,14}$ are found in the Khoi almost certainly as a result of gene flow, but the haplotype with universally the highest frequency in the San and in the probably racially least mixed of the Khoi populations is $Gm^{1,13,17}$. This, with $Gm^{1,21}$ is so typical of the San among Southern African peoples that it can on its own be used, and has been used (Jenkins et al. 1970; Jenkins 1972) as an indication of San or Khoisan admixture. Since it is either absent or very rare in Negroes outside of Southern Africa, it is a good indicator of the extent to which Southern African Negroes have been receptive of genetic contribution from earlier inhabitants of the region.

As mentioned in Chapter 4, gene flow between San and Khoi appears to have occurred mainly from the former to the latter; hence it is possible that a certain proportion of the San contribution to the Negroes has occurred via the Khoi. It is consequently not surprising to find $Gm^{1,5,6}$ and $Gm^{1,5,6,14}$ commoner in the Khoi than in the San, and the relative rarity in the latter of $Gm^{1,5,14}$, also considered a typically Negro allele, and its quite high frequency in the Khoi, may indicate that it is, in fact, a phenogroup of Khoi origin. Since it is found at appreciable frequencies in Negroes well beyond the bounds of Southern Africa, this may argue a more ancient contact of the Negroes with the forebears of the Khoi than with those of the San.

In a number of instances the tissue antigens, and especially the extremely polymorphic HLA-A and HLA-B loci, are even more informative than Gm. Here, however, they provide little effective evidence. High frequencies of HLA-Bw17 are strikingly common in Negro and Khoisan, while this antigen is found relatively less commonly in Caucasoids, but it is by no means certain that in these peoples HLA-Bw17 represents a single antigen. It is in any case a major component of the constellation of antigens in Melanesia as well, and this may be a further example of parallel selection. Negro and Khoisan tissue antigen patterns are similar not only in the frequencies of this antigen but also in including relatively high frequencies of HLA-Aw30 and low ones of HLA-A1, and a virtual absence of HLA-A11. This is in agreement with findings for Negroes in Zambia (Festenstein *et al.* 1973) but not with the results of Govaerts *et al.* (1973), working among Negroes in Zaïre, who found rather high frequencies of HLA-A1 and lower ones of HLA-Aw30 and HLA-Bw17 and detected HLA-A11 at appreciable frequencies. It is possible that the extensive gene flow from Khoisan to Negroes in Southern Africa, and even as far north as Zambia, has been responsible for such contrasts in pattern. Linkage disequilibrium of haplotypes is not informative here, either; the *A9, Bw14* Negro haplotype is present in Khoi but not in San, while the *Aw23, Bw17* Khoisan haplotype appears possibly to be present in the Negroes of Natal (Brain and Hammond 1973) but not in those of Zambia. Appearances such as these strongly suggest late gene flow. The distribution of the characteristically Khoisan, or more specifically Khoi, tissue antigen marker, HLA-Aw43 (Nurse *et al.* 1974), among Negroes is not yet clear.

Though in the red cell acid phosphatase system frequencies of $P^r$ reach their highest levels in the San, this variant is also very common in the Khoi, and is found in Negroes as far distant as the Yoruba (Ojikutu *et al.* 1977). The Caucasoid variant $P^c$, or one resembling it closely, occurs at low frequencies in the Dama and Ambo. There are fairly marked regional differences in the frequencies of $P^a$ and $P^b$; the rise in $P^r$ frequency in the Khoisan appears to be at the expense of $P^b$ but in the case of the Negroes at that of $P^a$. In the peptidase A and D systems the *Pep A*$^2$ and *Pep D*$^2$ alleles are characteristically Negro, and where they occur indicate Negro admixture. *Pep D*$^3$ is rarely found, principally in the San, occasionally in Negroes, and thus far not at all in the Khoi. Null or very low activity alleles of adenosine deaminase have been found in both San and Negroes (Jenkins *et al.* 1979), but it is far from certain that they are identical. There is very low polymorphism otherwise for adenosine deaminase in most non-Caucasoid Southern African populations.

Adenylate kinase is polymorphic in Caucasoids and Khoisan, but not at all or barely so in all the Negroes except the Dama, among whom $AK^2$ frequencies attain levels similar to those in the San. Other peculiarities of the Dama are their high frequencies of the $C_5+$ electrophoretic variant of pseudocholinesterase and of the so-called 'Richmond' variant of 6-phosphogluconate dehydrogenase. The genetics of the former is obscure; it occurs, however, in other Negro populations as well as in San and Caucasoids. Though the 'Richmond' variant was first described in a Caucasoid, its distribution is worldwide and it may possibly have spread as a result of the extensive slave-trade from southern Angola. The principal polymorphism of 6-phosphogluconate dehydrogenase everywhere, however, is for $PGD^c$ rather than $PGD^R$, and Negro frequencies pf $PGD^c$ are closer to Caucasoid levels than to those in the San, among whom it is often absent. Polymorphism at the second locus of phosphoglucomutase is found in both Negroes and Khoisan, with higher frequencies of $PGM_2^2$ on the whole in the latter, and not at all in Caucasoids. At the first locus the common variant $PGM_1^2$ has frequencies in Negroes similar to those in Caucasoids and is occasionally absent in Khoisan. It tends to be replaced in the Khoi by the much rarer $PGM_1^6$, which is occasionally also found in Negroes.

The three remaining commonly investigated polymorphisms are known to be concerned in the response to malaria. The boundary of Southern Africa coincides almost exactly with the furthest extension southwards of Haemoglobin S. This is found in the Ambo and Kavango Negroes at variable frequencies, very occasionally in the Venda south of the Limpopo, and not at all in the other Southern African Negroes or the Khoisan, though it is sometimes found in 'Coloureds' and even in Southern African Caucasoids. The thalassaemias, rather more typically Caucasoid than Khoisan, are present in the Negroes, but once again selection would have been more likely to conserve them in the presence of malaria. There are two glucose-6-phosphate dehydrogenase variants, $Gd^A$ and $Gd^{A-}$, which are found almost exclusively in persons with some Negro ancestry. Both are common in the Southern African Negroes, and this indicates with a high degree of probability that the mutations that produced them occurred before that for Haemoglobin S. The degree of protection against malaria afforded by $Gd^A$ is lower than that due to $Gd^{A-}$, and the much more frequent finding of $Gd^A$ than $Gd^{A-}$ in Khoisan as well as in the Negro probably indicates that the former represents the earlier mutation; its frequencies in them are too low, however, for it to provide much evidence of a common origin for Negroes and Khoisan. The almost universal absence of Duffy antigens in Negroes lends weight to the contention of Miller *et al.* (1976) that these antigens provide receptors for malarial parasites. The Khoisan, who throughout their history have almost certainly been much less exposed to malaria than the Negroes have, have only low frequencies of the Duffy null allele, $Fy$, which may, again, have been introduced among them by the Negro admixture. The significant frequencies of $Fy^a$ among the Negroes of Southern Africa probably reflect Khoisan assimilation.

## The origins and antiquity of the Negro

If we discount for a moment the early dates given to the Border Cave remains and consider only the later unequivocal Negro relics, we are faced with their probably much later evolution than either the Khoisan or the Caucasoids, and hence with the possibility that the Negroes may have evolved from one or other, or perhaps from hybridization between them. The weight of the serogenetic evidence here appears to incline towards the Khoisan. The present-day San, who are probably more typical of their ancestors than the modern Khoi since they have been less exposed to hybridization, manifest a lower degree of polymorphism than either Caucasoids or Negroes, and could represent an almost blank sheet upon which the Negro template was inscribed. It may well have been in an ancestral common stock that the mutations for acid phosphatase $P^r$, the Henshaw and V antigens and the $PGM_2^2$ variant at the second phosphoglucomutase locus occurred, while the major differences between the two races come from later, principally selectively maintained, mutations. It remains possible nonetheless that a Caucasoid Capsian strain may also have gone into the original ancestry of the Negroes and even that it was this very hybridization which made the evolution of a new race possible. It would be hard to explain otherwise how the E-bearing alleles of the Rhesus system, or $PGD^c$, or $PGM_1^2$, come to have such almost uniform frequencies in Negroes as opposed to Khoisan.

These conclusions are similar to those arrived at on somewhat different grounds by Tobias (1966) and discussed by him in more detail in his presidential address to the Royal Society of South

Africa (1972*a*). He emphasizes the morphological distinctiveness and uniformity of the Khoisan, and shows that the Khoi possess rather more morphological characters of Negro type than the San do. In fact, the evidences of greater morphological specialization in the putatively elder stock suggest that the proto-Negroes could have evolved from a sector of that stock which had undergone less specific differentiation. Indeed, it is likely that several of the most characteristically Khoisan morphological attributes would have been distinctly disadvantageous in tropical forest environments. The conformation of the eyes, for instance, so suitable under conditions of extreme brightness, would be a hindrance in dim shade; the striking steatopygia, easily managed in open country, would be cumbrous when meandering through undergrowth; and sparse peppercorn hair affords less protection against objects falling from above than does a thick and woolly growth. Once the ancestors of the Negroes had established themselves in a forest habitat their more specific morphological differentiation would have proceeded with comparative rapidity owing to the strength of selection for the more favourable recombinations of the genes governing morphology. The persistence of such characters once they had emerged from the forests may be ascribed to the relative weakness of the forces selecting against these traits in their more recent environments. Negro migrants would nevertheless tend to choose environments conforming as closely as possible to those to which they had been previously adapted. Following the abandonment of the deep forest some degree of selective adaptation to the savannah must have taken place, and it is probable that in Southern Africa particularly, and possibly in other parts of Africa as well, this process has been largely hastened and facilitated by hybridization with earlier inhabitants. In Southern Africa these would have been entirely or exclusively Khoisan; in northern and eastern Africa there would have been an appreciable Caucasoid component as well.

An intriguing possibility suggested by Cavalli-Sforza (1974) becomes rather more likely if we accept, on the evidence of Border Cave, the evolution of the Negro at least as long ago as the Khoisan and possibly appreciably before the Caucasoids. He points out that the Caucasoid gene marker array appears to lie intermediate between those of the Negro and the Mongoloid. From this we could either interpret the relationships between the stocks either cladistically, with a 'European' geographical stock on a 'branch' lying between those of the 'Africans' and 'Asians', or hypothesize that the Caucasoids may represent the product, after long *in situ* adaptation to the conditions of western Eurasia, of hybridization between Negroes and

Mongoloids. This might mean postulating that human speciation took place outside Africa, or even accepting the hypothesis of Coon (1963) that ultimate hominization occurred in a number of different places in several already geographically differentiated strains of *Homo erectus*. The latter is difficult to accept on present-day dogma, but is by no means inconceivable.

We remain, inclined, however, to conclude that the Negroes have evolved by selective adaptation to a forest environment from an original stock which was in the main the same as the original stock of the Khoisan. There may have been a Caucasoid element incorporated in the proto-Negroes, and it is not impossible that hybridization between early Khoisan and early African Caucasoids may have produced a strain which could adapt more easily to these new conditions. In response to specifically forest stresses certain mutations occurring in the Negroes have proved selectively advantageous and been conserved, and these variants have come to be included in the constellation of Negro traits.

## The expansion of the Negroes

If the earliest departures of the Negroes from their forest fastnesses resulted from expansion of populations beyond the limits which particular areas of forest would support, this would necessitate the acquisition in rather less productive environments of new technologies of food production. The most significant of these, of course, was agriculture, but even before that was learned fishing seems to have become an important means of subsistence for at least some Negroes. The fishing-camp site at Khartoum (Deevy 1949) associated with Negro relics may date from the fifth millennium B.C. or earlier. In the Nile valley and around the lakes which in the early Holocene still existed in the southern and central Sahara there remains ample evidence of waterside communities which fished and hunted hippopotamus and crocodile (Arkell 1966).

It is probable that the introduction of agriculture into Africa from the Middle East affected the non-Negro hunter–gatherers before it did the Negroes, and that when the Negroes began to expand they made contact with settled communities cultivating cereal crops and possessing sheep and goats. There are advantages, obvious to those who have been acquainted with famine, in producing a surplus of grain which can be stored and maintaining on the hoof an accessible source of meat. The appearance of food production in north Africa was relatively sudden, and coincided with a period of lowered rainfall which would have caused the limits of the forests to contract and would thereby have contributed to producing population pressures within them in the following generations. The hunting techniques of the open country would have necessitated a radical change in the hunting techniques which had been used in the forests. The edible wild plants of arid areas differ very considerably from those of the forests. It would be intellectually simpler and less wasteful of effort to learn how to hoe, sow and reap, and to care for flocks and herds, than to attempt to accommodate to the new ways demanded by persistence in hunting and gathering. There were doubtless elements among the emergent Negroes who did thus persist, and whose successors or descendants may be represented by Negroes who even today hunt and gather under desert or savannah conditions; but the vast majority of modern Negroes are agriculturalists. That it was as bearers of agriculture that they spread most considerably south and east, though debatable, seems fairly clear.

Agricultural techniques were eventually in addition carried back into the forests, and have come to be dominant among the cultures of many of the peoples dwelling there. Even the effort needed to slash and burn an area of heavy growth before sowing crops is worthwhile when the alternative is precarious dependence on a diminishing yield of wild fruits and on ever more elusive game. The original forest dwellers are today probably represented by the various and not necessarily closely interrelated populations of Pygmies. The larger build of the more widespread Negro strains would be made possible by an increasingly sedentary way of life and by increasing occupation of clearings, either artificial or naturally appearing, rather than the deeper ranges of the forest itself. The relative rarity of the remaining Pygmies is most probably less the result of their failure to thrive than of the contrasting extent to which their larger relatives have been enabled by technological adaptation to thrive even more. Extensive forest clearing would hardly have been possible before the introduction of iron tools, and the earliest gardens could have been planted along the banks of rivers. Metallurgy is of much later appearance than agriculture and some techniques may have been of indigenous development. The working of metals was known to the early Nubians, and was probably introduced to them from the Mediterranean or Middle East, but it does not seem to have spread from them in the same way as agriculture did. It is only from the middle of the first millennium B.C. that it appears among the Negroes of northern Nigeria who developed the Nok culture (Fagg 1969). It may even at that relatively late period have spread from the Meroitic civilization of Nubia, or indeed been carried across the Sahara by the indefatigable Garamantes; but its effects were sudden and widespread, and the centre of their radiation was certainly Nok. Iron-working makes its appearance almost simultaneously in Chad and eastern and southern Africa during the first few centuries A.D. There can be little doubt that its spread coincided with yet another expansion of the Negroes.

## The Bantu expansion

The languages of the great majority of the Negro inhabitants of Africa south of the Bight of Benin fall into a single large family known as Bantu (Fig. 6.1). Linguistic data provide a valuable measure of population movement in periods without recorded history, and in the case of Bantu this is especially so. All the Bantu languages are so closely related that it appears highly improbable that the depth of their historical divergence can be very great. Greenberg (1963) maintains that Bantu is part of a large Niger-Congo language family which had its origin in the area of the confluence of the Niger and Benue rivers, very close to the Nok culture area. Guthrie (1962) suggests that there was a nuclear centre in the southern part of the Congo Basin where a language ancestral to all the Bantu languages was spoken, and Oliver (1966) has amplified this suggestion and asserted that from this nexus the Bantu languages have spread in most directions though at varying times. Oliver's hypothesis would necessitate a certain amount of back migration along or across the routes which had brought proto-Bantu from west Africa. The rapidity of the movement away from this nuclear centre certainly does not lack archaeological support, though a dearth of evidence from the Congo basin itself is unfortunate.

The work of Hiernaux (1968) seems to indicate that in many biological respects the findings of the linguists can be upheld, and that a large part of the ancestral stock of the Bantu-speakers came from the savannah lands north of the rain-forest. From there groups crossed the forest until they reached a habitat similar to that to which they were previously adapted and, after a period of settlement, expanded widely into southern and eastern Africa. In the opinion of Clark (1970) the accumulated evidence suggests that 'the ancestors of the proto-Bantu speakers . . . made their way into the western and northern parts of the Congo Basin, perhaps in the second, but more likely in the first millennium B.C. . . .' (p. 214). Then, around the beginning of the present era, the introduction of metallurgy, and in particular iron technology, triggered off rapid and widespread movements from the heartland south of the forest. This would mean that the prior derivation of the stock from the vicinity of Nok was purely coincidental, though it could also be interpreted

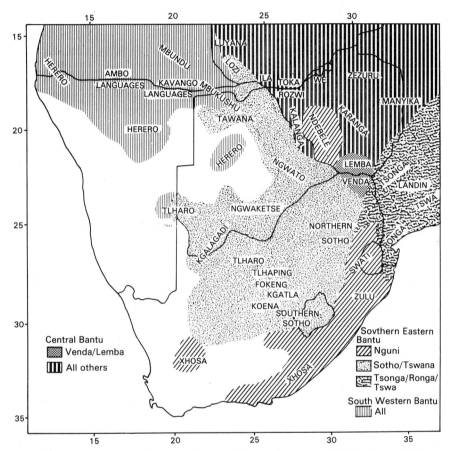

Fig. 6.1. Distribution of the speakers of Bantu languages in Southern Africa. Some attempt has been made to distinguish among the Western Sotho languages, but with the exception of the western outlier of Tlharo, and the regionally differentiated groups in Botswana, any fine distinctions would be nugatory, and the Tlhaping, Fokeng, Kgatla and Koena are placed with the rest of the Tlharo indiscriminately mingled in the area in which modern 'Bophuthatswana' is situated. Similar differentiation among the Cape Nguni has been disregarded, and all are classified as speakers of Xhosa. This map represents a modification of one appearing in Jenkins (1972).

as the flight of refugees who did not share in but rather were threatened by the new technology.

There have been a number of classifications of the Bantu languages, some at marked variance with others, but certain points of agreement do emerge. It is generally agreed that there are major divergences between the languages of the eastern and the western side of the continent, and that these divergences are carried south into the

Bantu languages of Southern Africa. In addition, the 'Proto-Bantu Homeland' at some time extended in a broad zone east and west of its Congo basin nucleus, with the result that even today there is a central African linguistic stratum centring on Luba and comprehending a variety of peoples who claim Luba ancestry. Moreover, Bantu languages have borrowed both lexicological and phonological elements quite extensively from the languages of non-Bantu-speaking peoples with whom they have come in contact, and some of the borrowings afford clues to the possible identity of Negroes other than Bantu-speakers who may have inhabited various areas previously (Nurse 1970).

We know that there were early Iron Age settlements possessing domesticated bovines and caprines in southern Zambia and in Zimbabwe by the third to sixth centuries A.D. and perhaps much earlier (Phillipson 1969), and that iron was worked on the Witwatersrand before the end of the first millennium (Mason 1962), but we cannot be sure that the people concerned spoke Bantu languages or even that they were Negroes. On the western side, there was a very substantial Iron Age settlement in southern Angola at least as early as the seventh to eighth centuries A.D. (Vansina 1966; Fagan 1965), but the same reservations must be expressed. It is not until the period of Portuguese expansion in the late fifteenth and early sixteenth centuries that we encounter any positive evidence of the presence of Bantu languages in the region.

By that time, however, Bantu-speakers had penetrated sufficiently far down the east coast to suggest that they had already occupied it for a considerable period. The earliest recorded words that we have from the peoples of the Transkei are indubitably Bantu, and they date from the sixteenth century. A hundred years later the Dutch at the Cape received information from the Khoi of a people in the interior known as the 'Bricqua', which is cognate with the modern Nama name for the Tswana. Our earliest knowledge of Bantu-speakers in Namibia dates only from the eighteenth century, but studies by Vedder (1934) of the genealogies of the Herero chiefs suggest that they had occupied the Kaokoveld at least two centuries earlier.

### Negro migrations into Southern Africa

In considering the movements of peoples into Southern Africa we need to return to an examination of the topography. This has been done in previous chapters in so far as the topography of the region affected the movements of the hunter–gatherer San and the pastoral immigrant Khoi, the one unburdened, the other restricted in choice of route by the needs of their livestock. In the case of the agricultural Negroes, the restrictions would have been somewhat more rigid. There are two ways in which a migrant party of agriculturalists can proceed. If they have an adequacy of beasts of burden, these can be laden with provisions for the journey, including the tools of agriculture and the seeds for cultivation. The extent of the migration is then limited by the necessity for cultivating a crop when the provisions are exhausted. Such a migration can consequently proceed only in comparatively short stages, most probably extending only from one harvest to the next sowing. The alternative is to revert for the period of migration to pastoralism and gathering, and depend on the natural increase of flocks and herds to augment a diet of game and wild vegetable foods. Where such a migrant party lost its livestock, whether by natural hazards or through theft, there would be no alternative to the adoption of outright hunting and gathering.

We know that for several of the migrations at the time of the *Mfecane* the pastoral alternative was chosen. In these migrations, those of the Ndebele away from Kwa Zulu and of refugees from Mantatisi's forces towards Barotseland, and those of the Maseko and Jele Ngoni to east central Africa and of Soshangane into Gazaland, the movement had to be swift and as far as possible unburdened. The Ndebele *izimpi*, lacking wives, did not even have women with them until they took wives from among the Sotho/Tswana. Much of this movement, too, was through settled country where the inhabitants could be terrorized by force of arms into providing other food as well as cattle. There is no reason to suppose that anything similar happened as the Negroes first moved south to occupy arable land tenanted until then only by sparse hunter–gatherers and herders. The archaeological record suggests a very slow pace of movement indeed, and it seems likely that not just a single season's crops, but whole generations, might elapse between stages of the journey.

This method of progression would restrict severely the routes which might be followed. Indeed, there is only one practicable way for the tardy agriculturalists to enter and subsequently spread out into Southern Africa, and that is across the Limpopo river. The point at which the river was crossed would depend on whether the migration was moving along the coastal plain or the plateau. There are three principal strains of Bantu-speaking Negro to the east of the Kalahari desert, the Sotho/Tswana, the Nguni and the Tsonga. The Sotho/Tswana are not now, and possibly never have been, a coastal people; though they certainly spread into Swaziland, and were probably represented in the interior of Natal before the *Mfecane*, it may be doubted whether they ever occupied much beyond parts of the inner fringe of the coastal plain. If they did reach the coast, the fact that they did not remain and flourish there suggests that the penetration would have involved small parties and perhaps exploration rather than long-term settlement. It is legitimate to suppose that they crossed the Limpopo in its upper reaches, below the northern edge of the Zoutpansberg. The first wave, represented by the modern Kgalakgadi, who live in the southern Kalahari, may even have moved up the Limpopo to its headwaters without ever crossing it. There is no record of their ever having been found on the right bank of the river. Almost all the peoples now resident from the Transvaal plateau basin westward into Botswana and southward across the highveld of the Orange Free State and the mountains of Lesotho and into the northern Cape Province and the uplands of the Transkei, are speakers of Sotho/Tswana languages. It does not seem likely that they were anteceded by any Negro people which they did not absorb.

The Nguni fill the coastal plain south of the Nkomati River, except for the area immediately north and south of the border between Moçambique and South Africa, which is inhabited by the southernmost of the Tsonga. The earliest known Iron Age sites in Southern Africa date from the fourth and fifth centuries A.D. and are all in areas either still or comparatively recently occupied by Sotho/Tswana (Klapwijk 1973, 1974; Prinsloo 1974; Mason 1973). It therefore appears that the Nguni are more recent comers. Virtually all the investigations until recently of the archaeology of the Natal, Transkei and eastern Cape coasts were carried out by Schofield in the 1930s, and his conclusions (1935, 1936, 1948), unsatisfactory as they were, were long current. He regarded the earliest potsherds as 'pre-Bantu' and ascribed the next oldest to the influence of the Fokeng, a Sotho/Tswana people, and, even more specifically and less credibly, to their Hurutshe division. A related style of pottery in the Transvaal has been dated to the eleventh century A.D., but no radiocarbon datings are available for the Natal finds (Fagan 1967b). Fagan suggested that they might date from the middle of the sixteenth century. Iron was certainly known to the peoples of this coast well before that time; the shipwrecked company of the Portuguese galleon *São João* were almost successful when they tried to buy a cow for half a dozen iron nails in 1553 (de Brito 1735). In 1970 Dutton reported a seventh-century iron smelting site at Ndumu near the Moçambique border, and Schofield himself (1948) described a ninth-century Early Iron Age settlement at Ntshekane in the Tugela basin. More recently Hall and Vogel (1978) have published a preliminary report on a fourth-century site at Enkwazini near St Lucia Bay. Their findings suggest that the styles of the pottery found there are related to those of the same period in the Transvaal. They contrast markedly with those of Schofield's NC3 phase as represented by Ntshekane, and may indeed be either 'pre-Bantu' or Sotho/Tswana.

The shipwrecked Portuguese, when they needed to cross rivers, went inland, and added considerable distances to their journeys in so doing. It is not clear why they did not take advantage of the sandy bars which obstruct the mouths of all the rivers; perhaps on their first landing they made for higher ground, and so were unaware of this facility. The attitude of the Nguni to the sea is an odd one: they are reluctant to eat fish, few of their villages are within sight of the sea, and they had few boats or rafts. It is impossible to say, since the relocation of peoples following the *Mfecane*, whether their migrations hugged the convenient shore and spread out between the rivers, or whether they, too, went inland to cross them. In either case, they almost certainly migrated into Southern Africa along the coastal plain. The ancestors of the Swati ascended into the foothills of the Drakensberg north of the Pongola River, and mingled with the 'Sotho' settled along the Usutu River. Some Nguni must have reached the plateau, since there are still groups among the Pedi known as Koni, the Sotho generic term for the Nguni peoples. Gardner's (1963) attempt to establish a Nguni presence at Mapungubwe is controversial and dubious. The most assuredly Nguni Iron Age remains are found in Schofield's NC3 sequence, which may show some affinities with Karanga artefacts from Zimbabwe. If this is the case, it fits in very nicely with a tradition recorded by Bryant (1929) that the Lala, one of the peoples who contributed to the formation of the Zulu, were of Karanga origin.

An even more likely route of spread of cultural styles would have been via the Tsonga peoples who pressed after the Nguni along the coastal plain from the north. The Tsonga comprise the third main division of the Bantu-speakers classified by Doke (1954) as south-eastern, and their southward spread probably resulted from the extension of Shona speakers to the coast. That extension itself may have been part of the rise of the Mwene Mutapa empire of the Karanga, or even of the earlier empire of the Rozwi, also centred on Zimbabwe. It has been claimed that the Rozwi were Sotho-speakers (Abraham 1964) and traditions collected in Malaŵi speak of Sotho-affiliated Nyai as the warrior-slaves of the Mwene Mutapa, whether Rozwi or Karanga (Nurse 1978).

The Tsonga have no very definite boundary today with the Nguni, and according to Bryant (1929) there has always been a considerable amount of interpenetration of the two peoples. Before the *Mfecane* the Natal peoples were singularly incapable, according to the same authority, of putting up any united or even effective resistance to invasions from any quarter. Though recent recorded history shows the Tsonga as relatively weaker, the early Portuguese records portray a subtle and determined people, an impression conveyed also by the thorough studies of Junod (1927). The Tsonga, unlike the Nguni, have established themselves in the escarpment of the northern Drakensberg and penetrated to the plateau, interdigitating there with the modern Sotho-speakers. The easy conquest of a section of northern Tsonga during the *Mfecane* by the absconding Zulu general Soshangane was deceptive, and his Landini have been virtually absorbed by the people they conquered. It remains uncertain to what extent the Nguni of Natal, as opposed to those of the Transkei, represent the outcome of hybridization with Tsonga.

All the immigrant Negro peoples into the eastern part of Southern Africa interbred to some extent with the Khoisan, and the Khoi component of the Xhosa has been estimated to be as high as 60 per cent (Jenkins *et al.* 1970). This estimate is tenable from historical evidence as well (Harinck 1969). The somewhat lower genetic contributions made by the Khoisan to other groups are not so well attested by tradition, and probably took place rather earlier. The principal fusions between Khoi and Xhosa occurred only in the eighteenth century. The sharing of relatively high frequencies of $PGM_1^6$ and $Gm^{1,5,14}$ by the modern Nama and the Kgalagadi suggests that the earliest of all the Negro immigrants might have been among those most ready to interbreed, and this suggestion is supported by the skeletal remains from Bambandyanalo in the northern Transvaal (Galloway 1959). There is, however, no main section of the south-eastern Bantu-speakers which lacks some evidence of Khoisan admixture, though the small Venda population of the extreme northern Transvaal, whose language may perhaps more properly be grouped with Shona, and the Chopi and giThonga-speakers of the coast near Inhambane, who have not been thoroughly investigated yet, may have less than the rest.

The giThonga-speaker, who name their languages but not themselves, and the Chopi, may conceivably be survivals of early Negro migrations which have left no other recognizable representatives. Their languages, which are distinct from each other, contain convincing evidences of having received Tsonga contributions comparatively recently, and it has been suggested (Bailey, personal communication) that their affinities lie at least in part with the Bantu spoken on the western side of the continent. Contact between the western peoples and the south-eastern Bantu-speakers could have occurred around the northern fringes of the Kalahari; biological evidence of this will be presented later. It is not inconceivable that the very earliest Negroes to enter Southern Africa may have come not directly across the Limpopo but from the flood-plains of the upper reaches of the Zambesi and its tributaries, and been overwhelmed and assimilated by the later waves of speakers of Sotho, Nguni and Tsonga.

The western Negro immigrants into Southern Africa all, with the exception of the Dama, have definite and fairly readily dateable traditions of immigration. The Herero came as pastoralists; their agriculture is rudimentary and unenthusiastic, and their cattle are of great ritual significance. Their arrival in the Kaokoveld of Namibia, probably at some time during the sixteenth century (Vedder 1934), may have been either from the east, across the less harsh northern part of the Kalahari, or, as seems more probable, from Angola across the Kunene or even, in advance of or together with the forebears of the Ambo, through the gap between the Kunene and Kavango rivers. The Yei of the Kavango delta in Botswana are said to be linguistically related to the Herero, however, and there are Herero still living in Botswana who are said not to be descended from the Herero parties who fled there after the defeat by the Germans early this century. The Herero have close linguistic and cultural ties with the Ambo and with some of the peoples of southern Angola. Brincker (1886) and Vedder (1928b) both record traditions of a common origin with the Ambo, and some names of patrilineal clans among the Herero may be of Ambo origin (Vivelo 1977). Both the linguistic and the cultural similarities may, however, have been acquired late.

The Ambo themselves certainly arrived in the area now inhabited at much the same time as the Herero entered the Kaokoveld. They afford a classic pattern of Negro migration. Their movements are still recent enough for fairly reliable accounts to persist in oral tradition; they acknowledge that they travelled as a number of separate but interrelated parties sharing language and clan affiliations but under autonomous leaders, and that they were consciously searching for land as similar as possible to that they had left. They brought with them seed, metal tools and livestock, but curiously enough, either no knowledge of metal-working or an inadequacy of smiths. They say that they depended on Dama and San for that technology, but it may, of course, have been taught to those autochthones by such immigrants as were adept in it. Some of the parties settled in southern Angola, one, the Kwanyama, straddling the modern border. Where they migrated from is remembered as resembling their present country, and as being somewhere to the north and east of it. Their cultural and linguistic affinities are with the peoples who have had links

with the Lunda and Luba empires (Vansina 1966), and it is likely
enough that their first homeland lay near the headwaters of the
Zambesi or perhaps the Kwando or Kasai.

The Kavango peoples who live to the east of, but do not abut on,
them, show similar affinities. They are much more recent arrivals,
having reached their present lands during the eighteenth and nineteenth centuries. They are usually classed together, though the
different peoples migrated at different times, and between the
Mbukushu and the other Kavango peoples there are marked linguistic
and some significant cultural and biological differences. All nonetheless state that they were originally Mashi, a people falling into the
same group as the Luyana and Lozi of Barotseland. There are, however, grounds for supposing that at least some elements among the
Mbukushu may have originated to the west, in the *planalto* of Angola.
Despite their being the easternmost of the Kavango peoples, they
are sometimes called Viko, or Westerners, and they maintained
strong trading links with the Luchazi and Chokwe of Angola, and,
through them, with the Portuguese on the coast. The Goba of the
Kavango delta may represent a group related to the Mbukushu, but
little or nothing is known for certain of them or other dwellers in
the swamps.

## The effect of the Kalahari on Negro migrations

At first sight it might appear as though the Kalahari Desert, where
agriculture is difficult and barely rewarding and where scarcity of
perennial water precludes the regular keeping of livestock, might be
responsible for the maintenance of the division between eastern and
western strains in the Negro migrations into Southern Africa. Yet to
the immediate north of it lie broad fertile flood-plains which have
been settled by iron-working peoples, presumably Negroes, for more
than a millennium and a half; and the affinities of the modern
inhabitants of the area with peoples lying to the west of them show
that no major topographical features hinder easy access between
Barotseland and the interior of Angola. On the other hand, the near
identity among the styles of all the rich assortment of early Iron Age
sites in Zambia and Zimbabwe attests the degree of contact, or the
closeness of common origin, of the earliest metal-using agriculturalists
in the area.

Indeed, even today most of the peoples of Zambia, north-western
Moçambique and Malaŵi, and more remotely, the Shona, have traditions of ancestral connections with the Luba/Lunda areas of Zaïre
and Angola. It is of course possible that these traditions represent

superimpositions dating from the period of greatest prestige of the empire of the Mwata Yamvu of the Lunda in the seventeenth century, reinforced by the outlying nucleus of his authority represented by the Kazembe court on the Luapula River. This is certainly the case with the 'Uluba' traditions of the Maravi (Nurse 1974, 1978), whose migrations into Malaŵi occurred far too early for the Mwata Yamvu to have had any part in their causation; there is nevertheless reason to suppose that such histories may simply have supplemented memories of an actual origin from the Luba country (Schoffeleers 1972; Linden 1972).

The speakers of the south-eastern Bantu languages have no such traditions, and neither do they speak of any early knowledge of the desert. If the Rozwi were Sotho-speakers, they are far more likely, considering the emphatically eastern affiliations of the Sotho/Tswana languages, to have entered the plateau of Mashonaland from the eastern coastal plain than by crossing the Zambesi or the northern Kalahari. This probability is strengthened by the relative paucity of relics of the early period of Rozwi dominance in north-western Zimbabwe whereas in the south-west and south they are abundant (Robinson 1966). The association of the modern Shona-speaking Rozwi with the north-west could represent the outcome of their displacement first by the Luba-connected Karanga (von Sicard 1954) and later by the Ndebele. Robinson (1959) is of the opinion that the closest affinities of the Rozwi were originally with the Venda, and Abraham (1964) has collected traditions of a crossing of the Zambesi southward by Karanga during the first half of the fourteenth century. The Venda, surrounded by the Shona-speaking peoples and the Sotho/Tswana, and not all that far removed from the Tsonga as well as athwart the path of at least one Nguni back-migration into Central Africa, have lain too open to influences from all directions for any except the cultural traits unique to them in the region to be legitimate clues to their affinities. Schapera (1937) counted them and the associated Lemba as an offshoot of the Karanga, and Guthrie (1967) sought unconvincingly to establish a Karanga derivation for all the other south-eastern Bantu languages as well as Venda.

The Ngoni of Malaŵi have a tradition that at some time during their migration from Mdwandwe in Kwa Zulu they fought a battle with the Lozi (Nurse 1973). They are not very circumstantial about it, so they were probably defeated. They do not claim to have crossed the Zambesi at all before their final crossing near Tete, and it is therefore probable that their opponents were not Lozi but Rozwi. The Nyai clan is said to have joined the Ngoni at that time, and the Nyai were associated with the Rozwi rather than the Lozi. The Lozi themselves were overcome by the Kololo, a Sotho/Tswana people, a division of the Fokeng, fleeing from the Tlokwa hordes led by Mantatisi. The dynasty of their king, Sebetoane, did not last long, but the line which preceded and replaced his had traditions of contact with the Rozwi and of origins from the Mwata Yamvu of the Lunda (Mainga 1966).

These are the sole available evidences of any movements of south-eastern Bantu-speaking peoples westward to the north of the Kalahari. None of these stories makes any claim to penetration beyond the tributaries of the upper Zambesi. On the other hand, the claim of a Luba/Lunda derivation for the Karanga at least is strong and persistent. It is also apparent that the Iron Age settlers who preceded and were presumably absorbed by the ancestors of the present inhabitants of southern Zambia, Zimbabwe and the eastern Caprivizipfel formed part of a tradition common to the whole plateau area, including that westwards into Angola, but not extending, it seems, to the east coast. It is consequently possible to suggest that any close biological affinities of south-eastern with south-western peoples, apart from those ascribable to their common Negro heritage, was more likely to have been due to westerners moving eastward than to migrations in the opposite direction.

The main effect of the Kalahari could therefore have been to deflect the migrations of agriculturalists eastward. Such migrations would almost certainly have been due to population pressures from expansion in numbers among the inhabitants of the rich agricultural lands of northern and central Angola. East coast migrants coming from the north would inevitably have been held up at least for a time by the broad and rarely fordable lower Zambesi. Even at its delta the Zambesi presents not a simple series of sand bars but an unhealthy complex of brackish morasses. People moving into unsettled lands on the northern bank would hardly be inclined to make the effort of crossing it in any numbers to settle on its swampy south bank until they were forced into doing so in response to the pressure of later comers. It is thus readily comprehensible why the plateau of Mashonaland should have been inhabited long before the coastal plain. The half millennium and more which separates the first Iron Age settlements in Zimbabwe from the earliest south of the Limpopo can be explained by the unattractiveness of the eastern Kalahari to migrants coming from the west, and to the lateness of the eastern migration, presumably comprising the ancestors of the Sotho/Tswana.

There is a strong possibility that the Sotho/Tswana may once have stretched much farther to the west than they do now. It is associated with a persistent and far from improbable tradition that the Herero

immigrated into Namibia from the east. Vedder (1928c) retails a
quite circumstantial account of the arrival of the Herero among, and
their good relations with, the Tswana, who eventually allotted them
lands in the far west of their own domains, that is, in the Kaokoveld.
There are Tawana bordering on the Mbukushu still, and Ngwaketse,
Hurutshe and Fokeng have penetrated the lands previously occupied
by the Kgalagadi in the south of the Kalahari; the so-called 'Tswana
Homeland in South West Africa', however, is populated by the
descendants of Tlharo, a division of the Hurutshe, who migrated
round the southern edge of the Kalahari in the late nineteenth century.
The earliest Herero immigrants were known as Mbandu, a name
preserved as Mbanderu by their modern descendants in Botswana
and eastern Namibia. The possibility of their eastern origin is supported by their practising a cattle-cult, not a feature of the western
Bantu-speaking peoples, and observing patrilineal as well as matrilineal descent, while the Ambo and their other neighbours are
exclusively matrilineal.

It is not easy to understand why, if there were population pressures
in central Angola at the beginning of the present era, it should not
have been until fifteen hundred years later that the Iron Age came
to Namibia. It is true that the Namibian archaeological sequence is
not yet by any means clear; but it is also true that the present
exsiccation of the area north of the Auas mountains is quite recent,
and much of it probably due to human occupation. The answer may
lie in the date of arrival of the Khoi. If the Khoi did, as has been
suggested in an earlier chapter, occupy the good grazing grounds
of the north of Namibia before moving on to the richer pastures
of the western Cape, their presence could have deterred Negro
immigration as long as fifteen hundred years ago. The first Negro
inhabitants of Namibia were undoubtedly the Dama; and in the
earliest accounts we have of them the Dama lived either as slaves
of the Khoi or as peculiarly miserable and, it is suggested, incompetent and fugitive hunter–gatherers. In the status of the Dama there
stood a warning to any prospective Negro immigrants not strong
enough, or too peaceable, to resist the Khoi. It is significant that
when the Khoi encountered the Herero their method of coping
with them was to attempt their enslavement.

## The problem of the Khoisan-speaking Negroes

The Dama enigma has been quite well publicized almost since these
people were first recognized as being separate from the Herero.
Separate they most certainly are; yet many of the earliest literate

travellers in Namibia accepted without question the Nama Khoi classification of the two peoples as identical except for certain insignificant cultural differences. To the Nama there were Dama who possessed cattle and other Dama who did not: the distinction was as simple as that. Little attention was paid to the fact that the people who did not have cattle did not have a language of their own, either, but spoke Nama, while the cattle-keepers spoke an obviously Bantu language. The morphological distinction between the slender fine-featured Herero and the robuster frame and coarser physiognomy of the Dama also seems not to have been apparent to the Khoi.

The Dama, lacking an iron technology or any except the most rudimentary agricultural knowledge, which may have been acquired late, and speaking not a Bantu but a Khoisan language, were at first thought to be unique among Negro peoples. It has been claimed, nonetheless, that the Kwadi of southern Angola, who also speak a click language (Westphal 1971) and of whom only a single small band remains, resemble the Dama (Estermann 1956; de Almeida 1964; Redinha 1974). True, the Dama are not the only Negro hunter-gatherers: besides the Kwadi, there are the Pygmies of Central Africa, who are numerous and have been known for a long time, and even in Namibia itself there are the Cimba of the Kaoko-veld. But the Cimba speak Herero, and the Kwisi, who live not very far away in Angola, speak another Bantu language, Kuvale; it was the language of the Dama which posed the greatest puzzle. Not only do they speak Nama, but it is Nama which is hardly distinguishable except in accent from the language spoken by the Nama them-selves. It is not a dialect, though it is said to include a few lexical items which are not used by the Khoi, do not appear to be Khoisan, and for which a 'Sudanic' derivation has been claimed (Vedder 1923). Most of the other hunter-gatherers of Namibia are San or possibly Khoi; the Dama even in outward appearance obviously are not, and their serogenetic profile shows them emphatically to be Negro. Their origins obviously afford a rich field for speculation.

The adoption of a language completely unrelated to one's own may come to pass for a variety of reasons. Slavery is only the most familiar of these. The vast majority of persons of Negro descent in the New World speak Indo-European languages, and the biologically Celtic Icelanders were probably speaking the Scandinavian language of the dominant minority even before the two strains merged. It would be easy to assume that slavery was responsible for the linguistic position of the Dama were it not that there are numerous Dama groups who deny that their ancestors ever were slaves of the Nama. Such denials are easy to make; they do not prove facts. Whether as slaves or as a freely negotiating people, the Dama abandoned their own language and assumed that of the dominant and aggressive Nama. The only tenable alternative explanation is that they descend from pre-Khoi Negro autochthones who lived in small groups all speaking different languages, and that with the arrival of the Khoi some form of cohesion was necessary (cf. Jenny 1967). For this they would need a *lingua franca*, and adopting the language of the new-comers would enable them to communicate not only among them-selves but with the more powerful people threatening them.

The recent recognition (Nurse and Jenkins 1977*a*) that the majority of the so-called 'Black Bushmen' of Southern Africa are not in fact unusually deeply pigmented San but biologically much closer to the Negro peoples may help to provide one possible answer to the Dama riddle. All the Khoisan-speaking Negroes so far studied are, or until recently have been, hunter–gatherers, and all speak languages which fall into the Khwe-Kovab (Westphal 1971) family. Westphal suggests that in the Tshu-Khwe sub-family, which includes all the languages except Nama, !Ora and Xiri, a substratum remains on which a 'Hottentot' superstructure has been imposed. Neither he nor Köhler (1961) makes any certain identification of the substratum, but it is suggested that it may derive from one or other Bush language family. It is not improbable that from one Tshu-Khwe language to another the substratum would differ.

The Khoisan-speaking Negro hunter–gatherers extend in a narrow belt from the Atlantic coast between the sixteenth and twentieth parallels of latitude south as far eastward as the western part of Zimbabwe. In addition to the Dama, they comprise the Kwengo or Barakwengo of the Caprivizipfel, south-eastern Angola and southern Zambia, the conglomeration of peoples known as 'Masarwa' in northern Botswana, and the 'Bushmen' of Matabeleland and the Wankie Reserve. If tradition and some quite recent travellers are to be believed, there were 'Bushmen' even in the eastern part of Zimbabwe in the last century, and those may have been either Negro or San. Morphological studies extending through the first seven decades of this century have repeatedly noted the extent to which these 'Black Bushmen' differ from the San, and ascribed it to Negro admixture; only the Dama were consistently recognized as being probably unmixed with San. Serogenetic studies on the Dama, Kwengo and two groups of 'Masarwa' have established that their genetic marker profiles are basically Negro, and that there has probably been no greater admixture of San or Khoi with them than with the Southern African Negroes who speak south-eastern Bantu. With the Dama there has probably been very little Khoisan admixture at all.

This does not mean, however, that genetic studies have revealed any significant overall similarities among the various Negro peoples who hunt and gather and speak Khoisan languages. The uniqueness of the Dama is unimpaired; the other peoples all resemble their neighbours more closely than they do each other or the Dama. The closest affinities of the 'Masarwa' of north-eastern Botswana are with two Kavango peoples, the Kwangali and the Gciriku, while the Kwengo are closest to their Mbukushu neighbours and an Ambo people, the Ndonga, who resemble the' Mbukushu closely. The Kwengo are less like the Dama or the 'Masarwa' than they are like any other Negro people apart from the highly aberrant Cimba, also hunter–gatherers, and the Nkolonkadhi division of the Ambo, who are also rather atypical. All three of these Khoisan-speaking Negro peoples do show some slight resemblance to the Nama Khoi of the Rehoboth Gebiet; but the Nama of Rehoboth are almost unique among the Khoi for having not been averse to intermarriage with the Dama. Their differences from the Khoi of Keetmanshoop are larger, but, except in the case of the Kwengo, not strikingly so; though they are very nearly as remote from the San as they are from Caucasoids.

The distribution of the Khoisan-speaking Negroes is quite as curious as their linguistic oddity, scattered as they are around the northern fringe of the Kalahari desert. They often claim to have been previously owners of cattle, but to have lost them, and this could indicate that they are the descendants of unsuccessful intending immigrants into Southern Africa. The infestation of the bush of the area with tsetse flies, and of the game with trypanosomes, would provide a very good indication of how this could have come to pass. Once their cattle were lost, in an area not readily amendable to systematic agriculture, they would have had no alternative but to hunt and gather. Indeed, the ancestors of the present 'Masarwa' and Kwengo may well have been the only successful survivors of such catastrophes, that segment of the migrants which by chance included enough of the genetic equipment of hunter–gatherers to be able to make a rapid cultural adaptation and reversion to the older way of life. Even so, such success as they have had has depended much on the comparative salubrity of these northern marches; Negro migrants who penetrated further into the desert before losing their cattle may have perished without trace. Their skills are not as finely developed as those of the San, and they are more given to entering cliency relationships with more fortunate Negroes. Today they are beginning to adopt, or re-adopt, agriculture.

The catastrophe theory of their derivation, however, fails to explain why they should today be speaking Khoisan languages. It has been hypothesized (Nurse and Jenkins 1977a) that the question could be one of appropriate semantics. A Bantu language oriented towards agriculture or cattle-keeping might simply have lacked the terms, and even the deep structures, which would be required for hunting and gathering. As the survivors precariously garnered the necessary skills from San or Khoi hunter–gatherers, so they found it increasingly necessary to learn the proper words for them, and the proper ways of using them. This explanation has not been accepted by the linguists. It would account, nevertheless, for the signs that the Tshu-Khwe languages consist of Hottentot upon an undefined substratum. It does not entirely account for the fact that these languages are Khwe-Kovab, Hottentot-related, and not more closely connected with the Ju language family, the Northern Bushman of Bleek (1927).

If we accept both the theory that all the Dama were originally brought to Namibia as slaves of the Khoi, and the suggested route of the Khoi immigration propounded on cultural grounds by Cooke (1965) and elaborated by Jenkins (1972) with biological supporting evidence, and described in an earlier chapter, the matter may appear more straightforward. The 'wild' Dama, the 'Berg' and 'Peak' Dama (Vedder 1923), representing the descendants of runaway slaves, show that it was possible for the slaves of the Khoi to escape. Once that is granted, there is no reason for assuming that it only became possible for them to escape once they had reached Namibia. It is true that the Khoi must have remained longer in Nambia than they are likely to have in the neighbourhood of the tsetse belt, but on the other hand the captives are less likely to have been particularly closely guarded during the ardours of the march. The runaways in northern Botswana and the Caprivizipfel would have had to take to hunting and gathering at once. The Khoi are known to have treated their slaves harshly; the desert life may have come as a welcome relief rather than a challenge. The imposed language would be retained, but undergo the natural course of divergence with time (Swadesh 1948). The Dama, more recently in contact with and subjected to the Khoi, speak modern Khoi. The Tshu-Khwe languages represent other lines of descent from the language ancestrally spoken.

None of the foregoing can be applied to the Kwadi; but the Kwadi have not been particularly thoroughly studied, and if they do become better known before they pass into extinction it is conceivable that they, too, will be able to be fitted into this scheme. Westphal (1971) certainly includes their language in the Khwe-Kovab family.

## Servitude and gene flow in Africa

During the concluding centuries of the slave trade by far the greater number of slaves sold in the world's markets were of African origin. It was well known that these people were not all newly enslaved; a large number were disposed of by other people whose right to dispose of them was recognized in the societies from which they came. For this reason slavery has been popularly regarded as an African institution. Nevertheless, it has been shown (Douglas 1964) that ownership of people, and even the right to trade in them, occurs in a number of African societies without the people sold necessarily being regarded as chattels or as lacking rights or status. One major stumbling-block to the understanding of African servitude by non-Africans, and of slavery elsewhere by Africans, has lain in this gradation of proprietorship exercised by some people over others in Africa.

This does not mean that all types of African slavery are necessarily benign. The Nama, and some West African societies, have been notoriously unsympathetic towards their slaves, and though the Yao of Malaŵi (Mitchell 1956b) were kind to their slaves they rarely accorded them any status. The position of the slave must be considered in the context of the society to which he belongs and his position compared with other members of his own society rather than with the free in Western society. Many African societies are characterized by their readiness and willingness to absorb outsiders. Nurse (1978) has shown that where certain categories of pseudo-kinship are important, the stranger who cannot be fitted into them can enter a society at first only as slave or as ruler. Although most are initially incorporated as slaves, the striking slavery-to-kinship continuum which exists ensures that within a very short period of time they become free members of the society. For example, among the Tawana of north-western Botswana a slave mother properly married to a free man gives birth to free children and thereby becomes free herself (Tlou 1977). This is, of course, similar to Islamic practice; and in Africa the influence of Islam ought never to be disregarded.

Turnbull (1965) has described the way in which Mbuti Pygmies become incorporated into the settlements and kinship system of the non-Pygmy villagers of Zaïre. The villagers claim to be the hereditary owners of the Mbuti. Relationships of actual incorporation involve a male villager taking a Mbuti wife, but never the reverse. The offspring of the marriage are considered by both Mbuti and villagers as in every respect villagers. Data on the number of such marriages are not available for the Mbuti but Cavalli-Sforza (1972) claims that marriages between Babinga Pygmies and non-Pygmy farmers in the Central Africa Republic are rare. In both cases the gene flow takes place almost exclusively in one direction: Pygmy to farmer or villager. Santachiara-Benerecetti *et al.* (1980) have used $PGM_2^{6\ \text{Pygmy}}$, which they consider to be a true private polymorphic allele of the Pygmies (present in all three Pygmy groups examined by them), to calculate the extent of Pygmy admixture in the neighbouring farmers and obtained a value of around 5 per cent.

Gene flow in one direction (Pygmy to villagers but not *vice versa*) such as Turnbull (1965) describes, and which gene marker studies confirm, is typical of the relationships which have existed between dominant and subject peoples in other parts of Africa. It is often assumed that slavery always involved sexual subjection and that the offspring of master-slave unions would be incorporated into the slave population. Classical studies in the United States by Glass and Li (1953) indicated that the American Negro population had a genetic constitution which was approximately 30 per cent Caucasoid in origin. Dyer (1976) has assembled evidence to demonstrate that, contrary to popular belief, much of the Caucasoid contribution came either from enslaved Caucasoids or from marriages between free persons of both races.

From the results of sero-genetic studies (Nurse *et al.* 1976) it can be deduced that the enslavement of the Dama by the Nama did not involve sexual subjection either. A number of high frequency genes in the Khoi are either present in very low frequency or altogether absent in the Dama. A number of these alleles are considered to be Khoisan markers and their significant frequencies in the Southern African Negro populations imply that considerable Khoisan gene flow into them has occurred. This is the case for $A^{\text{bantu}}$, $MSHe$, $Fy^a$, $Gm^{1,13}$, $Gm^{1,21}$, $Tf^{D1}$, $P^r$, $PGM_1^6$, and $PGM_2^2$. The reverse situation, namely high frequencies of certain Dama alleles showing significant frequencies in the Khoi (Rhesus $r$, $Gm^{1,5,6}$, $PGD^C$, $AK^2$ and $Gd^A$) would support the hypothesis that the gene flow has tended to be predominantly or even exclusively in one direction, from the servile to the dominant people.

We (Nurse *et al* 1976) have previously suggested that in many instances of servitude in Africa the situation appears to be similar to that existing in Rwanda where the Tutsi exercised hegemony over Hutu and Twa but did not interbreed with them (Maquet 1961). The gift of cattle by a Tutsi to a Hutu secured the latter's services but also conferred on him the protection of a patron, the so-called *buhake* contract. A similar state of affairs exists between certain San and the Kgalagadi, who are Bantu-speaking pastoralist/agriculturalists.

A Kgalagadi will describe himself as the *munyi* of a San, a word meaning 'owner' of a thing, but, when applied to a social relationship, this is better translated 'supervisor' or 'master'. Intermarriage is said to be rare and marriages between Kgalagadi women and San are most exceptional (Silberbauer and Kuper 1966). But even to the casual observer it is clearly evident that a great deal of intermarriage has taken place, and hard genetic data confirm this fact.

It seems extremely probable, therefore, that sexual exploitation of a subservient by a dominant population or individual occurs in Africa only as a preliminary to the incorporation of the lower-status person or group into the higher-status stratum. In consequence, gene flow is almost invariably socially upward, and gene marker evidence of descent from an 'inferior' population does not indicate social misrepresentation. There is abundant evidence to suggest that this phenomenon is not confined to Africa.

### Bantu-speaking hunter–gatherers in Southern Africa

The Negro hunter–gatherers who speak Bantu languages do not, like the Khoisan-speakers, form a single recognizably coherent group. In the far north-west of Namibia, and across the Kunene in Angola, live the Cimba, who speak not a dialect of Herero but Herero itself. According to van Warmelo (1962) they represent a sub-population which has had, on account of the loss of cattle from sickness or theft, to abandon the pastoralism of their Himba neighbours, with whom they are otherwise identical. McCalman and Grobbelaar (1965), however, regard them as quite distinct from the Himba, and so does van der Merwe (1969). This cultural sequence is also claimed for the Kwisi, who live among the Herero-related Kuvale of southern Angola and speak the Kuvale language (Estermann 1956). These peoples could consequently be two of the rather rare examples of reversion *in situ* to hunting and gathering after a period of settlement (cf. Lévi-Strauss 1952), and in this contrast with the assumption of hunting and gathering by the Khoisan-speaking Negroes in the course of migration. Altogether the Cimba probably number rather less than 100 and their population structure appears to be unusual owing to an intensity of inbreeding not found in other Herero-speaking peoples. The Kwisi are very much more numerous. Redinha (1974) states that in 1960 there were more than 5000 of them, some living around the Moçâmedes desert but others in forest country.

The Kgalagadi include agriculturalists in those rather sparse parts of the area that they inhabit which do permit agriculture, but are mainly pastoralists now. A few of them do still subsist by hunting and gathering. They speak a language which is Tswana-related but has undergone significant divergence from Tswana, and live mainly along the southern fringes of the Kalahari, to which, in fact, they have given their name. They are said traditionally to have retreated to those parts following the arrival of the second Negro immigrant wave, consisting of the ancestors of the modern Rolong and Tlhaping. Until quite recently their way of life greatly resembled that of the 'Masarwa' and the San, even to the extent of the assumption of patron-client relationships with settled Negro populations (Schapera 1953). Their numbers are hard to estimate, but they probably run into several thousands.

### Seaborne Negro immigration

No Negro people ever fully developed a maritime culture, though one was in the process of development among the Guinea peoples at the time of first contact with Europeans. The unprotected coasts of Africa, with their lack of good harbours, are an adequate reason for this, and it was the situation of the offshore islands of Guinea which made the position somewhat more hopeful there. But no Negroes of their own volition ever entered Southern Africa from the sea; all who immigrated by that route were slaves.

The slaves at the Cape were not entirely, or even mainly, of African origins. 'Never', wrote the Burgher Senate to Lieutenant-Governor Bourke in 1826, 'has a single individual of the numerous tribes of savages by whom we are surrounded, ever been enslaved by us, and even those who dwell within the confines of our colony are free'. The majority came from India and the East Indies, and it was principally at times of acute labour shortage that the East India Company sent slaving expeditions up the east coast and to Madagascar. Even then slaves were acquired not by raiding, but by the purchase of persons already enslaved. Trading in slaves from the west coast, the major source of the trade to the New World, was strictly forbidden; it was, under Dutch rule, the exclusive prerogative of the West India Company; and the single shipload of slaves from the coast of modern Ghana which was landed at the Cape, and remained there for ten years before being taken away and sold in Java, made its genetic contribution mainly to the 'Coloured' and Caucasoid communities, and not to the Negroes.

Only a very small proportion of the Negroes who arrived by sea left descendants who were incorporated into the Negro populations of Southern Africa. They interbred with other slaves, with Khoi and with Caucasoids, and their progeny melted in the main into the eventual 'Cape Coloureds'. Slaves who escaped tried, if they could, to do so by stowing away in or being taken aboard vessels sailing for Europe. Once they reached the Netherlands they were free; a humane convention which existed long before a court decision established a similar situation in England (de Kock 1963). Even those who did choose to flee inland would have had to travel a very long way before they met other Negroes. They were sheltered and assisted by the Khoi, and at least one Khoi population, the Xiri or Griqua, received substantial Negro gene flow in this way (Nurse and Jenkins 1974; Nurse 1975a, 1976). The only ones likely to have entered Negro communities were those who survived shipwrecks on the Natal or Transkei coasts, and they were mainly slaves of the Portuguese, not of the Dutch.

Emancipation, a frequent occurrence under the relatively humane conditions of slavery at the Cape, had a somewhat different genetic outcome. Freed slaves were permitted to return to their native lands, usually on payment of double the passage money; those who were too poor to pay were sometimes permitted to work their passages. Women who were freed by their masters in order to marry them and legitimize their children were mostly of Asian origin and their descendants are included in the present-day Caucasoid population. A stiff emancipation tax insured against the possibility that a freed slave might become a public burden, but did not curtail the practice. Freed slaves would in their turn purchase and manumit other slaves. Some returned to Moçambique or Madagascar, many remained to blend with the 'Coloureds'; there is no record of any tendency to choose life beyond the Karroo or the Great Fish River, and settle among the Sotho/Tswana or the Cape Nguni.

## The effect of Negro immigration on the environment

The life of the hunter–gatherer produces no depletion of resources, which are self-renewing, and almost always available in abundance. When they are not they can be sought with comparative ease elsewhere. Cultivation and the care of livestock, on the other hand, demand much more systematic effort and planning, and place their practitioners far more at the mercy of the elements. There is the need in the first place to choose suitable fields or pastures; the choice may turn out to have been a bad one. Then, in the case of the agriculturalist, there is the work of preparation of the land even before the sowing or planting, and for the pastoralist the need to scout widely for water adequate for his stock. The greater the gamble the more the exploiter of the land has to take out of the land in return. The agriculturalist will not readily abandon his gardens before they have reached the limits of their fertility, and the herds and flocks of the pastoralist can drink springs or wells dry before they move on to the next. In both instances replenishment only happens gradually, if at all. The springs and wells have to wait until the water table once more rises to the level at which they are functional; renewed wild vegetation in the gardens has to pass through several cycles of growth and decay before the soil is worth cultivating again. Sometimes inappropriate weeds supervene, and the arable potential of the soil is lost for ever, or the nutrients of the topsoil have been so exhausted that no growth of any kind can take hold.

Paradoxically, the need for fallowing is appreciated most where land is in shortest supply. It is more readily apparent to the agriculturalist than the pastoralist, who is tied less firmly to particular localities. The very large settlements of Sotho/Tswana along the eastern edge of the Kalahari could hardly remain in existence were not certain fields left fallow every season, while the much sparser Nguni/Tsonga inhabitants of north-eastern Kwa Zulu, who still slash and burn the primary vegetation of their sandy plain, each season reduce the capacity of their country to support their descendants. In a dry year the pastoralist may even come into conflict with the agriculturalist, who would eventually settle near and assume control over the most reliable spring or well he could find, and at the same time be less liable than the itinerant cattleman to exhaust it. It must be admitted, however, that there is no record of this having happened in Southern Africa except where the pastoralists involved were Caucasoids, perhaps because at one time the Caucasoid growth rate must have equalled or even exceeded that of the Negroes.

It does not appear that hunter–gatherers, even under conditions for them the most ideal, ever reproduce very rapidly. The possible reasons for this will be discussed in Chapter 10. The Khoi, the principal pure pastoralists of the region, are believed to have diminished drastically in numbers during the past few centuries, mainly by disease but also by assimilation into the 'Coloured' communities and into some Negro groups. They were probably never very numerous. The population growth among the Negroes, however, has been considerable and, in recent times, rapid. Where the western Cape, the most productive part of the country, never contained more Khoi than could graze their herds there without coming into conflict with one another very frequently, the Cape Nguni or Xhosa-speaking people even in the seventeenth century were clustered thickly in the eastern Cape and pressing across the Bashee river in search of more land to subject to slash-and-burn cultivation.

This type of agriculture, when practised by an expanding people with a limited amount of land at their disposal, is more likely to impoverish the environment than almost any other form of land usage. That the Nguni have continued to practise it up to the present day is probably the most eloquent evidence there is of their long separation from and consequent lack of recent cultural interaction with the Sotho/Tswana, who abandoned it long ago, though not until it had brought about a drastic reduction of the amount of arable land remaining to them. The total amount of damage done by it would of course depend on the numbers of people carrying it out. It is even possible that the *Mfecane* may have had some impetus lent to it by the consequences of overmuch slashing and burning in Natal. These would have been exacerbated by the fact that the south-eastern Bantu-speakers, in common with the Shona-speakers with whom their cultural contact has been moderately close, and unlike the peoples to the north of the Zambesi and to the west of its head-waters, are not only patrilineal but practise cattle-cults centring on the patrilineage. In consequence their herds tend to be large, as large as those of the pastoralists but not nearly as useful. Milk, usually fermented, is used as a food, but the cattle themselves are far too important to play any significant part as a source of meat. As game became sparser around the settlements, so would small stock, and especially goats, become more important food animals; and goats are especially destructive of the environment.

The cattle-cult of the Herero in Namibia and Botswana is reinforced by their dual descent system and suggests in them a strong cultural element derived, despite their western Bantu linguistic affinities, from the east. Their lands are even less suitable for agriculture than those of the south-eastern Bantu-speakers, and they were, prior to their coming under mission influence, much more dependent on hunting and gathering to augment the products of their flocks and herds than they were on agriculture. As a result, their principal damage to the environment has come about through over-grazing. This was probably not the case to nearly so great an extent as it is now, before the best part of their lands had been appropriated by Europeans; but it should not be forgotten that several of their nineteenth-century wars with the Nama were sparked off by grazing disputes. That they, like the Nama, were by no means numerous, is not particularly relevant when one considers the relative fragility of the vegetation of the lands that both peoples inhabited.

There are, and probably have been for a long time, many more Ambo than there are Herero or Khoi, and the development of large-scale European farming and industry in Namibia has attracted so much Ambo labour that it seems legitimate to conclude that even the fertile and well-watered lands of the Ambo are becoming inadequate to support them. The Ambo country is too flat to have suffered much from the exaggerated erosion and leaching of the topsoil caused by improper contouring of the fields of Negro agriculturalists in other parts of Southern Africa, a phenomenon which is accentuated where fallowing is either not practised or is inadequate. Poor contouring constitutes one of the most significant effects of Negro settlement on the environment, and though improvements are being introduced they will probably have come too late to conserve much of what is being lost.

By contrast, the more primitive slash-and-burn methods still being practised to some extent by the Kavango in their relatively fertile valley have not yet produced any noticeably adverse effects. As might be expected, this is because of their comparatively shallow depth of settlement and their low population density. It is probably only in western Kavango that any arable land at all remains uninhabited and unexploited in Southern Africa.

# 7

# THE SOUTHERN AFRICAN
# NEGROES TODAY

The taxonomy of the Southern African Negro peoples presents an intractable problem. For all of them there are current names, hallowed by long usage and the authority of observers who encountered or heard of them early. Unfortunately, few of the names were originally applied by the people themselves to themselves. The position is a little better than it is for the San, who use no group names for themselves at all, and perhaps a little worse than it used to be for the Khoi, who gave themselves nicknames.

The name 'Sotho' was first applied by the Mbo Nguni to peoples living along the Usutu River in Swaziland. The term is one of colour: it means blackish brown, and it is not certain whether it was applied to the river, or to the people, or to their clothing (Bryant 1929). Conversely, the Tswana and Tsonga first used their respective cognates, Koni and Ngoni, for the name Nguni which has since acquired such widespread usage. Mnguni is the *isitakazelo* or praise-name of address of the Nzimela clan, which, though not of great historical significance, is widely distributed, as well as of the Qwabe and Cunu clans, which are somewhat more prominent (Bryant 1929). Soga (1930) has argued that since Mnguni was the semi-mythical source of most important Xhosa genealogies, the name 'Nguni' is most properly applied exclusively to the Cape Nguni; but Bryant's evidence shows that it was also associated with Ndwandwe and the Negroes of Natal. 'Zulu' is also an eponym, that of the founder of the sister-clan of Qwabe, and only emerged into fame and eventual wide applicability during and after the *Mfecane*. 'Xhosa' is derived from a Khoi verb stem meaning 'to destroy' (Harinck 1969), and evidently demonstrates that periodical violence attended the early contacts between Cape Nguni and Khoi in the eastern Cape. Maingard (1934) states that they were known as //Khosa by the !Ora, and that the word signifies 'angry men'. This seems unlikely, as the word is not a masculine plural (or even dual), and there is no true cognate in Nama. In Nama, however, there is a verb //kho̅, to be rough, rude, coarse, or clumsy, but the Xhosa themselves were known as //Ko̅sakwa, and the difference in aspiration suggests that the two words are unconnected.

Junod (1927) has pointed out that the name Thonga, affricated at the initial consonant to Tsonga, is simply a convenient label for any group 'made up of populations of various origin which have invaded the country coming from different parts' (p. 32). The *thakazelo* of the Tembe, the dominant clan in the south, is, significantly, Nkalanga, and a number of sections of this people do in fact claim Karanga origins. The Tsonga are not the only people to be known by this name or its cognates: there are unrelated peoples called Tonga, unrelated to each other as well, in the middle and lower Zambesi valleys and on the western shores of Lake Malaŵi. All of them admit to having arisen from the fusion of elements from a number of other peoples. It would therefore be wrong to see the migrant wave which followed the Nguni down the east coast as anything like as coherent as its precursors. It probably consisted of a number of small migrations of east coast peoples, some of them Nguni or Sotho/Tswana who had been left behind, some augmented by contributions from the plateau depending on the fortunes of the various components of the state systems there, some, as we know, revertent due to the *Mfecane*. Nevertheless, the Tsonga language shows no sign of having undergone creolization, and the culture, if we exclude the nascent state formation in Gazaland, is remarkably uniform. There must have been some centralizing or dominant influence among the Tsonga to produce such standardization; but it can no longer be identified for certain.

Similarly, the name Ambo or Mbo is borne by a variety of peoples. A Mbo strain went into the Nguni; there are Mbo in Zambia and Malaŵi; and there is no evidence that any of these populations is at all closely related to any other. The radical is variously mildly or grossly pejorative, with connotations of worthlessness, ineffectuality or decay. Peoples known by such names rarely apply them to themselves. The Tswana, by contrast, have good-humouredly assumed the name, of unknown meaning, bestowed on them by the Xhosa. Prior to the mid-eighteenth century they were known to the Khoi and through them to the European settlers as the Birikwa, or goat-people, not pejoratively but in reference to their large flocks. It is even possible that the Khoi, though they possessed fat-tailed sheep very early (Cooke 1965), may not have had any goats until they got them from the Tswana. Various derivations have been proposed for the name Herero, which is used with some pride by the people who bear it.

Among the least current but most informative names of peoples in Southern Africa is the term Lala, used primarily of one of the constituent peoples of the Nguni. Soga (1930) defines it as having the connotation of 'autochthone' or 'earlier inhabitant', and *molala* is the Tswana word for serf. Soga also states that the Lala were skilled metal-workers. Clark (1955) has pointed out that the boats shown in 'Bushman' rock-painting of fishing resemble those made by the Lala–Lamba peoples of Zambia as well as those described by Junod (1927) as in use among the Tsonga. It is possible that this slender onomastic and artefactual link may indicate a substratum, derived from Central Africa, responsible for the earliest Southern African Iron Age settlements, and overlaid and absorbed by the south-eastern Bantu-speaking peoples. Archaeological evidence can be adduced both to support and to refute this possibility.

The Venda were so styled by the North Sotho; the meaning of the name is obscure, but it appears not to be pejorative and has been adopted by those to whom it is applied. As a linguistic and cultural entity intermediate between the Kalanga and the Sotho/Tswana, and the occupants or neighbours of archaeological sites of considerable antiquity, the Venda have been the subjects of much speculation, though hardly any of it either as imaginative or as well-based as that concerning the Lemba. The Lemba, a dispersed people living mainly among the Venda and the Karanga, show morphological and cultural features very suggestive of an East African and Arab-related origin (van Warmelo 1966). The derivation of their name is uncertain.

The foregoing will have supplemented Chapter 6 in indicating that it is well-nigh impossible to place the Southern African Negro peoples in satisfactorily discrete and incontrovertible categories. Facets of culture, fragments of historical tradition, and linguistic elements overlap between peoples, and the overlapping is not always consistent. Peoples who are seen from the outside as coherent may themselves be conscious of considerable internal diversity, sometimes on grounds which may appear trivial to the onlooker but which are nevertheless very real to them. No-one acquainted with them would consider the Kavango peoples, for instance, as an entity in any other than a geographical sense. Even the linguistic similarities of the Kwangali, Mbunja, Sambyu and Gciriku—but, significantly, not the Mbukushu—relate them more closely to a congeries of other linguistically similar Central African peoples than to one another. Peoples whose cultures or languages seem to set them significantly apart from one another may conversely possess traditions which lend them a real or spurious common historical identity. At the present time many have accepted for themselves taxonomies devised by often very well-meaning but sometimes ideologically or economically motivated outsiders. It is often in terms of these dubious feats of categorization that we as biologists are compelled to examine the differences and affinities of these peoples, and we need to remain aware that once natural or imposed splitting has occurred in a people the operation of genetic drift can frequently and rapidly produce an appearance of significant divergence where even a few generations previously the sections formed a unitary and homogeneous whole.

The biological study of the Negroes of Southern Africa started well after they had attracted the interests of investigators of their societies and cultures. Even today it is by no means complete, largely because it is difficult to be sure what the significance is of a set of findings in a sample whose precise affinities are indefinable owing to an inadequacy of other samples to complement it. In most of Southern Africa information on the biological nature of the Negroes is scrappy, imprecise and inexpert. The San, whose nature is perhaps of more fundamental theoretical interest, have lured most attention. Only in Namibia has there been anything like a representative sampling of all Negro groups, and even there a certain number of *lacunae*, a certain measure of uncertainty, remain.

### The curtailment of Negro expansion

As has been explained in Chapters 1-3, the agriculturist is necessarily more militant than people who can change their habitat without substantial loss or culture conversion. The first encounters in Southern Africa between agriculturists of different cultural traditions occurred when the advancing Negroes met the advancing Caucasoids. Both represented expanding populations; both saw the land in possession of the other as desirable. It was inevitable that clashes should occur along the frontier. The border of the European-administered Cape was pushed further eastward into the expanding coastal plain, the advance consolidated and maintained by the superiority of firearms over spears. Firearms, too, were being acquired by the Khoi: too late for them to affect the interaction of the coastal Xonakwa and the Xhosa, but in sufficient time and quantity to turn the !Ora of the valley of the Orange from simple herders into bandits, predators on the settled Tswana. It is not clear to what extent this may have provoked the campaigns of Mantatisi and other Sotho marauders like Moletsane and Tshane, directed not against the !Ora but against other Negroes; it seems likely that their fury and resolution were strengthened by being assailed from two directions, by both the !Ora and Xiri (Griqua) from the north and west and the Ndebele from the east. The Sotho/Tswana had already spread some way to the south, but the disturbances of the *Mfecane* curbed even this obvious direction of expansion. When the first Trekkers reached Sotho/Tswana lands, they thought them unoccupied, so severe had the recent devastation been.

Yet even had the Sotho succeeded in penetrating further southwards, they could not have got very far before meeting the vanguard of the equally land-hungry Caucasoids. The Great Trek is notable only for its accelerated pace; it represented the inevitable advance of the suppliers of the !Ora and Griqua to share less blatantly in the spoils laid open by these peoples. Like most suppliers of arms, the Caucasoids reserved for themselves the most potent concentrations of weaponry; but even with so disproportionate an advantage they felt themselves imperilled, and of all the peoples of the *Mfecane* only they made any attempt to consolidate themselves as they expanded. Even the Zulu power was content to remain within its lines of first conquest in Natal. Attempts to expand beyond them led, in any case, to fission. The Caucasoids, however, for all their military might, considered themselves as refugees, not primarily as conquerors, and derived a greater, even though not constantly operative, sense of unity from a perhaps exaggerated awareness of having suffered a common grievance. To remain together, or at least in contact, they had to fight just that much harder, and to occupy the land just that much more effectively. Theirs was the least advertent, but the most damaging, of all the interventions in the *Mfecane*. Without deliberately setting out to do so they put a stop to any possibility of continuing Negro migration on the old pattern. By taking permanent possession of land which they in many instances considered in all honesty to be open and unclaimed, they reduced in less than a generation the country available for occupation by Negroes to a mere fraction of its previous extent. In their new habitat they grew in numbers, and the prior tenants of the soil were penned into areas which constantly shrank.

The dominance of the Caucasoids was not due only to the superiority of their firepower. Their comparative wealth played a great part; they were able to barter tools as well as arms; their implements were of finer steel more finely fashioned, they owned gadgets which eased and simplified many of the procedures which they shared with the Negroes. The carpentry of the Bantu-speakers, for instance, was hardly worthy of the name; their clay pots could be replaced with pots of iron far more commodious and infrangible than any they could bake; an iron milling machine would grind more maize in an hour than a women with a pestle and mortar could in a day, and give it a softer and more even consistency. These things which were novelties to the Negroes the Caucasoids took as a matter of course, and were able, to some extent and at a price, to supply. A trading system emerged.

Trade was not, however, an important or immediate consequence of the Great Trek. Except in organization and arms the Trekkers were hardly more advanced technologically than the Negroes, and migrated with no intention of parting with their household goods. They benefited from the prior spread of firearms into the areas they were penetrating, but their sharing in the spoils was not advertent. Perhaps, as Livingstone (1856) claimed, they did indeed set up an arms trade themselves as a means of acquiring cattle; if so, they presumably simply exploited a pre-existent assured supply of ammunition. Their main acquisition was the ostensibly vacant land. The more intense routes of trading were along the coastal plains and the valleys just inland of them, between the Cape Nguni and the Caucasoids of what was regarded pre-eminently as the 'Frontier'.

For two hundred years the official policy, both Dutch and British, towards the frontier, was exclusionist. Periodically a theoretically inviolable government line dividing the settled farms of the Caucasoids from the Negro chiefdoms would be proclaimed. About 1780 this came to be fixed along the Great Fish River, across which all forms of trade or contact were strictly prohibited (Jeffreys 1928). But except in the dry gorges of its lower reaches the Great Fish River is easily fordable, and upstream it flows not in the expected direction of the frontier, from north to south, but for long stretches from west to east. The temptation of the rich pastures on the other side constantly beckoned covetous cattle-farmers. Furthermore, there were already Cape Nguni elements settled to the west of the river, with the Khoi acting in part as intermediaries between Negroes and Caucasoids and, in the process, undergoing limited assimilation by the former (Harinck 1969). Raiding and counter-raiding became endemic; the fissive tendencies of the Cape Nguni encouraged rapid population growth and accentuated the need for more land (Hammond-Tooke 1965) while the large families and rules of limited inheritance of the Caucasoid farmers militated against their containment within the boundaries of the colony. A British expedient which might have been effective had it been enforced was the negotiation of a neutral zone to the east of the Great Fish; but the negotiations were invalid under Xhosa law, and the purposes of the neutral territory were soon violated by the establishment of a town, the institution of trade fairs, and infiltration by elements from both directions (Benyon 1974). Further annexations ensued, facilitated by internal dissensions among the Nguni, by droughts, and above all by the Cattle Killing instigated by the prophets Mhlakaza and Nomnquase in 1857, perhaps the single grestest disaster ever suffered by the Xhosa.

As the frontier advanced, so did trade, and spilled across it. Each of the many 'Kaffir Wars' was followed by a greater availability of manufactured goods in the chiefdoms of the Cape Nguni. Further still to the east, well beyond the areas governed by frontier regulations, traders established themselves in Natal in the wake of the *Mfecane* long before the Trekkers crossed the Drakensberg. These latter, finding richer land than that on the plateau but equally unoccupied, set up the short-lived Natal Republic. The reaction of the British was rapid, and, partly to guarantee the interests of the traders, a colonial administration, at first dependent on the Cape, was established in 1845. An influx of Caucasoid colonists was added to the settlement of those Trekkers who preferred to remain, and by the time the Natal Nguni began to filter back from their dispersal by the *Mfecane* insufficient land remained unoccupied even to satisfy the needs of their diminished numbers. The Zulu kingdom retained its autonomy for a while longer, but once its military power was broken there began to be Caucasoid infiltration and settlement even across the Tukela. The development of the sugar industry, with the attendant immigration of Indian Caucasoids and semi-Caucasoids, imposed further constraints on Zulu expansion.

The main impetus of the Great Trek was directed through lands made vacant by the *Mfecane*, lands bounded in part by the unitary South Sotho state being created by Moshoeshoe I to the east and by the regrouped Tswana polities and the Griqua and !Ora to the west. The Tswana lands were hardly attractive, certainly no more so than the more easterly Highveld, and there was little conflict with the Tswana. The needs and ambitions of Moshoeshoe I were more considerable, and a series of small wars raged over the possession of the relatively fertile lands of the Caledon Valley until the Sotho sought and obtained British protection. The consequent treaties did not give them all they wanted or all they needed, but provided borders which have proved fairly stable ever since. To the north, the Trekkers displaced the still migrant Ndebele of Msilikazi and were confronted only with small weak chiefdoms until they encountered the remnants of the Pedi empire in the process of re-establishment by Sekwati, son of the great Thulare whose state had been smashed by the Ndebele.

This polity is known today as Sekukhuniland, after the son of Sekwati who succeeded in resisting the Caucasoid advance until 1879. Once his power was broken, only minor resistance was offered by the fragmented peoples of the northern and eastern Transvaal, most of whom succeeded by dint of submission and accident of relative inaccessibility in retaining much of their traditional lands. The singular gynaecocracy of the Lobedu (Krige and Krige 1943), for instance, persists almost undisturbed and undiminished to the present day.

Caucasoid influences were a great deal less important in producing restraints to the expansion of the Negroes of Namibia, except insofar as they furnished the tools of war. The Ambo until recently have had sufficient lands for their needs and desires, and though many Ambo now seek employment outside of them that is not due to any limits to the means of subsistence having been reached. The same applies to the Kavango peoples, who have fallen short as yet of anything approaching complete exploitation of the lands they occupy. The Herero, on the other hand, fell into dispute over grazing lands with the Khoi long before the effective presence of Caucasoids had been established, and had been cut off from all possibility of onward movement southwards by the time the German protectorate was proclaimed in 1885 (Bley 1971). It is true that it can be argued that the Khoi who resisted the Herero were forced into doing so by the displacement of the northern Cape peoples to the north of the Orange River consequent upon Caucasoid expansion into even those inhospitable regions; and the ultimate Herero war, against the Germans in 1905-6, resulted in the near-extinction of the Herero and the occupation of their best lands by Caucasoids; but the policies earlier followed by both contending parties were hardly any more humane, and aimed, in the long run, at much the same results as the Germans, however briefly, attained.

Awareness of the limitations and dangers posed by Caucasoid settlement was probably largely responsible for the directions taken by the main expansion of Southern African Negroes following the *Mfecane*. As may have been the case with many earlier and unrecorded migrations, these were movements of refugees; but they could no longer take place into unoccupied land, or land occupied only by small hunter-gatherer parties, since all that remained of such was unsuitable for agricultural occupation. Each expanding wave, of Sebetoane to the upper Zambesi, of Soshangane into Gazaland, Msilikazi into western Zimbabwe and of the Jele and Maseko across the lower Zambesi and into the hills above Lake Malaŵi and on into East Africa, was military; no pacific migrations could have prevailed among the already consolidated peoples to the north. These reactions, however, were in any case inevitable. Even without Caucasoid settlement, the coast would one day have been reached and occupied, and no more land would remain. The principal effect of the Caucasoids was to accelerate the curtailment of expansion, and to furnish some technological innovations which helped to reduce the severity of its consequences. The technological innovations were not in the first instance dramatic, and at no point have they constituted the ultimate answer to the problems of an expanding population. Nonetheless, their secondary consequences, in combination with the political effects of the Caucasoids, have produced a revolution in Negro development and been responsible for profound changes in the environment to which not only the Negroes but all the peoples of Southern Africa have to accommodate themselves.

### The growth of urbanization

Urbanization in Southern Africa was not a European introduction. Even leaving aside the archaeological evidence in the northern Transvaal, there was traditionally a tendency towards the formation of large towns among the Sotho/Tswana. As Sansom (1974) has pointed out, this was probably brought about by economic adaptation to the environment. Whereas the Nguni peoples and some of the Sotho of the northern Transvaal occupied environments which permitted dispersed self-sufficient settlements owing to the close association of available water, wood and arable land of specific types and aspects, and an accessible hinterland for grazing, hunting and gathering, on the Highveld there is much less concentration of a variety of resources within small configurations, and the Sotho/Tswana of that region and the adjacent semi-deserts were compelled to range relatively far afield to fulfil all their basic needs. This was accomplished by the selection of residential locations providing the best access to everything that was required; sometimes twenty-five miles would separate gardens containing two kinds of soils needed by a single Hurutshe cultivator in Botswana (Werbner 1970), and in such areas the grazing requirement per head of cattle is quite extensive. The sparse rains, also, are often geographically concentrated, so that a wide distribution of lands is necessary to increase the possibility of benefiting from them. There is, moreover, a complex interrelationship between rainfall and soil types, so that the risks to crops have to be carefully assayed. The Pedi recognize relatively infertile soils which nevertheless regularly produce sparse crops, while more productive soils demand more rainfall (Mönnig 1967), and it is considered desirable that a cultivator should have some of both.

Such concentrations of habitation promote trade, the exchange of surpluses for over-produced crops of a different kind, or for arte-facts, most particularly within the towns. Conditions for export would be more favourable than on a more dispersed settlement pattern; itinerant traders could make easier contact with a greater number of people. There is also a greater potential for the centraliza-tion of potential authority, and indeed the control of the Tswana chiefs over their headmen is traditionally stronger than is the case among the Zulu. Even among the Nguni, however, the chiefly kraals fulfilled to some degree the functions of territorial centralization, and afforded the sites for ritual, deliberative and judicial gatherings; but they were hardly urban, since few of the peoples who gathered for such occasions actually lived in them.

The towns of the Caucasoids were originally small and resembled in effect European towns as they were before the Industrial Revolu-tion; that is, they were the locations of markets, of fixed retail outlets, of artisans specializing in particular technologies, and of such relatively typical European professional men as the lawyer, the doctor and the clergyman. Like European towns, those in Southern Africa attracted labour by providing in exchange the cash needed for the acquisition of consumer goods. To some extent they have acted as places where produce could be sold by Negroes, but in parallel with their development there grew up an alternative system. The trade of Negroes with Caucasoids in rural areas passed principally through the trading-store operator, who might also act as banker, postal official, labour recruiter and even as the agent of European medical treatment. This effectively by-passed the need of the Negroes for the trade of the towns, but it accentuated the attractive-ness of cash and hence of urban labour.

Urbanization in Southern Africa accelerated mightily with the openings first of diamond, then of gold mines. Not only did these have an immediate effect on the neighbourhoods of the mines them-selves, but they also led to a considerable need for enlargement of the facilities of the ports which served them. At about the same time the payment of taxes by Negroes living in the territories controlled by Caucasoids began to be enforced. The fact that the money for tax payments could be raised by the sale of cattle showed that cattle had a cash value and led on to the idea of earning money to buy cattle for the payment of bride price. Cash, too, provided a con-servable resource by means of which the impact of food shortage could be lessened. As the Negro population grew, and more and more land was taken over by the Caucasoids, the probability of individual hunger even in relatively good years increased. According to Houghton and Walton (1952), in the still comparatively modest situation of overpopulation that existed in the Keiskammahoek 'reserve' of the Ciskei in the eastern Cape Province a quarter of a century ago, even a very good harvest in the area would only produce half the nutritional requirements of the population. In time of drought this fell as low as 5 per cent. The need for labour on the mines and in the towns and ports associated with them coincided with the need of the Negro agriculturists to earn money to supplement their subsistence productivity. The prosperity of the Caucasoids waxed, while the ability of Negroes to survive was at least somewhat improved.

In addition to economic motives, however, Dubb (1974) has indicated other reasons why Southern African Negroes move to the towns. One is the fairly obvious one of the superior availability there of such tools or implements as may improve the productivity of the land: but the purchase of these would not call for a prolonged stay. More important, probably, is sheer curiosity and the wish to verify or refute hearsay accounts of town life; more interesting is the function of town or mine work as part of a *rite de passage*, recorded for the Xhosa by Dubb and for the Ngoni of Malaŵi by Nurse (1980). It is amusing to find as a corollary to this that a sojourn in a town enhances the prestige of the migrant on his return home. A motive not mentioned by Dubb but which we have noticed is the wish of some young men who cannot find or afford wives under stringent tribal rules and in the face of polygamy, to make the simpler conquest of a deracinated town girl. Some few men of the perennially emergent middle class, too, find it easier to look for suitable educated and often more 'Westernized' wives in town than in the country.

The mines no longer play as dominant a part in the growth of towns as they once did. Manufacturing has become increasingly important in South Africa and Namibia, while in Swaziland as well as the other two countries cash crops of value have led to the development of urban centres. In general, the ports export more wool than minerals, as well as a good deal of timber and sugar. Parts of the region rely to a certain extent on tourism. Such non-mining industries continue to attract Negroes from the impoverished rural areas, even in the face of vigorous official attempts in South Africa to reverse the trend. And there is always a proportion, sometimes a major proportion, of immigrant Negroes who do not return to rural areas but add to urban congestion by raising families there.

## Changing disease patterns

Seftel (1977) points out that proper epidemiological surveys have not, to date, been carried out to ascertain the exact prevalences of most of the diseases which might show significant differences between urban and rural Negroes. It is often difficult to define urban and rural populations in Southern Africa clearly because so few are homogeneous in their composition. In many ostensibly rural populations one finds a preponderance of men who have spent a large part of their working lives in the cities or on the gold mines. Large numbers of women, too, have often made long sojourns in the cities. Often there are 'impressions' gained by clinicians like Seftel which may give us some leads in a field which is largely unexplored scientifically.

Mention has been made of the increasing incidence in the towns of metabolic diseases similar to those seen in the urban West. Diabetes is not unknown among rural Southern African Negroes; nor is obesity, although severe obesity has been rare among the Pedi of Sekukhuniland (Edginton *et al.* 1972). Hypertension was also uncommon among the rural Pedi, but it has been observed by one of us to have a fairly high prevalence among Tswana women in the northern Cape. All of these are nevertheless far commoner in the cities, where their rising frequency has been ascribed to changes in dietary habits and, perhaps, increased tempo of life and general stress. Diabetes is particularly common among urban Negroes and an interracial survey in Cape Town revealed a similar prevalence as in Caucasoids of European descent (Jackson 1978).

Seftel (1978), studying Johannesburg Negroes, and Seedat *et al.* (1978) studying urban Zulu in Natal, found hypertension extremely common. Among the Zulu the overall prevalence was 25 per cent (females 27 per cent, males 23 per cent) and in the age range 35–40 years women had a higher rate than men. The blood pressures were only slightly lower than for American Negro, West Indian and Nigerian populations but considerably higher than for most Caucasoid populations. Hypertension is the commonest natural cause of death in Negro adults in Johannesburg (Seftel 1978). An interesting fact to emerge from the study of Seedat *et al.* (1978) was the positive correlation between hypertension and years of residence in the city. Walker and Walker (1978) have shown that Negro schoolchildren aged ten to twelve years have lower blood pressures than those of Caucasoid (White) schoolchildren. Degenerative cardiac lesions other than the familiar idiopathic congestive cardiomyopathy, duodenal ulcers, dental caries, appendicitis and varicose veins are seen proportionately far more often in the Negroes of the cities than they are in rural areas. Robbs and Moshal (1979) found a twelve-fold increase in rate of admission to hospital of Negroes in Durban with duodenal ulceration. There was some evidence that change in diet from the traditional high-fibre one of unrefined carbohydrate in the form of maize, meat and milk to a diet consisting mainly of refined carbohydrate, meat and vegetables, might be responsible in those with ulcers. There was also evidence of increasing stresses and responsibilities accompanying better job opportunities. Conversely, there has been some decline in the prevalence of infectious diseases.

### Infectious disease as a contemporary problem

Although the urban communities enjoy higher living standards and better public and personal health services there still exist in the cities, as in the rural areas, unacceptably high prevalences of tuberculosis, veneral disease and infantile gastroenteritis; rheumatic heart disease is also extremely common (Seftel 1977). Malaria is the only infectious disease known with certainty to have been present in the Negro populations before they came into contact, direct or indirect, with Europeans or Arabs. It is unlikely, however, to have been the only one. Leprosy is an old and familiar disease of the East Coast of Africa and could have spread southwards with the south-eastern groups of migrants. The question of the introduction of the venereal diseases perhaps admits of more debate. Granuloma inguinale and lymphogranuloma venereum are probably of African origin, though their relative rarity in Southern Africa until recently may suggest their introduction or perhaps reinforcement by migrant labourers from Malaŵi, Tanzania, Angola and Moçambique. Gonorrhoea has been widespread throughout the ancient world for a long time, and in all probability entered Southern Africa by the east coast route; the types of promiscuity traditionally practised, however, would not be such as to promote its easy or rapid spread, and it could have retained a fairly circumscribed distribution until the breakdown in recent years of the more conservative patterns of sexual behaviour. Endemic non-venereal syphilis and yaws are both present in Botswana (Murray et al. 1956), but venereal treponematosis could first have entered the region from the East Coast or via the Cape. Trichomoniasis appears to be that much commoner in Caucasoids as to suggest that it has been introduced by them; but this may be due to the superior diagnostic facilities which they enjoy.

Of the remaining bacterial infectious diseases tuberculosis either is or until recently has been the most important. The pattern of response of the Southern African Negro to this disease implies a very low degree of prior immunity, and its introduction is almost certainly relatively recent. It could have come from East Africa, to which it appears to have been carried from the Middle East, but a more likely route of spread appears to have been outwards from the developing urban centres, particularly those which have grown up in association with the mines. Labouring in the mines has introduced its own special ancillary mode of spread; pneumoconiosis, in the form of silicosis, is a potent predisposing factor in the disease. In the form of asbestosis, found among the asbestos workers of the northern Cape, it may lead on not only to tuberculosis or pyogenic lung abscess but more seriously to malignant mesothelioma. Infectious respiratory disease of more acute type has probably always been found in Southern Africa, to judge from those rural communities, such as the migrant San, whose contact with later comers has been minimal, but the shortage of fuel has led to its being a particular scourge in recent times on the cold highveld. There have been serious outbreaks of diptheria and whooping cough during the earlier part of this century. Both of these would be of external origin, and both have now been brought under control. Tetanus, with a reservoir of the causative organism in the guts of cattle, is more likely to have been longer in the region. The dysenteries, again, are probably of ancient provenance, but typhoid fever and other salmonelloses probably arrived more recently, as did the European louse-born typhus, though tick-borne typhus, and tick-borne borreliosis, show so general a pattern of distribution throughout Africa as to suggest their long endemicity. Amoebiasis, commonest in Natal, may have been introduced by Indian immigrants of the late nineteenth century. Plague, however, enzoötic among the rodents and periodically going on to an epizoötic and sometimes, in northern Botswana and Namibia, to an epidemic, undoubtedly represent a relatively recent spread from Europe or the Middle East. Cholera has occurred mainly as sporadic outbreaks.

Of the viruses pathogenic to man a great number appear to be indigenous and to produce a reasonable level of immunity without great morbidity in the populations. We have no information about the spread of the smallpox epidemic of 1713 to, or its results in, the Negro peoples, but the disease is known to have occurred in Negroes both in the major and the attentuated minor form, and had probably been found among them for far longer than it had among the Khoisan. The other virus diseases which have produced major effects in Negroes although they are generally rather mild in Caucasoids, are varicella and most notably measles. Measles in Africa is frequently a killing disease, particularly when respiratory sequelae supervene on a severe febrile phase in children. Mumps and rubella are less serious, though it is possible that they, like measles, may be relatively recent arrivals. The Epstein–Barr virus, and infectious mononucleosis, appear to be of much longer antiquity in the region, and the association with malaria, Burkitt lymphoma, and childhood leukaemia has resulted in a fairly intensive ongoing study of this subject in the region.

The curbing of introduced infections has proceeded faster in Southern Africa than the management of the diseases which were originally there, but in many cases, and most notably that of tuberculosis, the measures taken have fallen far short of controlling the disease. The frequency of glucose-6-phosphate dehydrogenase deficiency, and of the Gd(A) variant, strongly indicate that malaria has been far more important in the past than it is at present, though the rarity of Haemoglobin S except in the Ambo and Kavango suggests that its greatest importance lay at a remove in time beyond the migration of the south-eastern people south of the Limpopo, and that even before the introduction of European medicine its prevalence in the south-east at least was no more than epidemic following the rainy season. Certainly its clinical severity when it strikes suggests a rather circumscribed acquired immunity in the modern Southern African Negro, except for the northernmost inhabitants of Botswana and Namibia. Malaria prophylaxis is not much practised, wisely enough considering that the disease is hardly endemic, but other control measures, such as the spraying of mosquito breeding sites, mass treatment during epidemics, and residual spraying of living areas at the appropriate times of the year, have reduced the morbidity and mortality that could have been expected without the intervention of European medicine. Trypanosomiasis is rarely seen clinically, though the fact that when it is it is usually in a visitor to the tsetse-infested areas, suggests that a certain measure of immunity may be present in those who live there and consequently that sub-clinical infections probably occur in them. It is, of course, uncertain whether this immunity is acquired or the result of a long period of selection for some inherited factors or factors which may confer protection. Similar protection against many arboviruses may also exist. Leishmaniasis has been found recently to occur in parts of Namibia; it is of a distinct species, with an animal reservoir in the hyrax (Grové and Ledger 1975), but there is no indication that it has ever been widespread in humans there, and indeed a proportionately higher number of infections have been found in Caucasoids than in Negroes, Khoi or San. Once again, the presence of inherited or acquired immunity to the disease is debatable.

We can be sure of the relatively recent extension of only one important type of helminthiasis. Schistosomiasis appears to be increasing and widening its area of prevalence, and even to be spreading to the Highveld as more dams are constructed. Snail control measures aimed at eradicating the intermediate host are being carried out, but it is too early yet to judge their effectiveness. Both of the species pathogenic to man elsewhere in Africa are found; *S. haematobium* is more prevalent than *S. mansoni* and appears to precede it in its spread. Both of the widely distributed hookworm species are present, too, though only *N. americanus* has so far been found among the desert-dwellers. Hydatid disease affects all races indiscriminately; it is becoming less common than it was earlier in the century, but it was certainly introduced into the region from outside with the development of sheep-rearing and the wool industry. It seems likely that ascariasis and infestations with threadworms and whipworms accompanied the early migrations of the Negroes, and though modern treatment has probably reduced the incidence of the more serious infestations of ascariasis, the other two are hardly debilitating and are often left untreated. There is sometimes vivid awareness of infestations with worms, and some of the herbalists' remedies appear to be effective.

## Genetic disease and congenital malformations

A moderately wide range of inherited diseases might be expected to occur in the Southern African Negroes, though generations of uncompensated selection against the most deleterious would keep the frequencies of the responsible genes low and result in a preponderance of new mutations among the cases of dominantly inherited conditions detected. Selection against recessively inherited diseases would be likely to operate much more slowly, though the pace of the elimination of the responsible genes would probably differ according to the degree of homozygosity promoted in different populations by different marriage patterns. Similarly, matrilocality and patrilocality would have a profound limiting or dispersing effect on genes for X-linked conditions.

The most notable inherited disease of the Southern African Negroes is oculocutaneous albinism, which has as conspicuously high a prevalence among them as it apparently has in most sub-Saharan Negro populations. Kromberg and Jenkins (1982a) report a frequency of cases of one in 3900 among the inhabitants of Soweto, the Negro-inhabited city adjoining Johannesburg. This gives a carrier rate for this recessively inherited condition of approximately one in 32.

Kromberg and Jenkins found that the rates varied in the different tribal or language groups, and suggest that the Southern Sotho case rate of one in 2041 might well reflect the high frequency of consanguineous marriage, and hence of homozygosity, known to prevail in this population. The forces leading to the high prevalence of oculocutaneous albinism in tropical populations have not yet been identified. It is not possible to choose with certainty between heterozygote advantage and random drift, but the fact that the phenotype occurs at a frequency too high for the gene or genes to be likely to be maintained in existence by recurrent mutation marginally favours the former possibility. Oettlé (1963b) suggested that the carrier may have a paler skin colour than the normal individual and consequently possess some social, and hence marital, advantage. If this is the case, it must argue a high degree of subliminal perception. Few obligate heterozygotes for the condition are obviously lighter in colour to the naked eye.

It has been claimed that the incidence of retinoblastoma among Southern African Negroes is, at one in every 10 000 live births, two to three times higher than in other populations, where the incidence tends to range between one in 17 000 and one in 32 000 (Freedman and Goldberg 1976). A subsequent report by Goldberg (1977) suggested that the rate might be rising, but the contention was based on too few cases from too limited a geographical distribution for its reliability to be evaluated easily.

Congenital bleeding disorders have been identified in a number of Southern African Negroes. As elsewhere in the world, classical haemophilia A is by far and away the most common. Its prevalence of 2.8 per 100 000 is approximately half that among the Caucasoids of South Africa (Lurie and Jenkins 1975). The other type of Factor VIII insufficiency, Von Willebrand's disease, has been reported in a Southern African Negro only once (Gomperts et al. 1969). A low prevalence of Huntington's chorea in the Negro peoples of Southern Africa has been noted by Hayden et al. (1980), who identified three cases out of a total population of nineteen million. There had been several earlier reports of cases in the region. The contrast with the high prevalence in the combined Caucasoid and hybrid populations has been highlighted by the estimate for the latter of a heterozygote frequency of 67 per million (Hayden and Beighton 1982).

The commonest birth defect found among nearly 30 000 infants born to Negro mothers at Baragwanath Hospital, Johannesburg, was post-axial polydactyly (10.4 per 1000 births), followed by talipes equinovarus (1.55 per 1000 births), hydrocephalus (1.3 per 1000 births), spina bifida and anencephaly (0.78 and 0.40 per 1000 births, respectively) and facial clefts (0.30 per 1000 births) (Kromberg and Jenkins 1982b). It seems, therefore, that whereas hydrocephalus is more common in South African Negroes when compared with Caucasoids, the other congenital malformations are significantly less common. The demonstration by Smithells *et al.* (1980) that the risk of neural tube defect can be considerably reduced by vitamin supplementation of the diet before and after conception has drawn attention to the importance of environmental factors in the aetiology of this group of birth defects. The monitoring of the prevalence of birth defects in developing populations as diets change might well give valuable clues to their aetiological mechanisms.

### Urbanization and the pattern of malignant disease

There are indications that the pattern of malignant disease, other than in children, showed an interesting geographical distribution even before urbanization began to have any effect on the biology of the Southern African Negroes. Today, in the rural areas, a significant negative association persists between the two commonest malignancies, primary hepatoma and carcinoma of the oesophagus. The distributions are obviously dependent on environmental factors, though these have not been identified with certainty (Harington *et al.* 1975). Whatever they are, they are probably not of recent development, except in the case of mesothelioma. An interesting change in one pattern of malignancy has been an alleged rise in the frequency of cervical carcinoma in women with the decline of the practice of circumcision as fewer and fewer men present themselves for the traditional initiation rites. There are no secure data about the incidence of industrial malignancies, except, once, more, mesothelioma. Gastro-intestinal malignancies are not yet common, but their incidence is rising, especially in the cities.

Studies of malignant disease in Negro men working on the gold mines of the Witwatersrand area, carried out by Berman (1935) and covering the period 1925–1933, gave an average annual crude occurrence rate per 100 000 of approximately 15, which was surprisingly high in view of the relatively young age of the subjects. A more recent study, initiated by the late Dr A. G. Oettlé in 1964 and completed by his co-workers in 1968 (Robertson *et al.* 1971a), found that the crude rate had increased over three-fold to 50.9 per 100 000 man-years in the 30 years. Berman found that liver and bladder cancers accounted for 90.5 per cent and 3.5 per cent respectively of all the cancers, and Robertson *et al.* (1971c) found that liver cancer was still the commonest although it now accounted for only 52.6 per cent of the total. The proportion of bladder cancer had risen slightly to 5 per cent.

Robertson *et al.* (1971c) investigated differences in cancer incidence according to the area of origin of the miners and the associations of those areas with certain specific tribes. Liver cancer is most frequent among miners from Moçambique, which at 78.1 per 100 000 has the highest crude cancer rate in the world. A high rate, 29.3 per 100 000, was also found among Zulu. This is in accordance with the high rate found in Zulu in Durban by Schonland and Bradshaw (1968). A high liver cancer rate was also found among Shangana/Tonga seen at Baragwanath Hospital near Johannesburg in another survey carried out in 1960-1964 by Robertson *et al.* (1971a), in spite of the fact that Shangana/Tonga of both sexes show very low overall cancer incidence.

More recently, a ten-year survey based on biopsy and autopsy histology at Baragwanath Hospital for the period 1966-1975 (Isaacson *et al.* 1978) showed the crude occurrence rate of liver cancer to be about 30 per 100 000. The authors pointed out, however, that this figure cannot be considered a reliable index of the incidence of the tumour in this population because, in recent years, extensive use has been made of serum alpha-foeto-protein levels in diagnosing the tumour. Liver biopsy is not invariably carried out. The male preponderance is apparent in this study (5.5:1), as it could not be in the studies on gold miners.

Although primary liver cancer is still extremely common in Negroes, there is evidence that its prevalence is declining. Harington *et al.* (1975) reported that over the eight-year period 1964-1971 the prevalence in miners from Moçambique showed a 44 per cent fall. For 1972-1974 Bradshaw and Harrington (1976) report that the rates dropped to 50 per cent of what they were in 1964. Over an eleven-year period this frequency of liver cancer in these miners from Moçambique consequently halved itself. The explanation for this steep fall in rates is not clear. The theory which postulates that the fungal toxin aflatoxin which may be ingested particularly in mouldy ground nuts is hepatotoxic (Oettlé 1965) has received some support. Van Rensburg *et al.* (1975) showed that in the area of Moçambique with the highest known liver cancer rate the diet contained the greatest amount of aflatoxin. Another theory incriminates the hepatitis B virus and rests on the observation of Kew *et al.* (1974) that it is found in the serum of 60 per cent of Negro patients with primary liver cancer, but in only 7 per cent of healthy controls.

Oesophageal cancer accounted for 13.0 per cent of all the cancers in the 1964-1968 survey of Robertson *et al.* (1971c). Berman

(1935) made no mention of cancer of this site in his 1925–1933 survey. This would suggest that cancer of the oesophagus has only made its appearance in South Africa since the Second World War. Burrell (1957) pointed out that oesophageal cancer was unusually common among Xhosa in the Transkei, and the more recent study of Robertson *et al.* (1971c) has confirmed a rate twice as high as that in the group with the next highest rate: 16.2 per 100 000 compared with 8.4 per 100 000 for miners originating in the Transvaal. There were lower rates among men from Natal (4.2) Botswana (3.0) and 'Northern Territories' (3.2) and there were no cases among men from the Orange Free State or Swaziland, although, admittedly, the numbers of workers from these areas were small. Among nearly half a million miners from Moçambique only six had cancer of the oesophagus, giving a crude rate of 1.4 per 100 000. In the same population there were 338 cases of liver cancer, giving the rate of 78.1 per 100 000 cited above.

Isaacson *et al.* (1978) have also drawn attention to the recent increase in incidence of oesophageal cancer in the Negro population of Southern Africa. From 1912 to 1927 no cases were diagnosed at The South African Institute for Medical Research in Johannesburg in spite of the fact that over 200 primary liver cancers were diagnosed in 1953–1955 alone. Higginson and Oettlé (1960) encountered 53 cases of the disease, giving a crude occurrence rate among males of 7.7 per 100 000. In 1958–1962 it was 15.1 per 100 000 males, and in the most recent period reviewed, 1966–1975, no fewer than 1331 cases of squamous carcinoma of the oesophagus were diagnosed, giving an average annual incidence in males of 23.68 per 100 000. The ratio of males to females is now around 5:1 whereas in the 1950–1954 period it was nearly 13:1 (Robertson *et al.* 1971a).

There is a high incidence of oesophageal cancer in both rural and urban populations. The cases occurring among urban residents of the Witwatersrand cannot be accounted for by recent immigration from rural areas. Oettlé (1963a) showed that the average duration of urbanization of Johannesburg men with oesophageal cancer was 19.5 years compared with 16.8 years among men with cancer of other sites, matched for age. Schonland and Bradshaw (1969) showed that oesophageal cancer is as frequent among the Zulu of Natal as it is among Xhosa of the Transkei. Johannesburg Negroes show a strikingly high incidence in the younger age group when compared with rates in other countries, and the striking preponderance of males over females sets the figures apart from those for several populations. Interestingly enough, in Californian 'Blacks', not all of whom are necessarily of Negro descent, the male to female ratio is 3.5:1 (Segi *et al.* 1979).

Cancer of the breast is significantly less frequent in Johannesburg Negroes than in Caucasoid populations in Denmark, England and Wales, but resembles that in Niger. According to Isaacson *et al.* (1978), it is 5.8 per 100 000. The histological types and grading do not vary markedly from other populations, contrary to the claim by Pearson (1936) from Nigeria that the tumour is more malignant in Negroes than in Caucasoids. The difference in prevalence rates is most likely due to prolonged breast feeding and greater numbers of pregnancies than is the case in developed countries.

The recent study of Isaacson *et al.* (1978) has cast serious doubt on the previously accepted view that among South African Negroes cancer of the cervix attains one of the highest incidences in the world. They report an average annual incidence of 19.3 per 100 000 in Soweto compared with 38 in Denmark, 20 in England and Wales, 24 in Californian 'Blacks' and 7 in Nigeria. The role played by circumcision of male partners in its aetiology is not clear: te Groen and Rose (1974) found that in the Negro population of the Eastern Cape Province, where circumcision is widely practiced, the rate of carcinoma of the cervix was extremely high. There is no evidence that there has been a rise in the frequency of cervical carcinoma in women with the decline of the practice of circumcision as fewer and fewer men present themselves for the traditional initiation rites. It must be pointed out, however, that these rites usually take place after puberty.

Late circumcision, as practised by some Southern African Negro peoples, can be expected to have less of a protective effect against penile cancer than does early circumcision. The rate in Johannesburg Negroes is 1.3 per 100 000, comparable with the high rate of 1.4 reported from Denmark. Approximately 4.2 per cent of men in Soweto have been circumcised.

Squamous carcinoma of the antrum occurs in Johannesburg Negro men at a rate close to that found in men in Western European countries and it is possible that snuff, to which various ingredients, especially burnt vegetable matter, have been added, is one of the aetiological agents. It is five times more common in men than women.

Carcinoma of the bronchus showed an elevated rate among Negro men in Johannesburg in the period 1958–1962 when compared with 1953–1955 (Robertson *et al.* 1971a) which is in keeping with the finding of Schonland and Bradshaw (1968) in Durban among the Zulu. These latter workers claimed that the incidence is approaching the level found in England and is becoming as great a problem as oesophageal cancer. Such a high incidence does not appear to exist on the Witwatersrand (Isaacson *et al.* 1978). An increase is undoubtedly taking place and is probably due to the fact that more persons are taking up the smoking of tobacco cigarettes rather than the more traditional hemp or pipe tobacco.

Isaacson *et al.* (1978) analysed data on large bowel cancer in Witwatersrand Negroes extending back over a twenty year period, but failed to demonstrate any significant change in the incidence. These findings are surprising in view of the presumed change in dietary habits among Negroes in the urban areas; but, perhaps, a longer 'latent' period is needed before a trend can be demonstrated. It is possible, of course, that the change to a mostly Western-type diet is not, in fact, taking place. Manning *et al.* (1974) showed that in Cape Town Negroes although there was an increasing taste for sophisticated town 'foods', the basic partiality for carbohydrate foods still prevailed. Economic factors might also inhibit any dramatic change to a Western diet.

## Changes in dietary practice

Data on the incidence and prevalence of the dietary diseases of deficiency and prosperity resulting from or associated with the changes in lifestyle of Negroes both in the towns and in the countryside are surprisingly sparse (Walker 1979). Such conditions bear relatively little on non-sedentary populations, and have not been dealt with at all in the foregoing chapters on the Khoi and San. When these practise their traditional modes of food procurement, they avoid both extremes. When they settle, their problems might be expected to resemble those of the neighbours whose patterns of settlement and techniques of food procurement they emulate. We may consequently expect the nutritional practices and problems of new-settled Khoi or San to mimic those of the nearest Negroes or 'Coloureds', and not to be specific to them. In the cases of most San, and most avowed Khoi, settlement will be rural, and what follows will most probably apply, with minor variations, to them as much as to the Negroes.

The diets of all three peoples tended traditionally to be high in fibre, and to include a relatively high proportion of fresh vegetables, many of them having an appreciable protein content. Energy-providing foods were mainly cereals for the agriculturists, but root vegetables for pastoralists and hunter–gatherers. All three modes of subsistence gave relatively regular access to meat, though this was probably consumed less often by the agriculturists. These last were particularly subject to the vagaries of season and climatic irregularities, being tied to the land. Marasmus and kwashiorkor in Southern Africa are virtually confined to the children of settled folk, and tend to be more common and more severe in the cities. Protein–calorie malnutrition in adults is also much commoner there than in the rural areas. The reasons for this may be found in the adoption of a cash economy and hence linked with the difficulty of procuring food in the cities except through the medium of cash, which in conditions of high unemployment is often in short supply. Associated with this is the tendency to eat a less traditional diet in the cities, to some extent on account of the standardization of available commodities. What can be brought is often more expensive and nutritionally less valuable than the rural foodstuff which it supplants; an example of this is the highly extracted maize meal which takes more effort to produce and has a lower protein and vitamin content than the crude variety.

A nutritional phenomenon which has occurred independently of the influence of the economic system and is more pronounced in the rural areas is the surprisingly high incidence of pellagra. Simultaneously with all the other changes which have overtaken them during the past few hundred years, the Southern African Negroes have been steadily abandoning their ancient grain staples, sorghum and bullrush or eleusine millet, and taking to cultivating maize, which is easier to till and produces bulkier yields but which is deficient in nicotinic acid. Provided that other sources of this substance, such as leaf vegetables and to a lesser extent meat, are available, this can be compensated for; but in a dry year, with few wild foods, or in poor circumstances in the city, maize may be the only food available, and the disease strikes, particularly at the women and children.

The economic differential can become even more marked with the adoption of a Western diet. Sometimes, what this amounts to is the consumption of as much of a Western range of foodstuffs as can be afforded. These are naturally the cheapest, mainly carbohydrate; most of the Western sources of protein and vitamins are too expensive, and the Western wheaten staples may be eaten in smaller quantities than the traditional maize would, so that occasionally even energy production suffers. A certain proportion of the cash expenditure, too, will be on costly items of limited nutritional value, such as tea and coffee, or on unfermented milk, which few Negroes can metabolize to full advantage.

These urban modifications, which owing to the influence of

trading-stores and of returning labour migrants are spreading into the countryside, go towards producing a diet which, though it may in most instances be adequate in energy and may contain enough gross protein, is low in animal protein and fat, low in calcium, frequently high in iron and borderline or low in most vitamins (Walker 1979). The fibre content is generally still high in the rural areas, but in towns its reduction probably plays a part in the increasing incidence of gut pathology.

## The limits of prior adaptation

Changes in the pattern of disease among Southern African Negroes can consequently be classified as those due to changes in the resource pattern, the two main types of resource involved being diet and the availability of effective therapy, and those due to changes in the environment, largely the effects of urbanization and the constriction of rural opportunities but also including a shifting distribution in the exposure to pathogenic agents. The two classes of course overlap: curtailment of traditional methods of food procurement by alteration in the availability of land leads to a need for new types of food, which may be better or worse nutritionally but are often in either case contributory to the pathogenesis of new diseases. Town life means exposure to a new range of infective and industrial hazards but also superior access to the means of combatting them. It may be at this level that the significance of prior adaptation recedes: nothing in the selective history of the Southern African Negroes could make their present position one which they themselves would have chosen.

# 8
# THE SEABORNE
# IMMIGRATIONS

The year 1648 saw the conclusions of both the Peace of Westphalia and the Treaty of Münster. The United Provinces of the Netherlands had been engaged for eighty years in an independence struggle with the Kings of Spain, and had profited more greatly thereby than most belligerent powers ever succeed in doing. Not only had independence been won and recognized, but the fortuitous union of Portugal with Spain had justified the Dutch in combatting the seaborne empire of the Portuguese, and taking over much of it. From the Sea Beggars of the 1570s had grown the struggling West India Company, and the immensely more powerful East India Company. While the rest of western Europe spent thirty years indulging in a convoluted effort either to establish the Elector Palatine on, or remove him from, the throne of Bohemia, the industrious and single-minded Dutch played their own game, profited where they could, and emerged as the greatest maritime power in the world.

The Treaty of Münster was conclusive; the Peace of Westphalia merely ended a war which had gone irrespectively badly for most of the combatants, leaving the Protestant factions dispirited and the Catholics obstinate and impoverished. Particularly in the Rhineland there remained large numbers of uprooted and destitute people to whom the contiguous prosperity of the Netherlands, more tolerant and hospitable by reputation than in fact, provided some gleam of hope and encouragement. Holland and Zeeland had become rich by exploiting the poverty of their seafarers until the seafarers themselves had slowly amassed enough wealth to leave the sea; the merchant companies faced an acute labour shortage which the Rhinelanders were providentially available to fill. From the viewpoint of the present, it might seem that the lower decks of Dutch ships became progressively less Dutch and more German, but the politics and boundaries of the seventeenth century were not those of today. Low Germany, in language and culture, extended to the Pas de Calais; the Spanish Netherlands were less uniformly Catholic, and the United Provinces more varied in religion, than they were to be three centuries later. The Flemings whom the Spaniards allowed to move northward and the Germans who flocked to the Maas and the Scheldt were not strictly foreigners any more than the citizens of one small German principality were in the eyes of those of the adjacent bishopric.

## The Caucasoids at the Cape

Consequently, at the time of the founding in 1652 of the Dutch settlement at Table Bay, the men who manned the ships of the East India Company originated to a very large extent from areas which would not now be considered to have very much in common with the Netherlands. These were the men from among whom the first Caucasoid settlers at the Cape were selected, men who were considered to have deserved land as a result of their faithful service to the Company and who were still sufficiently close to their agricultural origins to be able to run efficiently the smallholdings which were to provide fresh food for passing ships. Settlement, in fact, developed slowly; the first enterprise of the Company was the Company's gardens, and it was only when it became clear that simple shore-based service provided little incentive for efficient production that permanent settlement began to be encouraged.

No such sudden establishment of a colony as happened in New England came about here. Men, and even families, would stay a few years or a decade and then opt to return to Europe. Not all who stayed, or all who returned, were Dutch. There were people who, arriving with patronymics, dropped them and retained names to remind them of their places of origin: van Antwerpen, van Deventer; there were Germans who turned their surnames gradually over the ensuing years to Dutch forms, as Grünewald became Groenewald and Steinkampf, Steenkamp; there were Scots and Irish, luckier than many of their contemporaries, who had escaped from the 'pacifications' of Cromwell, so that today there are Gordons and others whose British connections are three centuries forgotten; there were adventurers from Scandinavia and the Baltic and, as the name Ferreira among others attests, even some from Iberia. But the majority by far were from Low Germany, Calvinist and of little social prominence.

To these were added, following the revocation of the Edict of

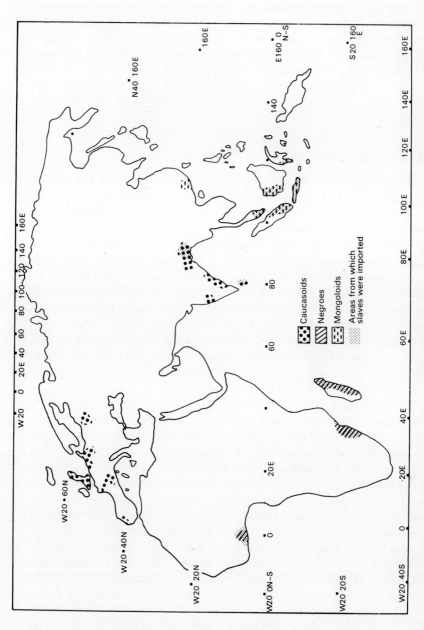

Fig. 8.1. Areas of origin of seaborne immigrants into Southern Africa. Migration of south-eastern Europeans principally post-dates the Second World War, while from the Baltic most immigrants were Jewish. This source became less important after the First World War.

Nantes in 1685, a sudden influx of French Huguenots of far greater social diversity. The seventeenth century French household was larger and more emphatically stratified than the Dutch one. Few of the earliest settlers at the Cape came from backgrounds in which an entourage of retainers were kept; for many of the leading French settlers it was a matter of course, and familiar servitors were as necessary to make them feel at home as familiar furniture was. Only rich or influential Huguenots, and those who depended on them, would have escaped forcible conversion and reached the Netherlands in the first place. There is some obscurity about their comportment and reputation in the United Provinces. Certainly, important reasons given for their further migration were their relative indigence, their industry and particularly their knowledge of wine farming. But the progress of their settlement at the Cape affords a comic and bloodless footnote to the history of the class war.

Emigration was lowest at either end of the social scale, with the unskilled and the high nobility least inclined to leave. Nevertheless, among the French settlers were Marais, who may have been among the cousins of the kings of France, who bore that name, du Plessis, sharing the surname of Richelieu, representatives of the great judicial family of Malesherbes, and several others of equivalent lineage. In the face of the bourgeois, peasant and artisan culture of those who had come before them, such people would wish to retain what they could of their own, and most blatantly their language. There were attempts to ensure that they would not all congregate in one area, that they would learn Dutch, and that they would not practise marital exclusion. Concessions were gradually made. Extending themselves geographically but maintaining a fair degree of cohesion proved economically advantageous to the French, and the economic benefits were consolidated by a concession on the language issue. By the middle of the eighteenth century the French language was no longer spoken at the Cape, but up to the present day there persists a tendency for persons of Huguenot descent to choose spouses from the same tradition. This is likely to be denied with some heat by many Afrikaners intent on celebrating the unity of their people, but may be easily, if somewhat tediously, substantiated by careful perusal of many of the family genealogical registers published during the past decades. That this reflects a trend of the eighteenth and nineteenth centuries as well can be shown similarly from the pages of the most general and respected of these compilations, the *Geslagsregisters* of C. C. de Villiers (Pama 1966).

The French immigration formed a single episode. The influx of Dutch and Germans coincided with and went on beyond it. As a means of rewarding Company employees, the allocation of land to which there was no other title but that of the Company, which could acquire it by simply staking a claim, was economical and satisfying to all parties. The Khoi perhaps did not find it very easy to understand how they came to be excluded from areas where they had previously grazed their cattle; but they were accustomed to giving way to other occupants of grazing, and if the occupants were now permanent there was nothing they could do about it. The colonists were at first compliant to the designs of the Company, which provided them with almost the only market for their produce; later, resentful of too strict controls, they would try to sell their provisions directly to the increasing numbers of foreign ships which frequented the roadstead. Men deserted from such ships, and in a community short of labour, and more particularly artisans, found employment. The links of the Orange family with the Stuarts deterred rather than attracted the Scots and Irish after the revolution of 1688, and the upheavals in 1715 and 1745 produced little change. A few mercenaries, some of them Slavic, Scandinavian or English but the bulk still German, with increasing numbers of Brandenburgers, swelled the numbers during the eighteenth century.

The question of the Jewish component in the early Afrikaners is not easily resolved. There were certainly Jews, many of them Christianized, among the eighteenth-century settlers, and the Marx family among the Afrikaners may have had a Jewish progenitor. It has tacitly been assumed that such Jews as came from Holland were necessarily Sephardic. Nonetheless, the persecution of the Jews in eastern Europe had already begun in the seventeenth century, and it is a matter of record that numbers of Ashkenazim sought asylum in Holland and in the Rhineland. Whether this contribution could be regarded as genetically much different from that brought by other Rhinelanders, is naturally uncertain, but the extent to which the Jews of Europe resemble their host communities, and the progressive secularization of Jewry in Germany during the eighteenth century, suggest that this is a subject which deserves only passing mention. It is certainly possible that a few characteristically 'Jewish' genes, such, perhaps, as those for Tay–Sachs and Gaucher's diseases, would have entered the Afrikaner at this period. It is interesting though hardly relevant, however, to note that the modern Afrikaner surname Gouws was originally the Huguenot Gaucher.

A less familiar contribution came from the Caucasoids of Ceylon and Bengal. One of the most steadily recurrent themes in the deliberations of the early councils at the Cape, and a subject repeatedly discussed by the Heren XVII, the governing body of the East India

Company, was the labour shortage. The Khoi had little incentive to take employment, for at least as long as they still had cattle, and in any case there were too few of them, particularly after the disastrous smallpox epidemic of 1713, to provide much of an answer to the problem. Slavery, indentured labour, or free labour were endlessly canvassed. The West India Company had a monopoly of the slave trade along the west coast of Africa, and on that the East India Company dare not trespass; though one shipload of slaves did indeed disembark at the Cape in the late seventeenth century, and remained for a decade before being transported to the east. This was long enough, it seems, for the introduction of the characteristically West African gene for Haemoglobin C into the Caucasoid gene pool, in which it today reaches appreciable but not polymorphic frequencies. But when at length it was decided to introduce slaves on a sufficiently large scale to provide the labour required, the first region from which they were actively imported was that with which the Company was primarily concerned. Overpopulated, underfed Bengal and Ceylon supplied intelligent and well-behaved slaves, but not the necessary husky labourers for the gardens and vineyards. After a time the closer shores of Madagascar and Moçambique began to be scoured. By then, however, the need for artisans and domestic servants was being filled by Bengalis, Sinhalese and some Malays, the majority of those whose descendents now make up the 'Cape Malay' community, but some of whose genes flowed into the forebears of the Afrikaners. Slavery at the Cape was never as gross or brutal as in the British colonies in North America, and several free burghers liberated and married Bengali or Sinhalese women, whose descendants were accepted as Dutch.

Non-Caucasoid gene flow into the early Caucasoid community seems to have been less regular. One of the earliest marriages cele-brated at the Cape was between a Dutchman and a Khoi woman. Some burghers married slave women not from Bengal or Ceylon but from the Malay Peninsula or the islands further east, and even some of the fairly numerous brides from Ceylon may have been Dravidian Tamil rather than Caucasoid Sinhala. The families of Chinese traders intermarried with the settlers, and are not forgotten in the genealogies. The evidence for Negro admixture, excepting the episode retailed above, is slight, but the origins of each and every slave bride are not stipulated, and some Negro genes may have been acquired at this time. It needs to be stressed nonetheless that there is very little, if any, evidence of gene flow from Southern African Negroes into the Cape Caucasoid community. All such would have been derived from rather remoter Negro peoples.

### Prior adaptation, selection, and genetic drift in the seventeenth, eighteenth, and nineteenth centuries

The original Caucasoid peopling of the Cape followed a deterministic pattern insofar as the predominant origin of the settlers was from Western Europe, and within that from low-lying parts of north-western Europe. The precise regions to which the French component may be said to have been adapted by long habitation were perhaps more varied. French Protestantism tended to be strongest in the south, with particularly recalcitrant small foci such as that in the Cevennes, though a number of towns and cities scattered throughout the country had sizeable Huguenot minorities, dependent often on the influence of regionally prominent families. That the principal new skill introduced by the French was viticulture, suggesting at least that many of them came from rural vine-growing regions, is of only minor relevance: vines are cultivated over a large area of France. It is nevertheless safe to assume that a significant proportion of them originated from parts with a Mediterranean climate, and were consequently already adapted to natural conditions closely resembling those of the Cape. This assumption is supported to a considerable extent by the researches of Boucher (1981).

There seems nonetheless to be no very good reason for counting the rest of the Caucasoid settlers as previously adapted to conditions at all grossly different from those of their new homeland. All came from regions described by the geographers who live there as temperate; and any grounds for so describing them would apply as well to the Cape Peninsula and the lands around it. Temperatures are perhaps not as extreme as they can be in Europe, but their means are very similar. Settlement at the Cape carried few biological hazards other than those of Europe, and the most familiar ones tended to be muted. There were better access to and greater availability of foodstuffs; the cold was rarely severe enough to be life-threatening, and even when it was, wood fuel was then still available in abundance; escape from the congestion of the towns meant lower exposure to and frequency of epidemic diseases (but when these struck they were often severe); and though medical attention was not always accessible, that lack may not have constituted an unmitigated curb on health.

It may be claimed, therefore, that in selecting the western Cape for settlement the Dutch East India Company was choosing for its employees an environment to which they had undergone prior adaptation; and for as long as the settlers remained within the shelter of the well-watered south-western valleys selection would operate on them no more harshly than it had in Europe. It was only with the slow development of pressure for more land which resulted from population growth during the eighteenth century that Caucasoids began to spread beyond an environment to which they were biologically fitted. In this extension into more arid lands with greater extremes of temperature the concomitant changes of altitude were probably of some biological importance; few of the original settlers came from any of the more mountainous parts of Europe. Furthermore, the familiar crops of the homeland and the new farms could not be grown further inland; there had to be a greater reliance on maize, and on meat and milk products. From agriculture the settlers of the fringe turned increasingly to pastoralism.

The Caucasoid drift beyond the temperate and well-watered west did not at first play much part in the evolution of the Caucasoid population. The persons who made the earliest moves in that direction were men, most of whom, where they left descendants, contributed their genes to the formation of one or other of the hybrid or 'Coloured' communities. From the beginning of the colonial period there was a shortage of Caucasoid women at the Cape, and though the Company periodically tried to alleviate this with parties of young orphan women from the Netherlands, the majority of new settlers during the eighteenth century were male. These men naturally were in competition with the male descendants of the first settlers for the hands of the Cape-born girls, and this led to the gradual displacement to the fringe of the settled area of those young men who had not succeeded in acquiring either brides or land. Eighteenth-century sibships were large; in economic circumstances superior to those enjoyed by most people in Europe, the survival potential of children was increased. Penetration of the hinterland happened slowly, but it was inevitable.

It is impossible today to reckon up the numbers of young men who migrated into the semi-deserts of the Boland and failed to leave descendants. The first wave of those who succeeded in so doing fathered their offspring on Khoi wives, and so pass out of the purview of this chapter and into that of the next; but they played their part in demonstrating that the land could be used by Caucasoids, and as the sex ratio at the Cape stabilized and population size continued to grow entirely Caucasoid families began to move into the area, displacing their hybrid cousins. It would have been on this and subsequent generations that selection for a changed environment would have operated most strongly and contributed most to the population structure of the Southern African Caucasoids today.

As the Caucasoids moved away from the apparatus of government and, more reluctantly, that of the churches, record-keeping declined.

It was, and is, the useful habit of the Afrikaner to record major family events on the fly-leaves of a Bible; but not all such Bibles survive. It would not be unreasonable to expect this migration, at least as its outset, to be attended with a rise in child mortality. Selective processes would operate with sudden sharpness against those moving into this new environment; moreover, the range of mating choice would be somewhat reduced, and over a few generations the degree of inbreeding could be expected to rise. For a while the pattern of settlement became more dispersed than it has been at the Cape. A reduced range of communication among families, among parties, lowered the potentiality for communal immunity against the communicable diseases of childhood, and these would have struck more severely than they had previously. Before, during and after the Great Trek the inland Afrikaners would have been subjected to forces of selection very similar to those which had moulded the Khoi. It is hardly surprising therefore, and not to be ascribed necessarily to miscegenation, that certain of the physical characters of the Afrikaner should so closely resemble those of the Khoi. It is not easy to define all of these characters, or relate them specifically to selective processes; but a certain sallowness of complexion in many individuals, a linearity of build found more frequently than it would be in the Low Countries, may be numbered among them.

The initial phase of this migration occurred during the eighteenth century, towards the end of which it had progressed enough for the first of the myriad of Boer republics to have been ephemerally founded at Swellendam and Graaff-Reinet (see Figure 8.2). The migrants nevertheless retained fairly close links with the centre of government at the Cape, and gene exchange continued to a greater extent than it would during the century which followed. But early in the nineteenth century the coherence of the Cape Caucasoids was beginning to break up, and this process was reinforced by the arrival of Caucasoids of similar broad provenance as the original settlers but forming different societies which remained to a greater or lesser extent distinct.

## Later European immigration

The British occupations of the Cape did not result in any immediate flow of British settlers into the country. Both occupations were initially opposed by the Dutch, but only half-heartedly. The British came ostensibly as partisans of a faction in Holland; perhaps not the faction most favoured by the Dutch at the Cape, but nevertheless one whose right to assert itself in the context of Dutch internal

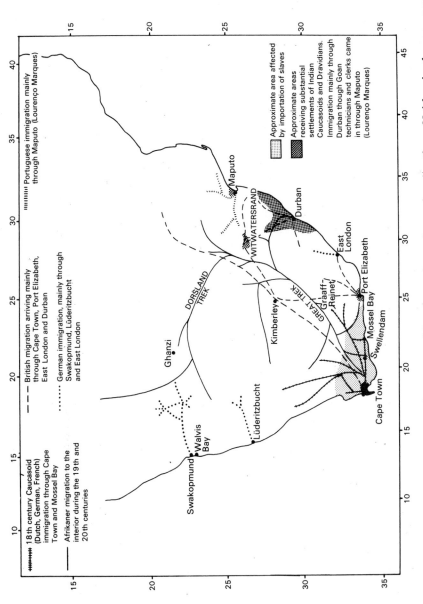

**Fig. 8.2.** The main paths of spread and settlement resulting from the seaborne immigrations. Neither in the map nor in the text is the very recent phenomenon of airborne immigration considered.

politics could be recognized. The second, ultimately permanent, occupation was admittedly directed against the party then ruling in Holland; but at the time it was intended as temporary, and it took the nine years up to the Congress of Vienna, and, perhaps more cogently, the exile of Napoleon I to St Helena, for the British to appreciate, or simply admit, that the Cape was strategically too important, especially in the context of India, to be returned to a Stadtholderate of diminished dignity and power. The House of Orange forfeited the Cape and a few other peripheral possessions, in exchange for the sour and fleeting tenure of Belgium.

Some British settlers arrived during the latter years of the Napoleonic wars and quite soon after their termination, but the most significant influx took place in the 1820s. The motives for this officially sponsored movement were fairly clear-cut. The establishment of a strictly unofficial British presence in the new colony, to counterbalance the Dutch numerical preponderance, was an obvious need, though probably not the main one. More pressing was the advisability of stationing a reliable military reserve as inexpensively as possible along the frontier with the restless Cape Nguni. Returned soldiers, a source, particularly in the economically uncertain post-war period, of potential unrest at home, could be granted land which they would then be expected to fight to defend. The lightly inhabited hinterland of Algoa Bay, where Dutch or Afrikaner settlement was sparse, was strategically crucial for this purpose, and such counter-claimants as there were to its possession could safely be disregarded. Land closer to the Cape itself was not available, but the propinquity of government was less necessary where loyalty could be assured.

The outcome was not quite as smooth as the government intended, but it functioned well enough. Unrest did occur among the settlers, but was expressed principally in the ineffectual columns of news-papers and lampoons hardly as blatant or as eloquent as those in London broadsheets of the time. There was a greater degree of accord and goodwill towards the Afrikaners than might have been expected, but little active support for the more positive aspects of Afrikaner discontent. Piet Retief, perhaps the most intelligent of the leaders of the Great Trek, was helped to some extent with the outfitting of his expedition, and sent off with an enthusiastic testi-monial. A few British settlers, mostly those of Calvinist persuasion, actually joined the Trek; but neither as an acquisition or as a loss were they particularly significant. Neither at this stage nor later was there any pronounced degree of intermarriage between these two main groups of Caucasoids.

This failure to blend is hardly as incomprehensible as it might appear at first sight. The British settlers had been selected deliberately from among sections of the population which were least likely to undergo social change when in contact with the unfamiliar. They were not linguists, any more than their descendants are: even today few English-speaking South Africans take the trouble to acquaint themselves adequately with the Afrikaans language or familiarize themselves with the idiosyncratic theology of the Afrikaner churches. British settler communities were moulded in part by the familiarities imposed on shipmates by the long sea passage; by the time they reached their future homes some measure of communal coherence had already formed, on which even the most cordial reception by other settlers similar in faith and language might scarcely have impinged. Certainly the relations between the Afrikaners and the unofficial Britons were good; but mutual respect and liking are not necessarily enough, and may indeed be inimical to, the exchange of mates between peoples.

What the eighteenth century had wrought biologically, and what the nineteenth century was to impose, on the Afrikaner, would be substantially bypassed by those British settlers and their descendants. The lands they were given were tolerably fertile, well-watered, at no great altitude; perhaps not very like the English midlands or the lowlands of Scotland, from which many of them came, but not all that unlike either. Escape from the Industrial Revolution, from crowding, poor hygiene, and the epidemic diseases of Europe, may have promoted somewhat better survival of infants, and compensatory dangers were rare enough for the communities to flourish. There were admittedly phases of economic discontent, due largely to market fluctuations. Few Britons moved onwards into the arid high interior before Afrikaner settlement there had worked to provide at least some amenities to mollify the effect of the environment on newcomers.

Much the same considerations held for the German settlers, many of them Hanoverians, who arrived while William IV still ruled over both Great Britain and Hanover. Many of these, too, were ex-soldiers, chosen to fulfil much the same functions as the British settlers. The Germans conformed perhaps a little more closely with what was expected of them and remained even more ethnocentric in their mating patterns. Intermarriage or interbreeding with Negroes and Khoi played little or no part in the formation of either community; once more, such miscegenation as there was went to swell the 'Coloured' population. These Germans seem not to have recognized in the Afrikaners the outcome of an essentially German immigration two centuries previously, and showed even less inclination than the British did to identify with them.

The European immigrations of the latter half of the nineteenth century were less organized and more sporadic. British, Germans and Dutch continued to arrive, and there was at least one British settlement scheme of modest size for Natal, to which some Germans also went. The effect of this in curbing the recovery of the Natal Nguni following the *Mfecane* has been considered in Chapter 7. The opening of the diamond mines at Kimberley, and subsequently the gold mines of the Witwatersrand, drew further settlers from Europe and were prime causes for the entry of Caucasoid elements other than Afrikaners into the interior. Many of the early inhabitants of Kimberley in particular were Jews from Poland and Lithuania; indeed, until the latter half of the nineteenth century there can scarcely be said to have been a Jewish community in Southern Africa, since most of the occasional Jewish immigrants, as mentioned above, had been absorbed into the Afrikaner people. Pogroms in the Russian Pale swelled this immigration, which continued until well after the major growth phase of the urban Witwatersrand and the abortive Decembrist uprising in Russia in 1905. By far the larger number of immigrant Jews were Ashkenazim, but after the first World War there was some Sephardic immigration following the fall of the Ottoman Empire. Most of the Sephardim went to the Rhodesias (Zimbabwe and Zambia), but there are a few in the Republic of South Africa still. Middle Eastern Christians began to arrive at about the same time; Lebanese, Syrians, Cypriots and a small number of Copts and Armenians, who tended, however, to assimilate with the Lebanese. Many Greek families which today maintain that their origins are in metropolitan Greece can be seen by accent, dialect and surname to be descended from Smyrniots, Phanariots or groups from the Anatolian interior.

From Southern Europe there was a constant steady trickle of migration, reaching a peak in the years following the Second World War, when the South African government, attempting to build up the White population in a hurry, relaxed the perhaps not explicit but nevertheless operative restrictions on Roman Catholic immigration. The propinquity of the Portuguese east and west African colonies, and the necessity of maintaining good relations with Portugal, led quite early to some Portuguese settlement in the Transvaal, and this grew as progressively more became disillusioned with the prospects of Angola and Moçambique. South African troops fought in the Italian campaign, and despite their being at first on opposite sides, good relations with the Italian people resulted. Italian immigration, sometimes involving the return of men who had first seen Southern Africa as prisoners of war, was for a time particularly favoured.

For the old lands, the lands with which Caucasoid South Africans had always felt bonds to be strong, the Netherlands and Britain, proved reluctant to encourage further migration to countries where the immigrants would immediately assume inordinate if precarious privilege. A people which had regarded itself as so morally unassailable as to be self-righteous was shocked to find itself condemned, and deprived of replenishment, on moral grounds. Later developments have not yet damped down the revulsion of many prospective immigrants from more enlightened lands, and though South Africa continues to attract some Caucasoid immigrants they are principally those who consider themselves refugees from one or other of the diverse kinds of socialism.

## The Indians of South Africa

The Indians of South Africa do not comprise only Caucasoids, and a good many of them might, by virtue of their descent from both the Caucasoid and the 'aboriginal' peoples of the sub-continent, be accounted hybrid. In truth, they can hardly be said to constitute in themselves a genetically or socially homogeneous population; yet we should prefer in our discussion of hybrid populations to confine ourselves to those in which one element of the hybridity at least is indigenously African. Hence it seems most fitting that we should describe these Indians by the heading under which the majority of their genes could be comprehended. The genetic profile of many Indian populations is that which could be expected from peoples of Caucasoid stock in whom the processes of selection and random drift operated a little otherwise than in Europe. In the Indians of Southern Africa we do not detect any marked deviations from what we should anticipate from this.

As with the Jews, the earliest immigrations of Indians happened effectively singly, and such descendants as they left were absorbed by either the Caucasoid or the hybrid populations. It was not until the middle of the nineteenth century, and once more in consequence of the *Mfecane*, that Indians arrived in any number and came to constitute a separate community. The Natal Nguni who began drifting back into their ancestral territories, or expanding to fill the lands they had once occupied, found that the country was no longer theirs but had been esteemed vacant and so taken over by Caucasoid settlers, in this instance, mostly British. Here the object was not simply subsistence, and provision for large families, as it most often was among the Afrikaners, but something more ambitious, vaguer, more pretentious and more openly exploitative. Colonialism was still respectable; more, it was praiseworthy; it brought industry to lands seen as idle, it promoted trade and the Christian religion, and it spread the then unquestioned benefits of Western civilization. If the Negroes could not utilize the country to what was thought to be its best potential, then it was not only the right, it was the duty, of the accomplished Westerner to do so. The Negro would not be neglected, of course; he would be taught to value and respect what his privileged supplanter and mentor thought worthy, and he would be helped, even encouraged to aspire to and, in some scarcely credible future, perhaps ultimately attain, similar goals. But the first stage would involve hard work, submission to the superior knowledge and skills of the immigrants, and the building up of capital. To none of these aims were the Natal Nguni by temperament or tradition inclined.

The settlers quite genuinely counted this as a grievance. To them it was not only morally wrong, it was an economic hindrance, that the sturdy and abundant labour that they saw around them, squatting on their own farms, drifting in idleness about the shrinking territories still open to Negro residence but already denominated 'Crown Lands', or half-heartedly trying to accommodate to the unfamiliarity of town life, should be so reluctant to be absorbed into a profitable economic system. It was wicked and wasteful; but fortunately Natal was not the only possible source of labour under the British flag. The Indians, after the suppression of what some termed their Mutiny and others, rather later, a Freedom War, were cowed, impoverished, starving and tractable. The Indians, or at least the underprivileged among them, were as ready to indenture themselves to the British abroad as some centuries previously they had been to sell one another to the Dutch. In the Madras Presidency and the dominions of the Nizam of Hyderabad labour was abundant and inexpensive; and thence the Natal settlers endeavoured to obtain it.

To its great credit, the India Office raised objections. Before the Tamil and Telugu poor could be permitted to dispose of themselves in servitude abroad, the India Office had to be satisfied that they, and not only their employers, would benefit. For some years the terms were thrashed out by the Government of Natal on the one hand and that of India on the other, until at last a compromise was reached. Indentured servants were to serve a fixed term, and then be given the option of repatriation or a plot of land in Natal. Free labourers could buy their own passages to Natal, and there make their own contracts, but would be responsible for their own arrangements when these terminated. Natal was to be opened to Indian

traders, who would be able to supply these new communities with
the commodities to which they were accustomed.

The industry which appeared to hold out the best prospects for
Natal was sugar, and it was to the canefields in the lowlands that the
Indian immigrants came, and for many years on the vicissitudes of
the sugar market that their fortunes depended. Certainly the treatment they received was, by and large, not as good as the Government
of India, but probably better than most ordinary plantation-owners,
intended. Disputes between these two parties continued long into the
twentieth century, long after Indian immigration had ceased and
more than one new generation of South African Indians had been
born. When a minority population is introduced as a solution, one
must not be surprised if eventually it comes to be regarded as a
problem. The industrious Indians, with a narrow range of requirements for subsistence, prospered; many of the trading families
became extremely wealthy. The less frugal British settlers were left
with a more restricted economic flexibility, and for some time they
feared that the acquisition of property by the Indians would lead to
their own displacement. The process was halted, not without difficulty, by the subsequently traditional South African expedient of
ethnically discriminatory legislation. An early attempt to combat
this was led by one of the most remarkable figures in Southern
African, and indeed recent world, history, the then rising lawyer,
M. K. Gandhi. In the Indian immigrants the earlier Caucasoid settlers
encountered a more equal match than they had in the less sophisticated peoples whose suppression had been relatively easy, and
though the Indians constitute a relatively small proportion of the
population of the region they have been more articulate than most
of the rest.

From the outside the Indians may look as though they form a
single entity, but more closely regarded they can be seen to be hardly
coherent either biologically or socially. The indentured labourers
had been drawn from the least privileged strata of what are now
Tamil Nadu and Andhra Pradesh. They were Hindu, but retained
much of the rural animistic culture of the South Indian countryside;
racially they were closer to the aboriginal dark Dravidian-speakers
than to the Caucasoids who had imposed a stable, perhaps rigid,
social system on the sub-continent. Few of them were literate in
any language, and hardly any were prepared or equipped to make
more than a simple geographical adaptation to their new setting.
The 'Free' settlers, though they included a number of Tamil and
Telugu-speakers, also included large numbers of speakers of the
Indo-European languages, Hindi, Bengali and Oriya, and many more

literate members. In both of these migrant categories were included sets of men coming from single villages, who when they prospered would send for wives from the same villages, but on the whole they represented a rather wider range of provenance than did the third category. The Vaiśya traders, though they adhered to the traditional caste affiliations, were principally Muslims from the Gujerat and Maharashtra, speakers mostly of Gujerati and Kacchi and related languages, and often arrived in very closely interrelated groups. They were, and are, given to cousin marriage and often simultaneously to sister exchange: they have carried over the highly inbred tendencies of their original villages into their present domiciles. Caste still regulates mating patterns for most South African Indians, irrespective of their religious beliefs; even those who have become Christians maintain their traditional preferences. There has been very little intermarriage among the descendants of the three categories, or with other population groups; though at one time it was common for traders, particularly, to take temporary Negro wives pending the arrival of their families from India, and there has been some slight gene flow from the Indians into the Natal Nguni.

Most of the posterity of the immigrant labourers remained in Natal, principally along the coast, which presents an environment not unlike that of South India. The traders, however, have spread widely; though debarred from residence in the Orange Free State, they flourish in the Transvaal and in some towns of the Cape and Transkei. They may be found in Lesotho and Swaziland but not, curiously enough, in Botswana, and in Namibia we have not seen any Indians at all. Not even all those in the Cape and Transvaal were originally part of the Natal migrations. Portuguese influence, and to a lesser extent official British colonial activities, drew in a few persons from other parts of India. There are or were speakers of Kannada and Panjabi, of Urdu and Mahratti, among the Southern African Indians.

The change of enviroment for the South Indians can hardly be said to have been significant. Their forebears had evolved under conditions climatically very similar to those of the Natal coast, and they brought with them the means of perpetuating their familiar diet. The Indians of more preponderantly Caucasoid stock, however, could be expected to derive more recently from ancestors adapted to colder conditions. It is admittedly some two and a half millennia since the time of Aśoka, but the Dravidian-speakers had been in the sub-continent far longer. Furthermore, the conditions of north-western India, with its harsh desert hinterland in Sind, its frequently extreme swings of temperature and its alternations of plain and plateau, are not strikingly dissimilar, except perhaps in population density, from those of the Southern African interior. The traders, in choosing to move away from the Natal coast, were making not only an economic choice, they were also moving into environments to which they were previously better adapted than the other Indian immigrants would be.

### The Mongoloid settlements

The slight but perceptible contributions of immigrants from the Malay lands and from China to the gene pool of the original Caucasoid settlers have been dealt with above. These Mongoloids were hardly settlers, though the 'Cape Malays' have retained an ostensible identity which will be discussed more fully in the next chapter. In the main, however, their descendants were absorbed into the amorphous hybridity of the 'Cape Coloured', and it was not until the twentieth century that any distinct Mongoloid settlement took place. Apart from the few Chinese merchants who established themselves independently, the first systematic attempt at Chinese settlement was bitterly opposed by the Rand Caucasoids and was ultimately abortive. The motives for the introduction of Chinese labourers into the gold-mines were similar to those which had originally inspired the Indian immigrations; but the Chinese proved less adaptable, more open to slanderous rumour, and possibly more unruly. They were in any case not a stratified coherent community, but simple labourers; they were ultimately deported in a storm of calumny, and only those traders who had served their needs, and, adroitly, some of the needs of the Caucasoids as well, remained to form the nucleus of a small and enterprising community which has since grown slowly, reinforced by occasional infusions from Hong Kong and South China in general.

The Chinese of Southern Africa are neither a large nor an important group, and though worthy of mention do not deserve any extended biological discussion.

### Physical characters: morphology

With so varied a range of origin, the physical characters of the descendents of the seaborne migrants may be expected to cover a wide morphological spectrum. Even those whose forebears came from Europe manifest their descent from those forebears in as many shapes, colours and sizes as there are in Europe. It has been suggested above, but hardly with any firm substantiation, that some Afrikaners, exposed through generations to an environment which probably moulded the Khoisan, may, without any Khoisan admixture, have begun to manifest some characteristics which are signs of adaptive selection and which resemble some Khoisan traits. On the whole, however, selection has been mitigated by technology, and morphologically the main body of South African Caucasoids resembles an anthology of European types.

The Indian-descended Caucasoids are of rather more restricted origin, and interbreeding between the two main stocks hardly occurs. The medium-sized round-headed straight-haired South Indian stock, narrow-waisted and round-hipped, with dark eyes, black hair and shiny dark-brown skin, shows very little internal morphological variation. Those whose ancestors came from the non-Dravidian north and west of India, on the other hand, and despite the prevalence among them of cousin marriage, reveal as broad a spread of physical traits as are found in their homelands. All are, admittedly, dark-skinned, and lightness of hair and eye, though it does occur, is very unusual; but facial features and body sizes and proportions vary considerably. This would suggest that in this community possible semi-isolates are kept in existence by inbreeding, and that there is still nothing approaching panmixis. Both Indian stocks show recent tendencies towards obesity, elevation of the blood-pressure, and maturity-onset diabetes, possibly due to dietary changes (Campbell 1963).

## Gene marker studies

Few studies of the gene markers of the Southern African Caucasoids have been made except in the course of investigations of other topics (Buckwalter et al. 1961: Brontë-Stewart et al. 1962), and most have been restricted to a narrow range of polymorphisms, mainly the blood groups (Hirsch 1958; Stoke 1966).

Botha (1972), in a masterly analysis of English-speaking and Afrikaans-speaking populations at the Cape, based, however, exclusively on the blood-groups, concluded that the differences between them, and the extent of their differences from their parent populations in Europe, could be ascribed to different patterns of gene flow from non-Caucasoid peoples. In the case of the English-speakers, the contributions came mostly from local, Southern African, Khoisan or Negro, sources, while in the Afrikaans-speakers they came from both these and East Asian sources. The proportion of non-Caucasoid genes was much the same, however, in both populations. He supported his findings with references to the genealogical investigations of Heese (1971) on the origins of the Afrikaners, but produced no historical facts to support his results for the English-speakers. Some were suggested by Jenkins and Nurse (1972).

Botha's study can be criticized on several grounds. His samples are likely to have included a fairly high proportion of persons of mixed ancestry 'passing for white', a common phenomenon in the environs of Cape Town, and consequently to represent a stage in the still ongoing formation of the South African Caucasoid populations rather than the finished product. Many such people adopt English as their home language in partial substantiation of their claims; Afrikaans is traditionally the language of the 'Cape Coloured'. English-speaking immigrants who settled at the Cape were not as representative of the forebears of the main body of English-speaking South Africans as were those who settled in the Eastern Province or Natal. The Afrikaners who moved into the interior originally can be divided into two well-demarcated strains, those whose genes went eventually to help form the hybrid communities and those who secured Caucasoid mates and practised ethnocentricity. There may even have been a modicum of social selection in the formation of the parties which made up the Great Trek. Botha's conclusions, interesting and revealing as they are, can hardly be extrapolated to the two largest South African Caucasoid groups as a whole.

This is illustrated in part by the work of Palmhert-Keller *et al.* (1983), in which small but representative samples of the Afrikaners and Germans of Namibia were investigated for variation in a wide range of genetic systems. While evidence of non-Caucasoid gene flow, probably from the Basters of Rehoboth, was detectable in the German sample, there was no evidence of it at all in the Afrikaner sample. Apart from such traces, the German sample was closely comparable with the range of published figures for gene frequencies in Germany; the Afrikaner one fitted in very well with the overall north-western European arrays. Much the same can be said for the results of investigations carried out in the laboratories of the Human Ecogenetics Research Unit in Johannesburg. Occasional evidence of non-Caucasoid genetic contributions, mainly genes of probably Khoisan or Negro derivation such as $A^{bantu}$ in the ABO system, the $Gd^{A-}$ and $Gd^A$ alleles of glucose-6-phosphate dehydrogenase, $PGM_2^2$ at the second phosphoglucomutase locus, the Rhesus allele $R^0$, and Haemoglobin C, were found in individual Afrikaners and a few English-speakers, but these did little to distort the characteristically north-western European picture. Similarly, the Jewish population studied by us (unpublished) showed no significant differences from other Ashkenazi populations elsewhere, and a small study of Indians on the Witwatersrand produced gene frequencies quite comparable with the stated origins of the ancestors of the members of the samples. Such findings are borne out by the data on tissue antigens among Witwatersrand Caucasoids kindly made available to us by Dr Gunthilt Meier, and those for the European-descended and Indian Caucasoids of Natal published by Brain and Hammond (1973). None of the earlier, more limited, studies shows anything different.

## Founder effect and seaborne immigrations

Seaborne immigrations possess several characteristics which demarcate them biologically and geographically fairly sharply from migrations which proceed overland. Perhaps their principal biological distinction is that migrants by sea rarely move as a preformed population, with the structure of the relationships among the migrants established in the country of origin, upheld though sometimes in abeyance during the voyage, and reasserted in the foundation of a settlement. It may be objected that there are historical instances of the movements of entire and coherent communities by sea: the peopling of New England is a familiar one, and analogous legends attach to many Pacific islands as well as to the establishment of the colonies of the Greek city-states in Magna Graecia. But these movements of peoples prove on closer examination probably not to have been necessarily representative of their parent populations. Moreover, in all the best-known instances landfall was made at good harbours, and ports were established, way-stations for the passage of goods, without much need to tame a hinterland. In Southern Africa the position was different. The first chapter has discussed the effects of the paucity of good harbours; of the three that exist, only one has ever been truly a channel of the products of the interior. Lourenço Marques, or Maputo, began as a trading station, but did not flourish until exploitation of the Transvaal by migrants coming from quite a different direction demanded the use of it as a route for the passage of goods inwards and outwards. Cape Town, a refreshment station, was meant to produce all that ships would need within a small but perfectly serviceable perimeter. Durban, it is true, was intended virtually from its foundation to serve for the export of produce from the lands beyond it; but it also operated as the portal of introduction of the migrants who were expected to provide the produce, and not as a means for the enrichment of the people whose title to the country anteceded theirs. All the other ports were, at least in their early stages, dangerous, exposed and unreliable, suitable only for rapid discharge and rapid loading.

A certain amount of introduced infectious disease certainly

entered the region through these portals of entry. As in the cases of the eighteenth-century smallpox epidemics, and the present tuberculosis prevalences, however, the principal effects are likely to have come through the introduction of such diseases into hitherto unaffected populations. The seaborne immigrants brought with them a certain measure of resistance to the diseases of their homelands, and were less immediately afflicted by them. Each seaborne migration, moreover, was followed by dispersion of the immigrants. Even where this was least extensive, the immigrants settled in areas much less densely populated, on the whole, than their areas of origin had been. The consequently reduced dangers of epidemics and other infectious diseases are attested by the survival to adulthood of many more members of the large sibships than are likely to have survived in the homelands. There must also, in most instances, have been a marked improvement in the diet, possibly leading eventually to an increased prevalence of diseases associated with excessive intake of food. These considerations apart, the diseases of the descendants of the seaborne immigrants are in the main the diseases of the stocks from which they originally came. In a few notable instances, concerned naturally with diseases having a strong hereditary component, founder effect has operated to modify the original patterns.

A large proportion of British and German settlement was made through Port Elizabeth and East London; Dutch settlement was more leisurely and gradual and took place only through Cape Town. Nevertheless, even the cumulative total of settlers under Dutch rule of a century and a half is not a great one. The Afrikaners of today are descended from a relatively small number of original immigrants, and their numbers reflect the favourable conditions for multiplication in their new environment rather than an abundance of original colonists. This, too, is a characteristic of seaborne migrations; that the numbers of the migrants are limited by the capacity of the craft to carry them. Such considerations still operated at the time of the principal early British and German settlements, but as the nineteenth century advanced ships became larger and calls at Southern African ports more frequent, so that predominantly British immigration could take place as a movement of quite large numbers of independent persons or families. By that time, however, the Afrikaners were already a distinct people, and nineteenth-century immigration produced very little effect on their genetic structure.

This mode of immigration is nevertheless of profound importance when considering all the Caucasoid immigrations into Southern Africa. The forebears of the Afrikaners arrived almost exclusively through Cape Town and radiated out gradually from that port. It was

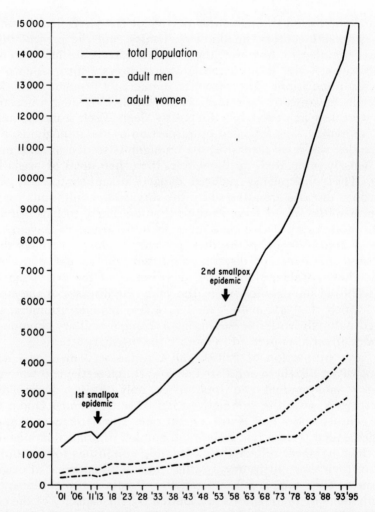

Fig. 8.3. The growth of the 'White' population of South Africa, 1701-95.

population pressure which brought about the slow cutting of ties with ships and the sea, in the first place, though the British occupation of the Cape finally sealed off the interchange and made the Afrikaners a continental people. The opening and settlement of the interior began before the Afrikaners had become distinct and while European contributions were still being received. We should consequently expect to find the Afrikaners prior to the Great Trek living in a temporal continuum of genetic association with Europe, with those nearest Cape Town the recipients of the most recent gene flow, this becoming progressively diluted and of older derivation as the distance from the port increased. The transition from the fertile valleys to the arid uplands would constitute a selective boundary, with the fertility and rapid growth in population size checked by the new environment and only the relative minority whose genetic constitution was favourable surviving to reproduce in those areas. Since it would have been the descendants of the earliest settlers who would be exposed first to this challenge, we may expect the genomes of the Caucasoids furthest from Cape Town to have included a higher proportion of genes inherited from the original families. Selection would, however, have operated on them the most strongly, and reduced the frequencies of, or perhaps eliminated, certain genes. But it is not only on selectively relevant genes that selection produces its effects. Every selective failure to reproduce would deterministically have removed from the population the genes in that individual which were concerned in that failure, but would have deprived the population stochastically also of the rest of the genes comprising the genome of the individual. A further component of random drift would be added to the founder effect dictated by the small numbers of the original settlers.

This bottleneck, probably exacerbated by a relatively high degree of inbreeding, must have been overcome to some extent by the Great Trek. Once more this represented a radiation outwards from the western Cape; but it also produced a wave effect, rolling the genes of the later comers northwards and eastwards across the sparsely settled uplands, gaining recruits diametrically across the progressive temporal dilutions of genetic material. The Trekkers were, of course, faced with the same challenges, and more, as the earliest Caucasoid settlers on the plateau; but larger numbers of people were involved, and the movement happened more rapidly, and by the time they and their descendants in their turn were settled enough to be exposed to the selective forces the advancing technology of the nineteenth and twentieth centuries had caught up with them. Once the tribulations of migration were over, the biological perils of the Afrikaners of the interior were already less than those which had confronted the pioneers of the previous centuries.

Similar historical considerations would have limited the impact of settlement on the British and later German settlers. Their ports of entry were the exposed, unsafe harbours of the eastern Cape, and the protected Durban Bay. At Port Elizabeth and East London the bulk of the immigrants arrived within circumscribed periods and proceeded at once to the lands allotted them, where for the greater part of the nineteenth century most of them remained. Immigration through Durban was more gradual, but there was little need or incentive for the settlers or their descendants to form themselves into a people distinct from that of their homelands. Indeed, they were less heterogeneous than the settlers under the Dutch East India Company had been, and the continued political connection with Britain made them more aware of being British than of being permanent expatriates. This did not, of course, apply to the Germans, who were a relative minority and who have preserved some measure of ethnocentricity even in South Africa, though this is more marked in Namibia.

The German settlement in Namibia was a stormy one. By the time Germany had extended her hegemony over the territory the best port on the coast, Walvis Bay, had already been claimed by Britain. The two possible harbours which remained were Swakopmund, just to the north of Walvis Bay, an exposed bight at the mouth of an inconstantly flowing river, troubled by periodical submarine volcanic activity, and Lüderitzbucht, with its broad hinterland of profitless desert. In the event Walvis Bay continued to be used for most purposes, while Swakopmund took on mainly military importance and Lüderitzbucht supplied the dry south as well as it could. It was in the centre of the country, however, that the most accessible and profitable farming lands lay. Slowly these were wrested from their previous holders, who had been weakened by the Herero–Nama wars that had partly prompted the German protectorate.

The settlers were preceded by mostly German missionaries and traders and some cattle-keeping Afrikaners from the Cape, who at first asked for British intervention to pacify the country. The British equivocated, and while they hesitated Bismarck moved; the first German forces entered the country in 1885 and imposed a military stamp on settler society which has lasted almost to the present day. Civilian settlers first trickled, and later slowly poured, into an environment only slightly less hostile in the centre, and considerably less inviting in the South, than the Karroo which had received the first Caucasoids to penetrate the Southern African interior. Many of these settlers were nevertheless only superficially civilian; farming land, as it became available, was allotted for preference to discharged members of the armed forces. In a political sense this practice gave more coherence and singleness of purpose to the settler community than the haphazard Caucasoid peopling of South Africa could ever have done; but biologically it served to accentuate the randomness of seaborne migration. The army of the new German Empire was drawn from a far wider selection of the German lands than the previous German immigrants into Southern Africa had been; and it was entirely and nationalistically German. In Namibia the Germans founded an amalgam such as would have been impossible in their recently united homeland. As at the Cape, however, there was a small inflow of genes from the earlier inhabitants, but not, in this case, directly from the Khoi. The Basters of Rehoboth produce comely daughters, many of them light enough in complexion to pass as Caucasoids. There seems to have been little or no contribution from the non-German Caucasoids already in the country. The impoverished and aggressive Afrikaner pastoralists who had filtered in from the south or crossed the Kalahari from the Transvaal were accorded toleration rather than cordiality.

When the first German settlers arrived they were greatly out-numbered by these Afrikaners. Those around Keetmanshoop and Warmbad in the south were hardly affected by German settlement until much later, but the farmers who had left the Transvaal in protest against the religious unorthodoxy of President Burgers and assayed an association with the Portuguese in Angola before setting up their own republic at Grootfontein were something of a barrier to German expansion northwards and may have saved the Ambo from the threat of colonization. It was not long, however, before they accepted German rule; and by that time it was becoming plain that the German settlements would be fully occupied in maintaining their position and would not need or be able to expand until rela-tions with the Herero and Nama had become less strained.

It is not our purpose to relate the unhappy history of the wars fought in the early twentieth century. Their impact was felt less by the Caucasoids than by the Negroes and Khoi. Their principal consequences for the Caucasoids were an increase in German immi-gration, once more and more pressingly of a military cast, and a relaxation of the earlier reluctance to admit more Afrikaners. The defeat of the Boer Republics led many of their more intransigent citizens to look to Germany as a possible ultimate deliverer, and those who were allowed to move to German territory settled mostly in the south and around Gobabis and bided their time.

The later Afrikaner settlements in Namibia, though they followed travel by land, possessed several of the characteristics of seaborne migrations. The Dorsland Trekkers formed a coherent party bound together only by the wish to flee Burgers and the need to survive. The immigrants after the Boer War were even more loosely grouped. Perhaps the most striking aspect of the Dorsland Trek, and one which may equally have operated during the more random passages of smaller earlier parties across the dry uplands, was the intensity of selection which the journey itself entailed. Super (1975) has suggested that the present high frequency of the gene for cystic fibrosis among the Afrikaners of Namibia may be because the high salt content of the sweat of the carriers of the gene acted as a deterrent to mosquitoes; so that proportionally more non-carriers died of malaria during the passage of the northern Kalahari. The mortality during the Trek was certainly extremely high, and many people died of malaria and some of starvation. This leads one to speculate on the selective effects of migration itself.

The high frequency of the cystic fibrosis gene in this population may thus be an instance of selection rather than random drift, and should not be regarded as an example of founder effect; nor can any of the 'ethnic' diseases now seen in South African Caucasoid minorities. There is no higher frequency of Tay–Sachs or Niemann–Pick or Gaucher's disease in South African Jews than in Ashkenazi Jews elsewhere. The frequency of $\beta$-thalassaemia in the Greek, Italian and Indian minorities is not elevated, and that of $\alpha$-thalassaemia may even be lowered. No single ancestral line can be blamed for the presence of any of these heritable diseases. There are, however, several conditions found at such high frequencies among the Afrikaner and so rarely in the rest of the world that founder effect has certainly been responsible for their prominence.

Founder effect produces its most striking consequences when a small number of migrants arriving as a fairly discrete group and attracting only minor later accretions to the group undergoes rapid increase in a relatively short time. It would therefore appear to be peculiarly likely to manifest itself in the progeny of small seaborne migrations. The most familiar instance of all, perhaps, is that of prophyria variegata, a dominantly inherited condition whose manifestations, before the advent of a number of new drugs which cause serious reactions in sufferers, were confined to an enhanced tendency for the skin to abrade and blister (Dean 1963). It is found in approximately 1 in 400 white South Africans, all of whom appear to be the descendants of a single Dutch couple who married at the Cape in the late seventeenth century. It has also been diagnosed in a number of 'Coloured' or hybrid patients, presumably descendants of the same couple.

There is less certainty as yet about the identity of the probably Jewish progenitor responsible for Gaucher's disease in the Afrikaner. This is not common, but is found relatively more often than it is in other Gentiles. Tay–Sachs disease is found in non-Jews, though much more rarely than in Jews, and it is possible that it may occasionally be confused with Sandhoff's disease. A rare recessive disease apparently confined to the Afrikaners is *dikkop* or sclerosteosis (Beighton *et al.* 1976), a progressive hypertrophy of the skull; whether it can be said to be due to founder effect or whether a new mutation is responsible, is uncertain. Similar uncertainty attaches to keratolytic winter erythema or 'Oudtshoorn Skin' (Findlay *et al.* 1977); though it appears likely that all the sufferers have a common ancestor, he or she has not been identified.

These conditions seem to indicate that the migrations inland of the Afrikaners have produced less panmixis among them than might have been expected. The highest concentration of porphyriacs is still to be found in the eastern Cape Province, and its name shows that all cases of 'Oudtshoorn Skin', wherever they may be living now, do have some connection with that small Cape town. Type II familial hypercholesterolaemia, also common in the Afrikaner, appears to be commonest in the Transvaal; genealogical enquiries suggest that it may have been carried thither in several members of one Great Trek party. It is difficult to postulate any selective advantage for dominantly inherited diseases. None of them is common enough to constitute a polymorphism; the commonest, porphyria variegata, has a gene frequency only in the region of 0.0013; and presumably in the absence of interference selection will in the long run extinguish them.

It is perhaps in the realm of the haemoglobins that founder effect can produce its most curious results. Haemoglobin S and Haemoglobin C, both of African origin, are both found at very low frequencies in the Afrikaners. The frequencies are not so low that the occasional homozygote does not occur (Lewis *et al.* 1957; Altmann 1945; Botha and van Zyl 1966). There have not, however, been any reports of Haemoglobin SC disease in Caucasoids, and this tends to substantiate the suggestion that the Afrikaners are not a panmitic population despite the forces of disruption and reassortment which have acted on them since the Great Trek.

## Inbreeding among Southern African Caucasoids

The rural Afrikaner, particularly in the Cape but also to some extent in the Orange Free State, the Transvaal and northern Natal, still tends to aggregate into communities which, if not precisely closed, are not very open to gene exchange with the exterior. Something similar is probably the case in some of the more tightly knit communities in the cities; even where the possibilities of increased mixing are there, factors such as family solidarity or religious exclusiveness enter into the situation. Assortative mating for Huguenot descent plays, on the evidence of the published genealogies (Pama 1966) some part in this, but is not likely to be the only, or even the most

important, influence on Afrikaner inbreeding. It is unfortunate
that no systematic study of this medically relevant but probably
politically delicate subject has yet been made. There are not even
data available on marriage migration.

Geographical closeness is probably the most important factor
in the choice of marriage partners among the Afrikaners as among
most peoples; but we have no hard data about the manner in which
it operates. There is certainly a great deal of mobility among government employees. The railways and the police shuttle young men all
over the country, and the considerable immigration of Afrikaners
into Namibia during the past thirty years has been brought about
largely through civil service transfers. During the same period there
has been a continuation of a trend, dating from the Depression of
the 1930s and possibly earlier, for Afrikaners to move into the cities,
sometimes purely in search of employment but increasingly to take
up assured employment secured for them by friends or relatives who
have moved there earlier. Increasingly industrialization has worked
to accelerate the trend, and the mines, once staffed largely by immigrants or men of British descent, are now predominantly staffed by
Afrikaners. Some of these migrations are made by families, but, as
one would expect, unattached young people, and no longer only
young men, migrate more readily. The population is growing, though
not as fast as the Negro population is, and it is largely the increase
which is moving into urban centres, while the rural population
remains static and probably continues to follow marriage choice
patterns which have not changed very much over the years.

Under such circumstances one would expect a certain amount of
inbreeding in the country areas, often following a pattern of reciprocal
mate exchange between lineages. Details would vary from district
to district, but such recessively inherited disease as occurred would
be more likely to be found in rural districts. Impressionistically, this
would appear to be the case. Whether the reduced panmixis of the
countryside is associated with a gradual reduction in fitness due to
reduced heterozygosity it is of course impossible to say, though
settlement in the northern interior has been of too short duration for
this to be likely to be so in the absence of other causes of inbreeding
such as rigidly observed assortative mating. The Afrikaner is, nonetheless, not notably more ethnocentric than other rural peoples;
during the investigation of 'Oudtshoorn Skin' it was found that the
principal affected lineage had intermarried with people of English
and Irish descent, and among the husbands of the women were a
Greek and a Croat.

The South Africans of British descent outwardly appear to be more mobile than the Afrikaners, even when they constitute rural populations. They also show more ethnocentricity in choice of marriage partners, though less attachment to their religious affiliations. Marriages between English-speaking and Afrikaans-speaking South Africans appeared at one time to occur most readily at the extremes of the social spectrum, with the offspring usually joining the Afrikaner moiety at the lower end and the English moiety at the upper; but a higher proportion of intermarriages now takes place in the intervening strata, and the affiliations of the children are not so regularly predictable. Even the most rurally-based of the South African English-speakers are city-oriented and contrive to spend a certain proportion of their lives in the cities. Many cling to ancestral links with the British Isles, and a good proportion choose their marriage partners there or in other English-speaking countries. No English-speaking community in Southern Africa can be called endogamous, with the inevitable exception of the Jews.

Some Jewish endogamy in South Africa must be inadvertent. There are well over a hundred thousand Jews in the country, the great majority Ashkenazic and descended from forebears who lived in a fairly small area around Vilna in southern Lithuania. The original Jewish settlers of the largest immigrations during the later nineteenth century came provided with surnames which they had only recently adopted, and which they retained or modified or changed completely almost at whim. Brothers arriving together would part, and when they met again would have surnames not recognizably similar. Their grandchildren more often than not would be unaware of any kinship at all. Concentrations of Jews formed on the Witwatersrand, at Kimberley, and eventually in all the large cities, while even the smallest towns now have some Jewish inhabitants. The most unfavourable consequences of inbreeding would appear to be avoided by dissipation of the rare deleterious recessive genes in this miniature diaspora, though some would be conserved and expressed through the unwitting alliances of kinsfolk. It should be remembered that the original immigrations, though large, came from a population which must already have been somewhat inbred and possibly thereby reduced in genetic variance. On the other hand the fact that the major deleterious recessive gene found among the Ashkenazim, that for Tay–Sachs disease, is not expressed in Southern African Jews significantly more often than it is in other Ashkenazic communities, suggests that inbreeding among them has not been considerable.

The smaller immigrations of Caucasoids present a different picture. The Italian community is hardly ethnocentric at all, and of such various origins in Italy that such endogamy as does occur is important only because of the possible expression of the common gene for β-thalassaemia. This is rather more important among the Greeks, who stem predominantly from rather circumscribed areas of Greece, Cyprus and Anatolia. They are perhaps the most ethnocentric of all the minorities in marriage, and the most resistant to genetic investigation. Endogamy among the various German communities is too recent for it to be of much importance, considering the wide spread of areas of origin of the Germans. The Portuguese immigrants, most of them recent, show a diametrically opposite pattern. Many of them are married to partners from the same area of Portugal as themselves, often from the same village, and there is a surprisingly high incidence of unusual recessively inherited abnormalities in their children. This may be expected to decline in future generations even if the Portuguese turn out to be as ethnocentric as the Greeks, since few lineages appears to have migrated with enough of their close kin for them to have much choice in future but to take marriage partners unrelated to them. Fortunately there is little or no thalassaemia in Portugal.

The tendency of the Indians to retain the mating patterns and preferences of their original villages has been mentioned earlier. Among them inbreeding is partly responsible for a moderate incidence of β-thalassaemia and may contribute to a much higher incidence of diabetes mellitus than is found in their areas of provenance. There is little doubt that this proceeds in part from their much larger consumption of food in Southern Africa than was possible for their forebears in India. The diabetes is of the maturity-onset type, which means that it does have a genetic basis and the high incidence may be due to the consequences of selection relaxation. Juvenile-onset diabetes is unusual among them, which is hardly surprising in the light of recent work showing an association between it and certain tissue antigen conformations, notably a D-locus antigen in linkage disequilibrium with HLA-B3 or HLA-B5 (Nerup *et al.* 1974). Brain and Hammond (1973) have shown that the Natal Indians have lower frequencies of these antigens combined than either Europeans or Negroes do. The incidence of recessively inherited disease other than β-thalassaemia is unexpectedly low in the Southern African Indians.

The population structure of the South African Caucasoids can therefore be seen to present a rather varied picture. In the case of those of British and Italian descent, it is somewhat looser than it is in the countries of origin, while the Greeks and Indians show structures formally not much different from those of the populations from which they originally came. The proximate ancestors of the

Ashkenazi Jews were part of an emigration which established Ashkenazic communities in several parts of the world, and their population structure resembles that of such sibling communities. The Germans and the Portuguese are at different stages of the same process, the formation of new population structures due to heterogeneity of the founding population. After three centuries of evolution the population structure of the Afrikaners is still far from stable, and there does not appear to be much prospect of its ever attaining uniformity. As long as agricultural and pastoral tenacity is counterbalanced by urbanization without reciprocal genetic contributions from the urban communities, and as long as small rural isolates continue to break up but reform, the Afrikaners will continue to show a high measure of heterogeneity and to be recalcitrant to overall genetic characterization. The numerous and often mutually contradictory genetic statements frequently made about them can consequently all be simultaneously true. The Afrikaner is a product of miscegenation, the last 'pure European', pathologically inbred and a manifestation of hybrid vigour, all at the same time.

# 9
# THE GROWTH OF
# HYBRID COMMUNITIES

The term 'Coloured' has long been in use in the English-speaking world to refer to people 'of colour', individuals with more melanin in their skins than that possessed by 'Whites', or people of European origin. American 'Negroes' used popularly to be referred to as 'Coloureds', while in Britain individuals of African or even East Indian origin were often given the same appellation. In South Africa the term has a somewhat more restricted use and is reserved to describe people of mixed racial origin. Legislators have even gone so far as to 'define' a Coloured person. In effect, the definitions embody the recognition of the emergence of a community of hybrids.

It will have been obvious from many of the foregoing chapters that the establishment of a particular ethnic category in Southern Africa is not generally followed by panmixis or uniformity even within that category alone. If anything, there is a tendency towards further fission; the Nguni tribes, which once must have been unitary, fall now into three main classes; the dichotomy between the Afrikaners whose ancestors trekked, and those whose forebears remained at the Cape, is as pronounced as ever; even the speakers of Southern Bush languages, before most subdivisions were extinguished, had time for multiple tiny groups and bands among them to be recorded.

The biological history of the peoples classified as 'Coloureds' in South Africa is even more diffuse, since their origins were not single but came about through a number of different processes. Their emergence or origin can be traced to the shores of Table Bay and a date soon after the Dutch East India Company had established a refreshment station there in 1652. Those close to the centre of Caucasoid contact, in Cape Town and its environs, result from a number of well-recognized and often contrasting historical and social trends, and to some extent preserve traditions reflecting their divided beginnings. Remote from the Cape, their origins are ever more patchy: simple large-scale geographically delimited episodes of miscegenation, even when one party is invariably Caucasoid, will differ in their parties of the second part. In South Africa a 'Coloured' is not simply someone of mixed ancestry, but someone into whose ancestry a Caucasoid element has entered. The other element may be Khoi, slave Negro, free tribal Negro or San, always ultimately a people originally African; the result is politically, and hence to some extent socially, identical.

### Origins of the hybrid peoples: the 'Cape Coloureds'

Despite the foregoing claim that the non-Caucasoid element in the 'Coloured' peoples is always ultimately, it is not necessarily uniquely African. It is almost solely at the Cape that a non-African, non-European strain has entered into these peoples; but it is at the Cape that there has been the best, even perhaps until relatively recently the only, opportunity for further blending of the mixed stocks, so that there is no longer any strain which stands out as a mixture of Caucasoid with Mongoloid or with the darker hybrid Caucasoids of India and Ceylon.

A possible exception to this might be the Cape Malays, whose ethnocentric persistence has been touched on in an earlier chapter. Their forebears were brought to the Cape as slaves, as convicts sentenced to terms of servitude, or as political exiles from the regions of Dutch control in east Asia. They were, as their name suggests, largely from the Malay peninsula and islands, including what is now Indonesia; but these were not their sole origins. Though there is probably no very great European Caucasoid element in their ancestry, they would certainly descend to some extent from the Caucasoid Muslims of Bengal, since their coherence comes not from ethnicity but from religion. During the nineteenth century and subsequently they received an additional Indian contribution from their co-religionists from the west of India, particularly the Gujerat, who came under British auspices to South Africa as traders. Even before emancipation the slave element among them led lives which were hardly servile; many had been manumitted either by their masters or by co-religionists who bought them for that purpose, and many of the rest were, despite their status, respected craftsmen whose relationship with their owners was one of convenient social stabilization, protection against general exploitation, rather than abjection. The slaves imported for heavy labour came mostly from East Africa and Madagascar, and were consequently either completely Negro or

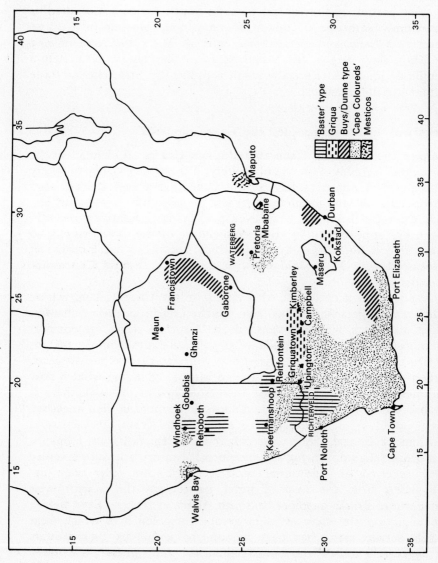

**Fig. 9.1.** Distribution of the principal types of Hybrid ('Coloured') population in Southern Africa.

Negro with a Malay or Indonesian admixture. They were mostly male, and would consequently have made to the eventual Cape Coloured people a contribution disproportionately small considering their relative numbers.

Until 1685 marriage at the Cape between European Caucasoids and persons of full colour of whatever provenance was legal, and the children of such marriages, even where one party was a slave, were absorbed into the European Caucasoid community. After that date slave-born children of European Caucasoids remained slaves until an age decided by the Dutch East India Company, and varying at different times. But even before that, unsanctified miscegenation was common; Theal (1909 (in 1907–10)) states that in the first twenty years of colonization 757 of the children born to slaves were hybrids. As the colonist community increased in numbers, however, pride of race became more evident and sexual relations between colonists and slave women decreased. Soldiers and sailors were not so readily deterred, less through any lack of ethnocentricity than because of the singular shortage of European Caucasoid women. Even slave women were in short supply, however, and extensive miscegenation took place between Caucasoids and Khoi.

Marriage between colonists and Khoi women was encouraged at first, and that between Van Meerhoff the explorer, and Eva the 'Hottentot woman', in 1664, was celebrated with a bridal feast at Government House. Nevertheless, such marriages were few, and the ordinary European Caucasoids eventually adopted a contemptuous attitude towards Khoi women and would not contemplate marriage even though sexual relations were commonplace. The offspring of such unions would remain with the mothers and became members of the newly emerging 'Coloured' community. In the eighteenth century the term Bastard was used to describe such offspring and was not at the time confined to the population which arose to the north west of the colony as the result of 'marriages', not solemnized by the Church or registered by the State, between the deprived younger sons of the colonists (*trekboere*) and the local Khoi women. We shall refer below in more detail to these Bastard or Baster communities.

One reason for the changing attitude towards the Khoi was the frequency with which miscegenation occurred between slave men and Khoi women. This also contributed to the creation of the 'Coloured' people and was particularly common at the Cape where, as late as 1708, the proportion of adult male slaves to adult females was six to one. Although this preponderance declined during the course of the eighteenth century, at the time of emancipation (1834) males still outnumbered females 21 613 to 17 408 (Theal 1910).

The hunter-gatherer San certainly inhabited the Cape at the time of its settlement and though Theal (1909 (in 1907-10)) implies that it was impossible for other people to live at peace with the San and, therefore, that they were not involved in the production of the 'Coloured' people, Marais (1939) has questioned this assertion. It seems certain that in the latter half of the eighteenth century many San were taken prisoner by civil commandos along the frontier areas and that they were distributed as apprentices, virtual slaves, among the farmers of the area. These San would presumably have been housed with the farmer's Khoi or hybrid servants and, in due course, would have been assimilated.

From the genetic point of view differences in contribution to the 'Coloured' gene pool by the Khoi and the San are probably more apparent than real. We have shown in earlier chapters that it was extremely difficult, if not impossible, to distinguish clearly between the two indigenous populations who inhabited the Cape at the time the Dutch arrived. Elphick (1977) postulates a cycle of Khoi-hunter interaction which had blurred the cultural differences between them; many people without cattle spoke the language of the Khoi. The relatively easy change from one culture to the other meant that gene flow occurred, and any genetic differences which might in past times have existed soon became minimized. This postulate, together with the likely fact that the immigrant Khoi and the 'settled' San might well have shared a common genome, makes it unnecessary, we believe, to consider separate Khoi and San contributions to the creation of the 'Coloured' people. The genetic constitution of the Khoi of the Cape at the time of van Riebeeck's landing is likely more closely to have resembled that of the present-day Central San of Botswana than that of the present-day Khoi of Namaqualand.

The social, and ultimately breeding, direction of the products of miscegenation covered a wide span. Legally begotten children of Europeans faced no difficulties in being recognized as members of the socially dominant group, especially at first. Even slave-born hybrids with one European parent entertained the certainty of liberation in adulthood, and though they could never become members of the Dutch community except by emigration to Holland, their daughters, where these were not born into slavery, being persons not of full colour, were eligible for European marriage.

The advantages of being light in colour are still intensely felt by the 'Coloureds' of the Cape, and the touching ambition to 'pass as white' has been a driving force among them ever since social discrimination against them began more than two centuries ago. When birth registration was, or is, not strictly enforced the chance reassortment of the genes for pigmentation could, or even today can, lead to the redistribution of a sibship among several of the social categories of the Cape. Within that category which came to be recognized as singularly and coherently 'Coloured' such genetic reassortment consequently took on a somewhat deterministic aspect, with those of lighter colour having the best prospects of mating and hence the highest chances of producing offspring, even though there was a strong possibility that those who were lightest of all would migrate socially upwards and deplete their parent community of opportunities to continue to produce the preferred phenotype.

Such assortative mating might be expected to have a small effect on the fitness both of the 'Coloured' and of the Caucasoid peoples; if anything, it should enrich the latter with increased genetic variance and heterozygosity, while the overall hybridity of the former would hardly be impaired. It is possible, in fact, that keeping the 'Coloureds' dark is, in an environment as subject to irradiation as that of Southern Africa, to their advantage. On the other hand, the transition from one ethnic category to the other would represent the flow into the Caucasoids of at least a few genes for pigmentation on the darker side of the Caucasoid range. There would therefore be a contribution to the Caucasoids of genes which would improve their adaptation to the environment. As this kind of transition has become steadily more difficult, so higher degrees of ingenuity and perseverance are progressively more necessary to those who make it. These selectively advantageous sociopsychological traits may be expected to be distributed in the community independently of the genes for colour, but each successful passage across the colour bar could be expected to represent as much a transfer of the genes for them as of those for skin colour. 'Passing for white' consequently operates marginally to the advantage of the Caucasoids and the disadvantage of the 'Coloureds'. The converse also occasionally happens with the passage of a dark 'Coloured' into the Negro or, less frequently now, the Khoi community. This is less likely to represent a voluntary social migration than a pragmatic accommodation to circumstances. Once again, the recipient population would receive the benefit of increased heterozygosity (cf. Findlay 1936).

During much of the time since it first came into existence, therefore, the 'Cape Coloured' people has represented less a coherent population than a genetic flux, absorbing and amalgamating all the genetic material entrapped in all episodes of hybridization at the Cape, occasionally putting out a flow in one direction or the other, into the Caucasoids, the Khoi or the Negroes, and at the same time receiving continued reciprocal flow from those and, more lately, from the variety of seagoing men and other travellers who call at the Cape. At the same time the products of the three types of hybrid process from which the first 'Cape Coloureds' derived, the matings between Caucasoid and slave, between Caucasoid and Khoi and between Khoi and slave, themselves have interbred and blended, so that except in the case of the Cape Malays there is no recognizable separate stock descended from any of them.

## Origins of the hybrid peoples: the Griqua

The Griqua enjoy less biological than social distinctness, despite the peculiar circumstances of their origins. The elements which have contributed to their genetic constitution are those which entered into the formation of the original 'Cape Coloureds', though the Khoi part may well have been the most considerable. The main reason for suggesting this is their retention until recently of a Khoi language, Xiri, and of an individual female *rite de passage* (the !*gabadas*) which is more likely to be Khoi than anything else (Nurse 1975a; Nurse and Jenkins 1975). Added to that is the evidence of their surnames, many of which indicate at least one Khoi partner in the marriages which contributed to their formation (Nurse 1976).

There seems little doubt that the primary circumstances in the formation of the Griqua were the coincidence of the breakdown of the Khoi tribe known as the ≠Karixurikwa, living about a hundred and fifty kilometres due north of Cape Town (see Fig. 9.2), as a result of the smallpox epidemics, and the manumission and arrival in the area towards the middle of the eighteenth century of an ex-slave called Adam Kok. Adam Kok seems to have been an extraordinary man; not only did he earn his liberty with such goodwill that his master equipped him with cattle and secured grazing and, somewhat mysteriously, burgher (citizenship) rights for him and his family, but he was accepted by the miserable remnant of the ≠Karixurikwa not as a rich foreigner on whose land they might squat, but as their rightful leader. His livestock proliferated and his family and followers grew in number. He was invested with official status by the East India Company and at the same time was sufficiently remote from the main centres of power to be able to furnish a refuge for persons who had failed to integrate into Cape society. Adam Kok's people attracted and absorbed runaway slaves, unemployed ex-slaves, deserters from ships and the services, dissident 'Coloureds' from the Cape and wandering parties of other extinct Khoi tribes. At the same

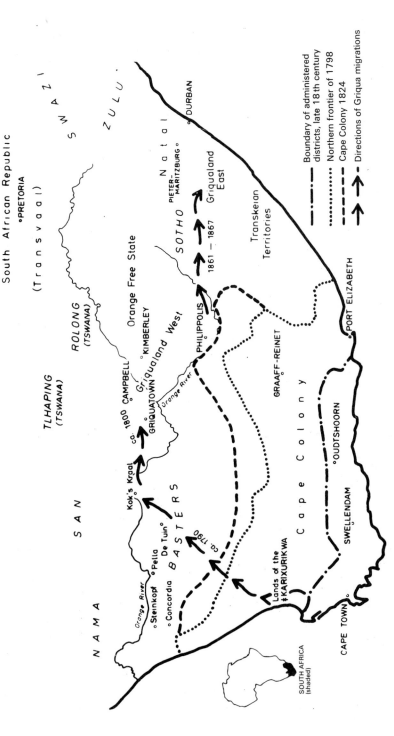

Fig. 9.2. Map of Southern Africa showing the migrations of the Griqua in the eighteenth and nineteenth centuries. (From Nurse and Jenks 1975, with permission of the American Journal of Physical Anthropology.)

time good relations and trade were maintained with the Cape. There was no reason at first, why Adam Kok's people should be at any disadvantage compared, if not with the urban Caucasoid population of Cape Town, at least with those who were beginning to penetrate beyond the fertile lowland valleys of the south west.

As so often, prosperity proved the downfall of this receptive and flourishing community. The more respectable Caucasoid population was exploding, overflowing beyond the limits to which Caucasoids were previously adapted and within which they had hitherto settled, becoming exigent for resources and demanding sole control over the springs of the accessible interior. Adam Kok's people, now sufficiently mictic to deserve the epithet Bastards (signifying, of course, not extramaritality but hybridity), were pushed northwards, into the harsher country of Little Namaqualand. There they encountered and, it seems, assimilated, other Bastards who had been earlier victims of the same process. As the Caucasoids moved after them they moved on, up to the very banks of the Orange River and then eastwards until they came to settle in the area still known as Griqualand West.

It was as Griquas that they came to be known. Their ignorant anglophone missionary (Campbell 1815) misconstrued the term Bastard; so they acquiescently hunted their history until they dredged up the memory of the ≠Karixurikwa and turned it to a Dutch-sounding name. Griqualand West, and the fertile Philippolis area to the east of it, were the scene of the greatest prosperity of the Griqua and of their earliest serious dissensions. Their principal wealth was in large and small cattle and in land; they were now well situated to be trade intermediaries between the Tswana of the Kalahari and the Caucasoids who were establishing themselves in the Karroo. They became transport riders and on occasion mercenary soldiers, supplied with firearms to keep the !Ora raiders from penetrating south. The Kok succession could not always bear its responsibilities; there were Waterboers and Barends who for a time or simultaneously ruled. The pastures of Philippolis were desirable and after the Great Trek the envy and slyness of the Trekkers contrived to possess them. Confined once more to the desert, the Griqua might have been safe, but the desert had diamonds among its pebbles, and so the British laid claim to and acquired it.

The last phase of the independence of the Griqua took place ironically in the richest land they were ever to possess. Displaced from their dry interior grazing, forced into a trek more anguished for them than it had been for the Caucasoid Trekkers, since they commanded fewer resources, they crossed the Drakensberg into a fertile area of coastal valleys lying between the colonists in Natal and the last independent Cape Nguni states. Here it is possible that their prior adaptation to the harsh uplands worked against them, and possibly against their livestock. We have noted the relatively small use made by the Khoi of the good pastures of the south-western Cape; with the Griqua we have a more recent and better documented account of the inability of a desert-bred people to cope with such. The fecklessness for which historians have blamed them was manifested particularly in their last homeland. Further-more, it was enthusiastically encouraged by outsiders who sought to profit from it. The official policies of the states adjoining theirs, too, were dispiriting. Moshoeshoe's Basutoland enjoyed British favour, Natal was a British colony, the Nguni states which survived were unlikely to do so for much longer, and prosperity turned out to be impermanent. Such wealth as there was provoked a succession dispute, and this was made an excuse for annexation to the Cape. The first confessedly racially hybrid state in Africa came to an end.

Griqualand East was annexed in 1874, but the dispersal of the Griquas had started long before that. At each change of settlement, some had been left behind. Improvident as they were popularly supposed to be in the conduct of their own affairs, they were possessed of certain skills which their neighbours could use. The Griqua spoke Dutch (or Afrikaans) and a Khoi language, and so could act as interpreters. They could lead the wild herdsmen of the interior to become tractable employees of the settling Caucasoids; as transport drivers they filled a niche peculiarly that of the hybrid peoples of the desert. As the diamond mines expanded at Kimberley, they were the foremen; on the fringes, they worked the diggings too unproductive for the attention of the white miners. The hybrid community in the northern Cape was growing as a result of the industry there, and into it many Griquas were absorbed. Still there were a few who kept their identity and their pride in their quondam independence. A few came back from Kokstad and took over what they could in the part of Griqualand West which they were still allowed to occupy. Others proceeded in an accelerated diaspora to establish small Griqua communities in the larger Cape and Transvaal towns, and to preserve a social if not a biological identity.

The Griquas above all others practised, and have continued to practise, the receptivity of the Khoi towards displaced persons. They increased in numbers not only through natural reproduction but by extending the prestige of their name to cover any who are prepared to invoke it. As it happens, despite their national destitution, the status of the Griquas, as the heirs of a once indepen-dent state, is marginally higher than that of the other 'Coloureds';

this is recognized among the convolutions of the South African laws on race. It is to the advantage of many non-Caucasoids to seek Griqua identity; and this has been extended not only to 'Coloureds' but to Khoi, San and Negroes (Nurse 1976). Parallel with the proliferation of race-based distinctions in the larger polity of South Africa, however, a previously indistinct social stratification has grown up among the Griqua, so that the 'old' families have more prestige, the 'newcomers' less; and this seems to operate more to the disadvantage of the darker than of the lighter recruits.

The Griqua, therefore, are no more of a stable community than the 'Coloureds' at the Cape are. Their survival as a political entity is not paralleled by any biological continuity. They will doubtless continue to expand, and contract, and maintain their relative numbers among the 'Coloured' peoples, as well as their identity.

### Origins of the hybrid peoples: the Basters

Ever since Fischer (1913) published his large study of them, the Basters have signified to anthropologists the centralized and coherent Rehoboth community; and in the context of contemporary Southern Africa that is who they are. Yet the name Basters was in the beginning more generally apportioned, though *sensu stricto* it referred to hybrids between Khoi and Caucasoids, as the Rehoboth people in the main are. Their origins can be found in one of the most interesting processes in Southern African biological history, and one to which there has been reference in earlier chapters.

As mentioned in Chapter 8, the large family sizes of the original colonists, coupled with continued Caucasoid male settlement, usually of discharged servicemen without wives, led to a continual surplus of Caucasoid males at the Cape. The newer arrivals, possibly because they were richer, possibly as a result of the greater attractiveness to females of males of exotic type, seem to have been preferred as husbands for the Cape-born women. The Cape-born men of low sibship rank, with less hope of inheriting land and with no resources besides a few head of cattle, could hardly compete. Even to secure grazing for their beasts they had gradually to move further and further afield and eventually up into the arid interior. There, doubtless, those less biologically equipped for such an environment failed to flourish, but there were survivors and, as Fischer points out, the survivors would be those whose genotypes best equipped them for their surroundings.

The land into which they expanded was thinly inhabited by Khoi, many of them remnants of the Cape Khoi, some of them indigenous to the area, but all of them the products of generations of selective adaptation to just that environment. Here the Caucasoid survivors of the traumatic transition to the semi-desert would possess the 'rare male advantage' which had made them losers in the search for mates among their own kind; they would be preferred over the more familiar Khoi mates when it came to securing mates. Moreover, they would command more economic resources; they would have guns, and wagons, and metal implements, and unfamiliar technical knowledge, and, poor as their cattle might seem under Western eyes, they would represent wealth to the Khoi.

The degree of formalization brought to the unions which resulted varied from one place to another. Calvinism was too rigid for the advantages to be over-exploited, or else they were just not considerable enough: there are few records of any Caucasoids taking simultaneously several wives. Most matings, even in the absence of the Church, were treated as marriages, and even today one hears the phrase *getroud maar ongeëg*, married but not consecrated, applied to a form of common-law marriage in the north-western Cape Province. When missionaries followed, as after a few generations they did, the position was regularized; but even before that the surnames of the founding fathers had been preserved and family pride maintained.

Early in the nineteenth century there were numerous Baster communities situated between the Orange River and the areas of Caucasoid settlement to the south; the Griqua, at the time, would have been counted among them. In the course of the century there was to be a great reassortment of these peoples. Caucasoid pressures were signalized at first principally in the seizure of water sources, and later in the provision of opportunities for paid employment and consequently the chance to buy desirable consumer goods. At first this displaced and impoverished many of the hybrids, and subsequently drew them into situations where they became indistinguishable from, and eventually intermarried with, 'Coloured' labour coming up from the Cape. Pockets of the original Khoi/Caucasoid Baster stock nevertheless survived, and the most considerable and best known of these became the Rehoboth Basters.

Once more, it was a concatenation of events which preserved this particular community. At the time of most considerable pressure from the advancing farmers the Basters around De Tuin in the far north were not only predisposed to organize themselves, but had living among them a missionary who was able to tell them how to do so. Instead of patiently starving, or seeking subservient positions, they rallied together and migrated northwards with their herds and flocks, up to and eventually across the Orange River, through the inhospitable southern part of what is now Namibia and on to the richer lands just to the south of the Auas Mountains. There they took possession of an abandoned mission station and divided the country around it among themselves.

Coincidence seems to have entered again into their being able to accomplish all this peacefully in an era and a region racked by the bitter Nama-Herero wars. The acquisition of the land appears to have been quite regular, by treaty with the Khoi tribe which claimed to possess it, though for many years afterwards there were disputes about whether the Basters had carried out the terms of the treaty properly. The Khoi tribes, many of whom had received an admixture of Caucasoid genes themselves, seem to have regarded the Basters as Khoi and hence potential recruits against the Herero; when they discovered their mistake they tried to redress it by military means. The struggle was not pursued for long, however, and when it was abandoned the position of the Basters of Rehoboth had been consolidated.

The Basters were seen as allies also by the next power to intervene in the region. The German settlement in Namibia has been described in the preceding chapter. The Germans seem to have been more aware of the Basters as semi-Caucasoids than as semi-Khoi. Indeed, Baster culture is selfconsciously Afrikaner, and preserves the values and virtues of the rustic Boers of the last century far more tenaciously than their sophisticated descendents do today. Under German influence, however, anteceding by a long time the German political presence in Namibia, they are Lutherans, not Calvinists, and this, too, may have made the Germans consider them as natural allies. For as long as the German mission in Namibia appeared as one of pacification, the Basters collaborated with it. When they saw the autonomy of the surrounding peoples crumble and give way to the type of Caucasoid settlement which they had fled to avoid, however, they knew themselves once again threatened. The First World War supervened; they appealed to the South African Prime Minister to rescue them; and the arguments about whether their little state is, or is not, autonomous have gone on ever since.

Unlike the Griqua, the Basters have cherished a marked pride of race. The children of a Baster who married a non-Baster hardly ever acquired Baster status. Even among the founding families, Fischer (1913) detected hierarchy, and ascribed it to the differing proportions of Caucasoid descent in the various families when they moved north out of the Cape. He suggested, too, that some covert miscegenation had taken place and been accepted at the price of lowered social status, even after the arrival at Rehoboth. During fieldwork in the area, we were made aware of a popular belief that the Basters of the eastern part of the area were closer to the Khoi and at the same time less respected, less privileged, less easily accorded their rights on the common lands, while the rich families of the south-west were able to command pretty nearly whatever they wished in the way of local resources. This division was reflected in the political field as well, with the privileged seeking more local autonomy and the poorer people aspiring to equal integration in an essentially independent Namibia.

The social hierarchy operated, and still operates, at several different levels. It is possible for an immigrant to apply for Baster status, but the criteria for granting it are fairly strict. It is hardly ever given to anyone other than men married to women from one or other of the most influential families. Each application is debated in detail by the Baster council, and the level of education of the applicant, his earning capacity and his capital are all taken into account. Some whole families get by through claiming that they are Basters who were left behind in the Cape at the time of the original trek, and providing genealogical information in support; the Basters are enthusiastic genealogists. The fact that most applications are approved does not indicate any relaxation of criteria. Only those most confident of success are likely to apply, since the future in the area of those whose applications are rejected is unlikely to be prosperous.

Movement out of the Baster ranks has been going on for some time, and more particularly since the arrival of the Germans. The children of Basters who marry non-Basters of colour tend to be assimilated into the Khoi or, in the larger towns of Namibia, the 'Coloured' communities. The Basters themselves are too proud and self-assured to make any attempt to 'pass for white', though many of them are indistinguishable from Caucasoids. On the other hand, a number of their daughters have married Caucasoid men. The converse type of marriage, of a Baster man with a Caucasoid woman, is much rarer, but not unknown. Though the children would rank as Basters, whereas in the former type they count as Caucasoids, most such marriages take place outside Namibia and their results are lost to the community.

### Origins of the hybrid peoples: the smaller local groups

In the foregoing sections we have dealt with the origins of the very large 'Cape Coloured' community and those of the smaller but distinct Griqua and Baster peoples. We have shown that each of these is in its own way coherent. The coherence of the Griquas and the Basters is political, that of the 'Cape Coloureds' and the Basters biological. We have indicated also that the three categories differ in the types of gene flow which characterize them. In this section we have to deal with a rather amorphous assembly of often sharply contrasting groups about which little is known.

One of the more distinct of these is the Baster remnant in the northern Cape, of which the Richtersveld community has been subjected to fairly detailed sociological (Carstens 1966) and biological (Jenkins 1972) study (Fig. 9.3). This community probably shows fairly exactly what is likely to happen to such people in the semi-desert environment when they are left to develop on their own, virtually without interference but in close interaction with a people constituting one half of their own ancestry. The Baster identity here has been virtually lost, and the people have merged biologically with the Khoi, though their culture continues to be one of synthesis, with retention of Khoi practices where these are dictated by the surroundings, or by material poverty, and the adoption of Western methods where these convey either prestige or advantage.

Despite the ostensible precision with which the gene marker profile of the Richtersveld people can be identified as predominantly Khoi, the question of adaptive selection of course remains to some extent open. Even in the short period of their evolution, there has been time for them to shed such Caucasoid genes as were disadvantageous in their environment and to conserve and increase the frequency of advantageous Khoi ones. Against this possibility we must put the fact that the Rehoboth Baster gene marker profile shows that people, at least in its representative original families, to embody almost precisely equal quantities of Khoi and Caucasoid genetic material (Stellmacher et al. 1976). Selective pressures have perhaps operated less rigorously on the Rehoboth than on the Richtersveld people, but the differential is unlikely to have been so considerable. The increase in the proportion of Khoi genes in the Richtersveld must be due to admixture.

The Richtersveld process is probably paradigmatic of what has happened over a great deal of the Cape interior. Wandering, relatively privileged parties of Basters have been welcomed by the disorganized remnants of the Khoi tribes and by intermarriage have facilitated the recognition of the latter as relatively high-status 'Coloureds' rather than as the more lowly regarded 'Hottentots'. The proportionate contributions to the admixtures will have varied widely from one place to another; there will occasionally have been sporadic

(a)

(b)

Fig. 9.3a, b. !Kuboes, a remote village in the Richtersveld, northern Cape Province, inhabited by Baster-like, hybrid people. (a) General view of the village;
(b) a typical Khoikhoi hut covered with straw mats behind which is the *komb-
uis* or kitchen which is covered in this case with hessian. The donkey cart has
replaced the ox-wagon and pack oxen of former years.

further gene flow from the Caucasoid side or even, as we have noted
in Riemvasmaak (Nurse and Jenkins 1978) from Negroes of surprisingly wide provenance. Caucasoid settlement and the growth of

towns have provided opportunities for paid employment and the acquisition of new skills and the farms and towns have also promoted further intermarriage and a greater uniformity of the genetic mixture. Nevertheless local differences do remain, and the relative sedentariness of the contemporary 'Coloured' communities of the interior will probably help to accentuate them.

There are, of course, parts of the Cape to which the hybrid strains resulting from the processes of Baster genesis never penetrated. Miscegenation in the Eastern Province resulted to some extent from the extension of Caucasoid settlement through the Little Karroo and to some extent from the opening up of Port Elizabeth. Since relatively few slaves were brought to this area, the proportion of East Asian genes in the 'Coloureds' of the Eastern Province must be low, while the proximity of the Cape Nguni, and the miscegenation which had already taken place between them and the Khoi to produce the Xonakwa, means that these 'Coloureds' are not a blend of equal quantities of Caucasoid and Khoi genetic material but comprise also a variable proportion of Negro genes. There has, nonetheless, been migration to and from the Cape, and in the larger towns there may be some approach to relative 'Coloured' homogeneity.

Rather apart from these communities stand those which have arisen outside the areas of Khoi habitation and as a result of miscegenation between Caucasoids and Negroes. Two examples of these are especially well known: the Buys 'Coloured' community of the Zoutpansberg, where the admixture was between Boer and Sotho-Tswana, and the Dunnes of Natal, the outcome of the marriages of one British adventurer to several Zulu brides. These stocks have tended to preserve their identities by marrying only mates of similar provenance; their aspirations, however, seem to be quite different, and whereas the Buyses strive for acceptance as 'Whites', the Dunnes constitute a recognized and respected clan among the Zulu. There are other, less well-known, instances of local miscegenations which have produced the Coloured communities of numerous small South African towns. Such communities would then provide centres for the migration to the towns of rootless persons of mixed descent from the surrounding countryside or from farther afield. Similar processes appear to have taken place, on a very much smaller scale, in Namibia, Swaziland, Lesotho and Botswana.

### The deracinated Khoisan

It has been mentioned in Chapter 4 and above that one reason for the sharp decline in the numbers of Khoi during the nineteenth and twentieth centuries has been the relatively higher social and political status of 'Coloureds'. It has been to the advantage of the Khoi to claim that status, and at the Cape it was generally easy for them to do so provided they assumed a reasonably Western way of life and acquired proficiency in the Dutch, or later the Afrikaans, language. Mission stations set up among the Khoi have subsequently been found to be populated entirely by 'Coloureds', without the intervention of Caucasoid miscegenators. Even the San have been able to take advantage of this to some extent, sometimes by first blending, as they so easily can, into a Khoi community, and sometimes by simply settling down, speaking Afrikaans and claiming to be 'Coloureds'. The rapid and otherwise almost inexplicable disappearance of most of the speakers of Southern Bush languages can probably be ascribed to this cause. Though no study of gene markers has, so far as we know, been carried out on the 'Coloured' peoples of the country around Knysna and Oudtshoorn, casual observation of them inclines us to believe that they, and probably several other 'Coloured' communities, are of substantially unmixed Khoisan stock.

### Origins of the hybrid peoples: the mestiços

The mestiços of southern Moçambique are included here simply for the sake of completeness. They have resulted from a far less socially regulated process of miscegenation between Portuguese and, in the main, Tsonga/Ronga-speaking Negroes. They are hardly conspicuous, since no social distinction has been made between them and the privileged classes in Moçambique; we cannot speak, however, about their position now that the country is independent. Since they have played little or no part in the formation of those hybrid communities which either elect or are forced to flaunt their hybrid identity, and since they seem not to exist as a community in themselves, they are of rather marginal importance to the study of the adaptation of the Southern African peoples to their environments, except insofar as they have in the past functioned as a channel of gene flow between one group of Negroes and a particular Caucasoid group which intermarried very little with the other Caucasoids of the region.

The share of Negro genes in the metropolitan Portuguese could in any case be expected to be high compared with the rest of Europe. There was a proportionately more extensive importation of Negro slaves into Portugal during the Renaissance and later, than into any other European country. For a time the labour force of the Alemtejo and the Algarve was almost entirely Negro. Emigration from Portugal to Moçambique, however, seems to have occurred more considerably from the north  and centre than from the south, and for a time at least to have included mainly men of privileged background setting out to make their fortunes (Lobato 1954). There could have been some flow of genes from West to Southern Africa by this circuitous route, but it is unlikely. What is most interesting about the mestiços of Moçambique is their origin in a more tropical and humid environment by a process analogous to that which produced the Basters: an excess of technologically relatively advanced and wealthy Caucasoid males adjacent to previously undisturbed indigenous communities.

### Physical characters and miscegenation

In earlier chapters we have, albeit with some hestitation, given rather typological descriptions of the gross morphology and gene marker profiles of the peoples we are describing. It will be seen from the preceeding parts of this chapter that for the hybrid communities it would be unusually absurd even to attempt to construct a typology, despite the pioneering and fruitful work of Fischer (1913). Fischer was fortunate in being able to deal with a coherent and well demarcated hybrid community whose very hybridity seemed likely to be informative about the biological demands of a particular colonial environment on its human inhabitants. It would be possible, with varying general relevance, to carry out studies similar to his on several of the other hybrid peoples, many of them less numerous and most of them less coherent than the Basters. In each case the morphology and the genetic profile would depend on the proportionate sources, on founder effect produced by the individual representatives of those sources, on the size of the original party constituting the community and on drift. Each case would need a study more considerable than that of Fischer. Only after all had been completed would we be entitled to generalize about the physical traits of Southern African hybrids; and even then only if we found that the phenomenon of hybridity had indeed generated such facts as could be interpreted as generalities.

### Health and the hybrid peoples

The high concentrations of 'Coloureds' in the western Cape and on the Witwatersrand, and their proximity to the main centres of medical expertise and research in Southern Africa, have meant that a considerable corpus of information about the health of these communities has become available. That the information is none-theless necessarily uneven and unsuited for overmuch generalization will have been apparent from the foregoing accounts of the ways in which the communities have been formed. The prevalences of the wide range of disease to which the hybrid peoples are subject depend on their varying economic and geographical situations, on social status and on the extent to which they are varyingly liable to the diseases of their ancestral populations. Only in the context of in-herited disease, and then particularly when founder effect can legitimately be invoked, can we talk about specific rates of disease.

Even here it is possible only to give rates for not necessarily homo-geneous populations. Haemophilia A is reported to have a prevalence of 2.9 per 100 000 'Coloured' people living in South Africa (Lurie and Jenkins 1975). Three cases of Von Willebrand's disease have been reported in hybrid patients (Gomperts et al. 1969). Hayden et al. (1980) ascertained thirty-three individuals with Huntington's chorea among approximately three-quarters of a million probably not panmitic 'Coloureds' in the western Cape Province. Whether this finding of a higher frequency than in other parts of South Africa is due to more complete ascertainment, as the authors claim, or whether it is due to real differences in geographical distribution, remains to be established. Hayden and Beighton (1982) claim for it the same high prevalence, 67 per million, among hybrids as among Caucasoids.

An unexpected and unusual finding by Hayden et al. (1982) was that 15.7 per cent of all the Huntington's chorea patients were juvenile. This is a figure higher than any previously reported. The usual preponderance of paternally derived disease was evident. No satisfactory explanation for this finding has yet been offered.

One of the most notable instances of inherited disease in a hybrid community is undoubtedly due to founder effect. Lipoid proteinosis is a rare autosomal recessive disease (Fig. 9.4). Of the 200 cases in the world literature no less than sixty have been reported from South Africa, twenty-six of these from a small inbred hybrid com-munity in Little Namaqualand (Gordon et al. 1969). The remainder are found among the Caucasoid Afrikaners. The ancestry of all cases can be traced back to a couple who married at the Cape within a few years of the original settlement (Gordon et al. 1971; Heyl 1970).

Unlike the Southern African Negroes, the hybrids show a very low incidence of polydactyly. According to Stevenson et al. (1966) this is about 0.98 per 1000 births, but once again one would expect figures to vary from one part of the country to another. The inci-dence of hydrocephalus, stated by Horner and Lanzkowsky (1966)

Fig. 9.4. A man from !Kuboes in the Richtersveld showing the signs of lipoid proteinosis: coarsening of the skin of the face with pitting resembling smallpox scars and marked thickening and induration of the lower lip.

to be approximately one in every 2000 births, appears also to be lower than in Negroes, given the same reservations. Conversely, neural tube defects are said by Sineer *et al.* (1979) to have an incidence similar to that in Negroes and probably lower than that in Caucasoids.

## Social mobility and social selection

Among the main 'Coloured' communities, that is, those whose formation and maintenance depend on random genetic accretions and losses, there is, additionally to the occasional social loss upward into the more privileged Caucasoid community, the equivalent but rarer occasional relegation of a member to a less privileged category. This can even be momentarily advantageous; there are a few employment opportunities in South Africa and Namibia open to educated Negroes but not to any 'Coloured', and in the newly independent states of Lesotho and Swaziland, though to a lesser degree in Botswana and Moçambique, mixed ancestry may be or come to be politically and socially detrimental. Nevertheless, in those parts of the region still dominated by Caucasoids, the hybrid persons who move into the Negro community most generally do so on account of their physical appearance, and forfeit some social and economic advantages in so doing.

These trends contrast fairly sharply with the ideological concomitants of being a Griqua. The Griqua, as has been pointed out earlier, are extraordinarily receptive to recruits, and though there are gradations of prestige among them, these do not limit the choice of marriage partners except possibly for the ruling Kok and Stenekamp families. No stress at all is laid on physical ability or appearance, on educational attainments or on dexterities or the possession of a craft. Anyone who has a sufficiently strong wish to become a Griqua can eventually become one; but for many Griqua the motives for remaining Griqua are sometimes less than overwhelming, and though there are institutions of Griqua solidarity, particularly churches, throughout South Africa, there are still Griqua who manage to use a change of residence as an opportunity to transfer into another community. Unlike the Basters, the Griqua do not find it easy to enter Caucasoid society. Their most usual outlet is into the broader 'Coloured' society, and though this entails some loss of prestige and even slightly of political standing, it can be economically beneficial.

Being a Baster has rather more considerable economic advantages than being a Griqua, and the status is the goal of only those relatively few who can be confident of attaining it. Similarly, despite the comparative ease with which an ambitious and able Baster can acquire acceptance from not the least influential segment of the Namibian Caucasoids, such defections are not common. The Basters traditionally show much more national assertiveness and cohesion than the Griqua do. For all their free enterprise society, the achievements or the disgrace of a Baster are shared by the whole nation, whereas a Griqua would receive sympathy or applause rather than involvement and support. The Basters, in short, and probably all the similar, Dunne-like or Buys-like, hybrid communities, present a social rigidity which may be their present shield but could be an important future weakness. The policies of the Griqua tend to their own extinction; those of the Basters could, if carried to an extreme, provoke their eventual extermination.

The extremes of colour-based stratification which have formed the traditional framework of social structure in many parts of Southern Africa are disappearing fast. There have never been very many Caucasoids permanently domiciled in Lesotho or Swaziland, and the hybrid descendants they left behind there derived no particular social advantage from their descent, though some inherited money enough to retain for them some privilege. In Botswana many of the

Caucasoids who took non-Caucasoid mates were recognized even locally as proletarian and not especially privileged: their descendents can be socially mobile, though more by their own wits than by favour of their skin colour. This applies to the more recently formed mestiço families of Moçambique as well, though not to those anciently established. Marriage laws have operated at various times to curb inter-racial matings in all the countries except Moçambique, but at the time of writing are current only in South Africa. While they have been operative, they have, as shown above, worked both to curb and to promote inter-racial gene flow. Where they have been enforced effectively, they have given to some of the hybrid communities boundaries which otherwise might never have arisen; where they have been circumvented they have sometimes led to some degree of genetic impoverishment of the hybrid community, though genetic replenishment could have been as easy and as frequent.

# 10
# ADAPTATION AND SELECTION
# IN SOUTHERN AFRICA

At the time of the earliest effective contact with Caucasoids, the indigenous peoples of Southern Africa were practising the three main varieties of simple subsistence economy. In different proportions, all three of these persist today among the peoples who practised them then, though they have been greatly eroded by the cultural influences of the West. This lack of uniformity of subsistence economy derives to some extent, of course, from regional and local variation. It reflects the diversity of ecological niche afforded by the range of climate and terrain, and the changes both the climate and the terrain have undergone and are undergoing. Upon these variations has been imprinted the impact of technological change, leading to the use of some ecosystems in ways other than those biologically the most suitable. Such uses are often dictated by, and lead to, short-term advantages to the users. They have been responsible for a range of socio-economic divisions among the various groups of human inhabitants greatly exceeding those which might have been expected from the simple meetings and minglings of migrants and autochthones.

The position of the Caucasoids in this socio-economic spectrum is particularly interesting. Mention has been made in earlier chapters of the primarily agricultural objectives with which the earliest Caucasoid settlements were made, and the extent to which a high sex ratio enforced the adoption of pastoralism immediately before the intermixture with Khoi which produced the major hybrid peoples of the interior. Pastoralism, too, was the mode of life adopted by all the Trekkers, though their involvement in a cash economy enabled them to have access to more basic agricultural products, such as dried cereals, than are generally available to pastoralists. Reversion to hunting and gathering, such as has occasionally happened on the fringes of Southern African Caucasoid society, has again been modified by the possibility of barter or sale of the products at least of the hunt. Curiously enough, the only 'Coloureds', or hybrids, of whom any record of hunting and gathering may be gleaned, have been those forced by poverty into such pursuits at the urban edges. The 'Coloureds' of most of the rural Cape Province and some of Natal are settled agriculturists; those of both Namaqualands and part of the Cape interior are pastoralists; most of the rest are urban dwellers involved essentially in contemporary Western life styles.

The extension of the agricultural Kalanga into north-eastern Botswana has encouraged some of the Khoisan-speaking Negro hunter–gatherers to emulate their success and attempt even to raise staple crops along the margins, or in the beds, of dry rivers. Both cattle-owning and cultivation are practised by many Bantu-speaking peoples, such as the Xhosa and Nguni, while the Tsonga fish intensively in river estuaries and lagoons. The Herero perhaps present the purest picture of pastoralism in the Southern African Negro, while the adjacent and probably biologically related Ambo and Kavango come closest to being the most exclusively agricultural.

There is unfortunately a large dearth, and not only for Southern Africa, of case studies of the anthropological ecology of these different modes and varieties of livelihood. One needs to rely on a combination of piecemeal evidence on different aspects to produce what can only be regarded as simplistic and schematic ecological syntheses. As a basis for particular habitat analyses, the adaptive responses to the demands of the Southern African biome should first be considered.

## Adaptation and the Southern African environment

Successful biological adaptation rests primarily on the physiological responsiveness of the human organism to a wide range of needs and stresses. At the same time, there may be particular genetically determined morphological modifications which through local selection confer particular advantages. Amongst Southern African peoples the rôles of both these attributes are clearly demonstrable. Technological achievements can be judged, like biological responses, by survival success, in the face of the whole array of environmental problems attending food getting, disease, shelter and security. In the Southern African milieu the ecological equilibrium attained by different groups, even with similar technologies, ranges as elsewhere from the highly productive and secure to the precarious and impermanent. Since the economies range through simple fishing and collecting, shifting cultivation, pastoral herding, mixed agriculture, mechanized farming, village and westernized urbanization, to advanced industrialization, dependence on biological and socio-cultural adjustments necessarily varies with the degree of techno-logical mastery of the environment.

The physical environment does not present the great extremes of continuous cold or of severe heat and humidity, great water scarcity or high altitude anoxia which affect human groups in other parts of the earth's surface. Nevertheless, there are tangible and often severe stresses associated with seasonally high and low air temperatures, high solar radiation, periods of drought, soils of marginal fertility, infectious disease affecting plants, domestic stock and man, and the peculiar hazards of industry, above all of deep level mining on the Rand and in the Free State. Hard physical work remains inseparable equally from the pursuit of food-getting and industrial occupations.

### Anatomical and physiological factors in adaptation

The most obvious adaptations tend to be morphological; one of them in particular, pigmentation, is discussed in more detail below. Mor-phological adaptations appear in the majority of cases to occur with a rapidity which suggests the preferential accumulation of particular arrays of genes rather than any kind of saltation. The outward phenotype depends not only on the interactions of discrete genes, but on the combined operations of a variety of factors. Where it is important for environmental accommodation and for reproduction in a new environment, those individuals whose outward phenotypes approximate most closely to the ideal for the conditions will naturally survive better and produce more offspring. Monogenically deter-mined traits, with the exception of balanced polymorphisms, tend towards the extinction of unfavourable alleles. Balanced polymorph-isms rely on heterozygote advantage, and are consequently most likely to represent either the outcome of mutations which occur after the environment which favours the heterozygote has been occupied, or the chance conservation of alleles which in the previous environment were adverse, and whose course to extinction happens to have been interrupted by the change of environment. The latter alternative, which would be an instance of pre-adaptation, involves so many imponderables that it seems improbable. It is more likely that identifiably balanced polymorphisms should arise most com-monly where new mutations result in selectively advantageous heterozygotes.

Selection for morphological characters, however, depends to

some extent on the patterns of reassortment of genes, often those already in the genome. With the involvement of a number of loci, the range of alternative combinations of genes becomes quite wide. Selection for the most advantageous of these will consequently happen relatively quickly, since elimination of deleterious phenotypes will mean the simultaneous loss of disadvantageous alleles at several loci.

It is probable, too, that in parallel processes of selection it is not always the same genes, or even the same loci, that are involved. Ford (1964), in a discussion of mimicry in butterflies, suggests that even in a two-locus system recombination at the level of a modifier locus can produce selectively favoured phenocopies further modifiable by selection to perfect mimicry of an established species. This is exemplified in the simple but definite and long-established ecological rules which govern the variation of animals according to their environments. In 1847 Bergmann pointed out that the smaller the body the greater the heat lost per unit of volume; consequently body size tends to be smaller in warmer regions. Allen (1877) amplified this rule in the light of the fact that the closer the body approximates to a sphere the lower its surface area, so that animals of the same species have longer extremities in warm than in cold regions. Gloger's (1833) rule states that animals living inside forests are likely to be black or red in colour, and Rensch (1947) demonstrated that animals in deserts do not have their subcutaneous fat evenly distributed but carry it in special localized depots.

Compliance with these rules would not necessarily involve loci homologous between species. Even in a single species, it is quite likely that in unrelated populations different arrays of loci might be involved in closely similar processes. There is no reason to suppose that similarities of outward shape and colour in different human populations betoken a common origin. They are more likely to imply that morphological mimicry has occurred on account of close likeness rather than identity of environments during the major part of their evolution, possibly from distinctly different original stocks.

The Southern African environment, for all its lack of climatic extremes, nevertheless imposes on its human inhabitants the necessity to make physiological adjustments to it. The human animal, like other living organisms, is sensitive in different ways to the different elements which go to make up the climates of its habitats; both the sensitivity, and the responsiveness which accompanies it, arise from the need to maintain homoeostasis, that is, to keep the whole organism functioning at an overall level which represents the greatest amount of thermodynamic efficiency consistent with the optimum survival, development and reproductive capacity of that organism.

An important constituent of adaptation to a climate is the ability to maintain homoeostasis in the context of the heat and energy balance. Where homoeothermy is disturbed there tend to be effects on the body's efficiency. Human adaptive responses to changes in the thermal environment may be of immediate or long-term physiological type, anatomical or social and cultural. The efficiency of the responses determines bodily comfort, the ability to perform physical work with the minimum of fatigue, the attainment of skill and dexterity and the patterns of growth and development. The body cannot tolerate any great departure beyond the 2°C diurnal range of the deep body temperature, regulated by a mechanism in the hypothalamus sensitive to rises and falls in the amounts of heat added to or removed from the body.

On continued or repeated exposure to heat there can be a marked improvement in the ability to endure heat stress. This is shown in a complex of signs, the principal cause of which is a faster response and more copious output of the sweat glands. Other, more localized adaptational responses occur with repeated exposure to cold. This acclimatizing ability is shown by peoples of all races, including those which have evolved in cold or temperate climates. Imposed on this physiological plasticity are the various anatomical characters for which selection has taken place in diverse climates. For instance, differences in thickness of body fat regulate the ability to tolerate heat and cold, and to some extent the selective 'rules' enumerated above are responsible for other anatomical differences among the races of man. The demands made on both anatomical and physiological resources vary, however, according to the mode of subsistence and the nature of the food supply, and to exposure to disease.

The spread of man across a large series of contrasting climatic zones has been made possible by the supplementation of biological adaptation by technological ingenuity. The ability to light, or to conserve, a fire; the development of various kinds of protective clothing and housing, and in hot climates of ventilation; have been as important in recent human evolution as anatomical and physiological changes were at earlier stages. Indeed, the question of heat exchange is too closely bound up with that not merely of ambient exposure and the strategies used to deal with it, but also of energy production in the organism, for any of these matters to be dealt with in isolation. As might be expected, different modes of subsistence provide different examples of types of accommodation to thermal stresses and of energy utilization.

### Ecology of the hunter–gatherers: energy flow cycle, climate, and work

Bearing in mind that patterns of hunting and gathering are not uniform, we need to examine the ecological complex that constitutes the total adaptive response of this form of livelihood. As in all ecosystems the nutritional process is the basic dynamic system; the efficacy of its functioning conditions and is influenced by the responses to climatic, disease and population pressure.

As an outcome particularly of the work of Marshall, of Tanaka and of Lee and DeVore, the ways of life of the Kalahari San, and in particular of the !Kung and of the G/wi and G//ana, have tended in recent years to be presented as a paradigm of the human ecology of man's earliest and most prevalent subsistence method. The analyses by Lee and DeVore (1968*b*) of the ecology of the northern Kalahari way of life are of significance for elucidating the strength and limitations of the hunter–collector economy. But even within Southern Africa the !Kung community which they studied represents only a variant, however important, of this way of life. Remains of San, as we have seen, are found scattered over the whole subcontinent, and the studies of Tanaka (1969, 1976) show that even among the G/wi and G//ana, exposed to much the same recent influences as the !Kung, there are significant divergences. It follows that ecological circumstances must have differed quite widely; some environments may have been even more favourable to the San mode of subsistence than the present-day Kalahari. The relative dependence on hunted animal food as against collected vegetable food must have been different under other ecological conditions. The associated material culture would have shown tangible regional variation. This would accord with the archaeological record.

According to Lee and DeVore (1968*b*) the San of the Central Kalahari obtain on a yearly basis more calories from plant than from animal sources. They are much more secondary than tertiary consumers in the food chain. Both sources are not exploited to capacity, partly because the total energy requirements are not as large as those of more complex economies, where the input energy needs to be relatively high and the yield is needed for many more purposes than the San aspire to. The low family size of the San group also reduces the energy each producer needs to extract. Lee and DeVore's oft-quoted ratio that the equivalent of two man-days work per week will suffice for survival for seven man-days provides the justification for claiming hunter–gathering as the original affluent society (cf. Sahlins 1968). But the output/input energy ratio is undoubtedly less favourable than that suggested by a simple assessment of the time costs of food procurement. It is easy to under-estimate the cost of providing sustenance if one confines the measurement to primary food getting, actual gathering and hunting, and if one does not use the real energy costs of the essential ancillary activities, collection of water for cooking, collection of fuel and the work of preparation of the raw food. Nevertheless the !Kung maintain a fairly secure margin, since family size is small. Small family size is probably demanded by the nomadic form of life.

That the high heat loads which prevail for a large part of the year must exert substantial limitations on sustained hard work in the Kalahari and to a lesser extent in other parts of the South African plateau is certainly true. There is a reasonably constant limit to thermal tolerance above which moderately hard work becomes physiologically stressful, that is, the body core temperature rises progressively and the evaporative and circulatory thermal heat loss potentials approach their full capacity. The level of heat tolerance will depend on the intensity and duration of the combined heat load, environmental and metabolic, which the population experiences. San of different regions not only show variability in their level of acclimatization, but even the highest levels attained are by no means particularly high. Nevertheless San do exhibit some degree of acclimatization. The fact that it is not specially great must mean that they reduce their exposure to high heat levels very significantly, principally by technological adaptations (Strydom 1974).

## The ecology of the pastoralists and their energy flow cycle

Pastoralism in Southern Africa has constituted the main form of subsistence traditionally for the Khoi and for some hybrids, but has hardly ever been adopted by the San. The Negroes and the Caucasoids have taken to unalloyed pastoralism mainly as a means of sustaining migration, and abandoned it when given the means for a settled agricultural life, though we must probably continue to consider the sheep-farmers of the Karroo as pastoralists, even though they are not nomads. Good agricultural land was being used by the Khoi as pasture when the first Caucasoids arrived. It is probable that the lands later occupied by the Cape Nguni and cultivated by them may earlier have been grazed by the herds of some of the Khoi they had displaced or assimilated.

Owing largely to the disposal of the land in Namibia by the German colonial power, the lands with the greatest concentrations of pastoralists today are hot, arid and over-grazed. The large cattle favoured by the Herero and Nama, and the goats of the latter, have been allowed to breed beyond the capacity of the land to sustain them. Lands of similar or only marginally more favourable type support the prosperous farmers of merino sheep in the Karroo and karakul sheep in southern Namibia; but they are no longer over-grazed, and boreholes and pumps keep them adequately supplied with water. The contrast is to some extent one of capitalization, but arises also from the number of the units of land per individual. Nevertheless, the introduction of ranching into northern Botswana, and its careful supervision by a responsible government, holds out promise for a profitable persistence of pastoralism in the region.

The abandonment of nomadism due to the constriction of open lands and the spread of agriculture has virtually put an end to all forms of pastoralism which cannot be maintained through technological adjustments. Prior to the arrival of the Caucasoids at the Cape, the exuberance of the pastures would have reduced the necessity for large migration cycles and improved the quality of the yield that could be expected from the herds. The milk which, soured, formed a major part of the diet of the Khoi would have been richer in nutrients and more abundant than that which the thin grazing of the interior provided. The goats and the fat-tailed sheep, principal sources of animal food, would have been fatter, and the yield of wild fruits might possibly have been higher.

On the other hand, it has been suggested recently by Harpending (unpublished personal communication) that since the nutriments in food plants are most concentrated in their reproductive organs, and since the highest yield of these occurs often in the wild as a response to adverse conditions and represents a physiological profusion to ensure survival, plants are probably not as valuable as foodstuffs for man in good years and fertile areas, where they put out more leaves and fewer seeds and are consequently of most value for their availability to grazers and browsers. The supply of animal protein, from herds and from hunting, however, would almost certainly have been higher. Nevertheless, we cannot be certain of the proportion of the Khoi who actually took advantage of the rich conditions of the western Cape. The suspicion remains that many may have opted for, and felt themselves more comfortable in, the arid interior.

The Khoi are still hunting, and the Basters and the Herero, with similar life styles, are too, irrespective of the wealth tied up in their herds and their present familiarity with the cash economy. Their gathering has probably increased since their relegation to the desert, and the abandonment of nomadism has permitted the limited adoption of agriculture. The present-day diet of all three of these types of pastoralist almost certainly contains more carbohydrate and less protein and fat than the traditional one. Being relatively sedentary not only reduces the range and variety of feed available to the live-stock; it means also that the pastoralists themselves are no longer able to follow any geographical sequence of wild products and are rather more dependent on local variations in rainfall. To compensate in part for this, the actual physical work carried out by the men and boys has been curtailed, as, though to a lesser extent, has that of the women. In addition, the presence of markets, including a labour market, permits the accumulation of cash resources to help cushion the impact of poor years and seasonal shortages.

## Climate, work, and energy flow in the agricultural ecology

The late emergence of agriculture in tropical Africa and southwards may well have been due to the secure food base which the savannah offered. This affords an environment so rich in game and vegetable resources that food-gathering communities there could maintain themselves comfortably without recourse to the great deal of extra labour involved in agriculture or stockbreeding (Clark 1959). Accord-ing to Clark, the record suggests that population pressure eventually forced one or more localized groups of Negro hunter–gatherers to start cultivating on a considerable scale. Such indigenous grasses as sorghum, African rice and millets, together with pulses, yams and other roots and fruits were probably the first crops. Population pressure may have been generated less by the food-gathering com-munities themselves than by the southwards migration of herders of sheep, goats and cattle under the influence of climatic deteriora-tion in the Sahara. It must have been a southward migration at least of ideas and techniques that introduced agriculture to the continent in the first place. But the opportunities in Southern Africa for successful cultivation were not over-plentiful. It has been calculated that no more than 10 per cent of the land area from the Zambesi to the south is cultivable by the simple systems used (Allan 1967). Displacement of the indigenous hunter–gatherers (whether San or Negro) by Iron Age or possibly Neolithic farmers must have been a relatively gradual but inevitable process. The modes of resource domestication appropriate to a predominantly savannah ecology were shifting cultivation or animal husbandry or both. Agriculture and metallurgy had reached the savannah woodlands and much of Central and Southern Africa by the end of the first millennium A.D.

The early agricultural societies were smaller, less technically

advanced and materially poorer than many of the later Iron Age
peoples, some of whom reached a high level of culture, technology
and trading practices. Early Iron Age villages consisted of wattle and
daub huts with thatched roofs situated on the edges of clearings in
the savannah, near fertile soil, good grazing grass and water supplies.
Cultivation was of cereal crops such as sorghum and millet; cattle,
small stock, and dogs and cheetahs were kept. But the villagers still
relied for part of their diet on hunting and gathering (Fagan 1965).
These provided the foundations of rural life; very similar practices
have persisted to the present day. The shifting cultivation system as
well as nomadic pastoralism are the necessary outcome of the low
rainfall pattern with its long dry seasons. The carrying capacity for
cattle is generally low. Estimates for Zimbabwe and South Africa
range from ten acres per beast in the better areas up to forty acres
per beast in the driest grazing conditions (Allan 1967).

The only well worked analysis of the energy flow cycle for an
African cultivator system is that by Fox (1958). This Gambian
model has received substantial confirmation from less comprehensive
but nevertheless valuable studies in many African savannah communities (cf. Weiner 1980). The cycle is practically open since trade
may be a component of the food-getting process.

The essential factor in the energy equilibrium of the Gambian
model is the existence of a 'hunger season', due of course to the
characteristic rainfall pattern. This imposes a severe limitation on
the overall productivity. The long dry period in Fox's analysis is also
causally connected with the work patterns imposed on the labour
force. Periods of intensive agricultural activity are interrupted by the
dry poor growing season when agricultural work is slowed down and
the energy stores available have dwindled. Even when the work of
harvesting ensues the overall activity is restricted by the low food
reserves, the loss of body mass resulting from the hunger season, and
the prevailing high temperatures.

That rainfall is capricious as well as seasonal in Southern Africa is
a major fact of life. Prolonged drought is a danger and catastrophe
that African peoples have to contend with everywhere. As a result
recurrent energy crises are endemic in African agricultural societies.
When a 'hunger period' strikes a Southern African farming community
the biological consequences will not be very different from the
Gambian case.

Another strong indicator of the precarious equilibrium of the
Southern African cultivation system is the fact that game hunting
and wild food collecting have remained part of the economy right up
to the present. The archaeological record testifies to this and is

closely documented by Fagan (1965). He provides evidence for this state of affairs for Later Stone Age farmers on the middle Zambesi, Iron Age farmers on the Zambesi during the seventh century A.D., and the community at Leopard Kopje at the end of the millennium, where there is evidence of trade, and at Kalomo in Zambia. The mixed economy of the Inyanga people is quite comparable with contemporary practices of Negro agriculturists in Southern Africa, and of an ongoing instability. A present-day example is reported from Malaŵi by Nurse (1975b). He has shown a clear connection between the deprivation of the hungry season, the loss of body weight, the recourse to hunting and the restoration of body weight as a combined result of game utilization and the new harvest. Land in relation to populations is plentiful among the Bemba according to Richards (1939), yet they certainly have hunger months during which they resort to hunting. Only by an appreciation of the energy equilibrium of a shifting cultivation system can one understand the necessary conservatism of the economy.

During most of what we may for convenience refer to as the European colonial period, extending from the establishment of trading posts south of the Sahara by the Portuguese in the fifteenth century until the present day, the study of energy generation and utilization in indigenous African communities has been bedevilled by the conviction of many representatives of the colonial powers that the African, especially the agriculturalist, was essentially idle. This curious, almost obsessive, belief was applied to a number of other inhabitants of the tropics as well, and was based on superficial and unanalytic observations which coupled the relative fertility and productiveness of some tropical ecosystems with the essentially seasonal nature of the main work carried out by the men.

The Venda of the Northern Transvaal were the subject of an exhaustive ethnographic study by Stayt (1931). Despite the detail of his narrative we cannot tell how much time they spent in food-getting activities or in leisure pursuits. For all the wealth of information he provides we cannot derive an accurate input/output analysis. It appears nonetheless possible to make a sufficiently reliable one against which we can judge his assertion (p. 34) that 'In spite of these natural advantages (of soil fertility, water for irrigation), the slothful character of the average Muvenda prevents him from doing more than the minimum amount of work necessary to maintain life and enjoy occasional festivities'.

What Allan (1967) described as the 'seeming indolence' of the African can only be understood properly in its ecological context. As Fox (1958) pointed out in the Gambian case, the agricultural system imposes 'indolence' not only at times when no work can or needs to be done but when calorie shortage makes a high work level difficult to attain and then only at the cost of semi-starvation. The level of work input also falls in the immediate post-harvest period when the energy required for food producing activities is not pressingly high. At this period energy goes more into non-sustenance activities. And it must be remembered that when the agricultural cycle demands it the active input levels can reach high values, though again activity may be restricted when the heat level rises and in individual cases when endemic disease overtakes the farmer or his wife. What the uninformed and at times prejudiced observer sees as indolence or sloth, therefore, are the necessary consequences of the changing input/output ratios inseparable from the ecological forces of semi-arid savannah biotopes.

## Reproductive physiology and the Southern African ecosystems

Among the physiological traits enabling human populations to establish homoeostatic equilibria with their environments are the reproductive variables. These include the age at puberty, the proportion of conceptions which do not proceed to the production of a viable infant, the mortality rate before the attainment of reproductive age, and the duration of lactation infertility. At different technological levels these may be supplemented by a number of artificial aids to or precautions against conception, to which we may add the practice of premature termination of pregnancy. The natural mechanisms are all female-oriented; since women invest the greatest amount of energy in reproduction, they are the limiting resource (Short 1976).

Natural selection operates to maximize the reproductive potential not only by promoting fecundity but also by limiting the production of offspring to numbers which, given the available resources, will be able with the greatest possible conservation of energy to survive and themselves reproduce. Human technology on the whole has operated to overcome the limitations without always amplifying the resources, with the result that in the majority of societies the natural checks and balances on population growth have ceased to be effective. Perhaps nowhere else is there as wide a contrast between different methods of biological control of population size as in Southern Africa. The San have never been proved to use any artificial method of birth control, though their neighbours are all convinced that they secretly use herbs whose powers only they know. Much the same can be said of the transhumant Khoi, while the settled Khoi are exposed to the same forces as the rural Negroes and 'Coloureds'. The Caucasoids, as well as the more prosperous segments of the urban 'Coloureds' and Negroes, have access to the full modern range of mechanical and hormonal control of conception. In addition, superimposed on both the most natural and the most sophisticated circumstances we find social practices not necessarily concerned directly with reproduction but influencing it greatly.

According to Frisch (1978), all four of the variables enumerated above are influenced by socioeconomic deprivation. Poor nutrition, hard physical work and exposure to thermal stress and communicable disease all delay menarche, promote miscarriage, favour infant and childhood mortality and mean that lactation leads to a net loss of nutrients, not immediately replaced by intake, to the mother. Frisch claims that the fact the undernourished peoples are less fecund than well-nourished ones can be regarded as an ecological adaptation to reduced food supplies, and hence advantageous to the population. Population reduction occurs by prevention of conception, or by voiding of the foetus before the attainment of viability, or by infanticide or other forms of death in the lower age groups, in whom less resources have been invested yet. The mechanism is consequently less wasteful that the regulation of overpopulation by mortality at later ages.

Curbs on population growth, brought about by undernutrition, might consequently be expected to be operating in Southern Africa in ecosystems where reduced food intake is common: the urban slums, newly unproductive or overcrowded agricultural areas, and among pastoralists suffering on account of overgrazing. Nevertheless, such areas are showing rising population figures, sometimes but not invariably due to government-regulated resettlement, though their populations are not growing nearly as fast as those in economically more favoured situations. One reason for this might be the reduction in the period of lactational amenorrhoea brought about, especially in the towns, by bottle-feeding, sometimes reinforced by the need of the mother to be the breadwinner. Short (1976) claims that more births everywhere are prevented by lactation than by all other forms of contraception put together. Abandonment or curtailment of breast-feeding in communities which do not possess or often use artificial methods of contraception may lead to increasing birth rates independently of the effect of nutritional deprivation on the mother. This deprivation is most likely to be passed on to the child at weaning, with a consequently higher infant mortality (Knodel 1977).

The foregoing considerations apply to the greater number of the

inhabitants of Southern Africa, but not to all. Nutritional and lactational effects are in them modified in different ways by social and
cultural practices, and by technology. Education, and in particular
adult education, regulates to some extent the rate of conception
among the literate. Religious beliefs in one direction encourage
family limitation, in another stipulate the unrestrained production
of offspring. Such considerations transect at all levels the Negro,
Coloured, Caucasoid and to a large extent the Khoi populations. To
the San hunter–gatherers they apply hardly at all.

Recent work, reviewed by May (1978), has indicated that population regulation in hunter–gatherer societies may have been operated
predominantly by physiological control mechanisms resembling
those found in other mammals. It has been shown that in the !Kung
(Howell 1979; Kolata 1974) the overall fecundity is lower than in
almost all human societies practising other subsistence methods; and
it appears to be possible to apply the !Kung results to the other San
peoples, though it seems a little premature to do as May does, and
hold that they are valid for all hunter–gatherers. The reduced fecundity appears to be due to a number of factors, involving mainly two
of the four variables and one other. There is a later age at menarche;
prolonged breast-feeding leads to suppression of ovulation and hence
to lengthy intervals between births; and there is a constant and
definite interval of adolescent infertility, such that however frequently sexual intercourse takes place during the period following
menarche, the first conception does not occur until some years later.
The reason for this seems to be that viable ovulation does not com-
mence, in the majority of cases, until several years after the onset
of menstruation.

The period of adolescent infertility is reduced or abolished by the
accelerated puberty and consequent early maturation found in
societies which have undergone Western-style development. The
responsibility for this lies most probably with Western food habits.
In many non-Western societies the adolescent growth spurt is not
accompanied by an extra allocation of foodstuffs to the growing
adolescent, who consequently tends to become more linear and
slender. Once the ingested nutrients are no longer needed to support
the accretion of bony and muscular tissue, any surplus can once
more be stored as fat, as it was in childhood. In modern Western
societies, however, adolescents, particularly at the time of their
growth spurt, are often encouraged to overeat. Frisch and McArthur
(1974) have suggested that there may be a critical weight-height
ratio for menarche, and a somewhat higher one necessary for the
maintenance of menstrual cycles. Where food intake is marginally

adequate or not increased during adolescence, both menarche and the ability to reproduce would tend to be postponed. The crucial category of foodstuffs would appear to be that providing energy. Where malnutrition is present but the intake of calories is nevertheless high, as happens in some deprived Southern African societies, the effect on reproduction would be likely to be less, and to function principally after conception.

This suggestion is supported to some extent by investigations of the endocrine status of the San carried out by Van der Walt *et al.* (1977, 1978). In 1936 Marrett had suggested that there might be a hormonal basis for the characteristic morphology of the San. This suggestion was later taken up by Tobias (1957, 1966), who reported that the oestrogen level in San urine was markedly higher than the levels in either Negroes or Caucasoids. He speculated that diet might be responsible for this over-oestrogenization, and that small stature and infantile morphology might result from the possible feminizing effects of these oestrogens. The rather more refined techniques available to van der Walt and his co-workers revealed that circulating hormone levels in the male San were not significantly different from those in male Negroes and Caucasoids. Female San, however, were found to have suppressed circulating levels of gonadal steroids and little corpus luteum function, while gonadotrophins were in the normal range. They account for these findings, and for the relatively low fecundity of the San, by postulating a mechanism by which the ovaries may be suppressed, thus limiting conception to times of high nutrition. The San are not at most times conspicuously undernourished, though their diet is somewhat restricted as far as energy-rich foodstuffs are concerned. The endocrine findings consequently do go some way towards supporting the hypothesis that high calorie intake promotes fertility.

### Pigmentation and adaptation: some paradoxes

Some morphological characters are immediately obvious to even the most casual observer as being geographically variable. Of all of these the one taken most seriously by lay persons coming from a Western background is coloration. The reasons for this are intriguing and mainly non-biological; a surprisingly similar symbology of colour stretches across a number of probably discrete cultural systems. In Southern Africa the range of human pigmentation is probably wider than almost anywhere else, and its cultural significance is greater. The range is a function of the historically conditioned migration patterns, and except in the cases of the peoples most anciently settled there is not yet of obvious adaptive significance. There are nonetheless indications that it may be; and in any case it is a powerful agent of social selection. It consequently deserves some particular attention here.

Two pigments, melanin, which is present statically in many tissues of ectodermal origin, and particularly the skin, and haemoglobin, the circulating and optically almost invariable pigment of the red cells, between them are responsible for the colour complexes presented externally by the body. Of these, hair colour appears to be of little or no adaptive significance, and eye colour is important only insofar as it is relevant to eye function. Most of what is physiologically important about pigmentation is included in the role of the skin.

Melanin absorbs ultraviolet radiation, which is damaging to tissues. It might consequently be expected that dark pigmentation would indicate that a stock had evolved in regions where exposure to ultraviolet radiation was more intense, and that light skin would suggest less of such exposure during the evolutionary period. It is certainly the case that all the most deeply pigmented human races at present live, or have historically recent origins from, equatorial or at least tropical regions. Not all such regions, however, are exposed to intense ultraviolet irradiation; South India and the other regions of provenance of very dark Asians were originally thickly forested, and much of Melanesia still is, while in earlier chapters we have produced evidence for the evolution of the Negroes in forested areas of Africa. It may be that pigmentation is governed to a greater extent than is admitted by Gloger's rule, for which there is little or no selective explanation except camouflage.

Certainly the irradiation hypothesis is beset with numerous objections. The San and Khoi, who almost certainly evolved in highly irradiated areas of Africa, are relatively lightly pigmented; though it can be objected to this that the San skin is often encrusted with an adequately protective layer of grime. The relatively dark Mongoloids, particularly those of northern Asia, and the Eskimos, have had it suggested for them that they have only migrated to their present areas of origin too recently for selection relaxation, if that were indeed likely to take place, to have had much effect on their pigmentation. In fact, the polar regions can be subjected to irradiation quite as intense as any in the tropics, compounded and exacerbated by reflection from the generally white surfaces of the surroundings. In any case, the biological and technological adaptation of the Eskimos to their own environment is so good that they probably have inhabited it for an evolutionarily significant length of time.

It has been suggested that reduced pigmentation represents not selection relaxation but a state for which there has been active selection due to the action of ultraviolet light in converting skin steroids to vitamin D, a reaction promoted by lower melanin levels. Darker persons in sunny environments would consequently be protected against overproduction of a substance which in excess can be toxic, while lighter persons in climates less sunny would receive enough irradiation to be protected against rickets. On the other hand, the absorption of visible light by dark skins leads to increased heat-intake due to its liberation as heat, but it is unlikely that the extra load on the temperature-regulating mechanisms is significant, even in strong sunlight. Radiation of heat from the skin is virtually independent of skin colour.

In Southern Africa the selective action of pigmentation might at first be thought to have operated *in situ* almost exclusively on the Khoisan peoples, but there is evidence, albeit unqualified and impressionistic, that the Caucasoids have also been affected. Mention has been made in Chapter 8 of the extent to which the Afrikaner physical 'type' seems to have begun to drift towards that of the Khoi, without equivalent monogenic evidences of Khoi gene flow. It must be re-emphasized here that the emergence of the Afrikaners from the western Cape, to which they were biologically previously adapted, into the vastly different interior, was almost certainly accompanied by intense selection, and this would have operated as much, or more, on pigmentation, as on other characters.

It may or may not be significant that the Bantu-speaking Negroes of Southern Africa are on the whole less deeply pigmented than the Negroes to the north of them. Whether this is due to Khoisan admixture, as is generally supposed, and as is to some extent indicated by gene marker studies, or whether it represents a stage in selective adaptation to a somewhat more temperate climate, remains obscure. Certainly Southern African Negroes are not exposed to any lower levels of ultraviolet irradiation than their forebears were; if anything the levels are higher on account of the residence of many of them at rather greater altitudes than most Negro populations.

## Selection and the monogenic markers

Few monogenic markers have been demonstrated as yet to have much bearing on Darwinian selection, and the majority of those that have, and certainly those for which the demonstrations are most convincing, are concerned with inherited defences against malaria. These are found in Africa principally in Negroes, and have been discussed in earlier chapters. One of them has marginal, and the other appreciable, relevance among Caucasoids. Haemoglobin S has been incriminated as a possible but rare cause of haemolytic anaemia in the Afrikaner, while $\beta$-thalassaemia is being recognized more and more frequently among Southern European immigrants and their descendants, and among South African Indians. Selection in their favour will be relaxed in the absence of malaria, and they will become detrimental and decline in frequency, though this may take a very long time (Splaine *et al.* 1971).

Various selective associations have been proposed for a variety of other markers; a number of these are discussed in the Southern African environmental context by Nurse and Jenkins (1977*b*) and do not need more than a passing mention here. The role of haemolytic disease of the newborn in the maintenance or decline of various blood group polymorphisms is unlikely to vary much with the environment, whose precise effects on such polymorphisms is uncertain. Real environmental effects are likely to lead quite rapidly either to fixation of the favoured allele, as appears to have been the case with Duffy negativity in West Africa, or to a balanced polymorphism where there is even a very small heterozygote advantage. The transferrin heterozygote $CD_1$ appears to have a greater affinity for iron than either homozygote, but it is not clear how this could be a selective advantage (Jenkins and Dunn 1981). The more rapid depletion of haptoglobin type 2 may help to account for the elevated frequency of the $Hp^2$ gene in the San and this could slow down the development of tropical splenomegaly in the presence of malarial haemolysis through compensating for the rise in IgM by depressing the level of the circulating $a$-2 globulins, but this is just speculation. Caucasoids also tend to have higher $Hp^2$ than $Hp^1$ levels, but malaria is hardly likely to have been involved in the evolution of the proximate ancestors of the largest divisions of the South African Caucasoids. It is rather more likely (cf. Race and Sanger 1968) that reductions in the $P^1$ gene frequency have come about in the Khoisan and Negroes through their exposure to pathogenic helminths with surface antigens resembling $P^1$ and the consequent slight selective advantage conferred by the capacity to manufacture antibodies against the worms.

The question of the adaptive significance of the ability to absorb lactose remains obscure (Simoons 1970, 1978; Flatz and Rotthauwe 1977). The recent suggestion (Cook and Nurse 1980) that either the allele for absorption or a gene in linkage disequilibrium with it, but more probably the two in concert, may be concerned with fluid conservation, would help to account for the finding of the allele in the San though not in the Negroes, who are thought to have evolved in a much wetter environment. Variations in type of liver acetyl-transferase among the San have been thought to be due possibly to a superior ability of one or other phenotype to utilize, or detoxify harmful substances in, unidentified plant foods. Such foods have also been blamed for or credited with the variations in the ability to taste PTC; non-tasting either brings plants bitter with thioureas into the pale of the edible, and so confers an adaptive advantage, or leads to overconsumption of such plants, with consequent diminution of thyroid activity and a selective disadvantage. More recognizably relevant to survival is the colour-blindness polymorphism, which Post (1962) has suggested might be important in the context of food procurement by hunting, where ability to distinguish colour could help in the location of prey.

It is possible that eventually the monogenic markers most closely associated with selection will turn out to be those connected with the immune system. Variations in some complement components have been found in both Caucasoids and Negroes, and reduced adenosine deaminase in an Ambo family and a San population. Probably more important than either of these in the interaction of the genotype with the environment is the complex of histocompatibility systems known as HLA. Clarification of the distribution of the D locus antigens in various populations is likely to be especially revealing.

## Culture, technology, and adaptation

Medawar (1975) pointed out that 'in human beings, exogenetic heredity—the transfer of information through non-genetic channels —has become more important for our biological success than any-thing programmed in DNA'. The acquisition of speech, or at any rate a system of signals capable of conveying complex information, would have been one of the most important milestones in the hominization process. All the most effective technological adaptations which followed would have been impossible without it, since it enabled a learned trait or ability to spread far faster than any inherited one could, and to be conveyed horizontally, to non-kin, as well as verti-cally, to kin. It could have produced selectively a saltative effect, so rapid would have been the competitive elimination of those unable to acquire the trait.

Other cultural and technological adaptations to the environment fall into two main types. In one, the more ancient, the adaptation operates by changing the microenvironment while leaving the macro-environment as it was. The utilization of fire for warmth or cooking, and clothing and ventilation, are probably the most obvious examples of this type, which generally consists of the adoption of cultural devices related to aspects of the environment other than food procurement. The earliest techniques which produced drastic changes in the macroenvironment seem all to have been related to the search for or cultivation of food; the use of fire in hunting, the denudation of tidal rocks, and slash-and-burn, or even simple clearing, in the preparation of gardens. As long as human populations remain small, such techniques interfere hardly at all with ecological regeneration; but once the populations grow and the resources are exploited beyond possible replenishment, the populations are once more exposed to the rigours of Darwinian selection, this time favouring principally those individuals whose chance assortment of genotype confers some pre-adaptive advantage in the new setting.

Technology has hitherto kept pace with, or in advance of, the exhaustion of resources. In Southern Africa advanced technology has been controlled rather jealously by the Caucasoid minority responsible for its introduction, so that though Darwinian processes have continued or recommenced among the Negroes and Khoisan, and to a lesser extent the hybrids, the Caucasoid populations are still actually or potentially shielded from their effects by the possesion of a protective technology. That technology is, however, itself dependent on resources the procurement of which is becoming steadily more difficult; this subject has been discussed in earlier chapters. What is less obvious, but not less cogent, is that technology, while reducing the challenges of the environment, is something towards which some degree of adaptation is also necessary.

### Adaptation to technology

Modern man has been described as *homo faber,* and one of his possible ancestors classified as *Homo habilis.* The adoption of each tool or technique as cultural evolution proceeds places a gap between those able and those unable to manipulate it. Selection in a technological society is not purely biological; it is inevitably also ergonomic. Some crude popular realization of this comes over with particular strength in the context of warfare, where possession no longer of biological might but of better weaponry is naively felt to be a property of the just and righteous: God is on the side of the big battalions. Yet not only the possession of but also the capacity to use a new device is involved. Were walnuts the only staple, selection would be for strong teeth; with the invention of nutcrackers, edentulous people with the full complement of fingers would be much more likely to survive than those less able to manipulate the instrument.

It is as necessary to adapt to a technology as it is to devise a techno-logy as part of the adaptive path.

The question of the ability of the different categories of popula-tion in Southern Africa to adapt to advancing technology is a vexed and politically delicate one. Among the Caucasoids there has already been selection for types of dexterity quite different from those which have helped to fit the Khoisan and Negroes into their environ-ment; the San have abilities which the Caucasoids and Negroes find it hard to emulate; Negroes and Khoi probably have the pastoral skills to survive and maintain their herds where the majority of San or Caucasoids would not know how to proceed. At the present it is the technology of Western Europe, which has evolved *pari passu* with the evolution of Western Europeans, which is dominant. A slight change in the availability of resources could make some other technology superior, or necessitate the invention of a technology wholly new, and a further adaptive process could lead to the success of some individuals and the failure and perhaps even elimination of others unable to grasp or to utilize another set of tools and ideas.

# 11
# HUMAN AFFINITIES IN A CHANGING ENVIRONMENT

Like a number of other societies which include a colonial element, the Southern African complex of populations appears to hold out the promise of being particularly revealing about the nature of human selective adaptation. Unfortunately, aspects of the situation which one might expect to be especially informative often turn out to be able to tell us nothing at all. The oldest settlement of a non-African people in the sub-continent is that of the Portuguese in Moçambique, yet in Moçambique there is not now, nor for a long time has been, any endogamous Caucasoid stock tracing back its coherent existence through the nearly five centuries of occupation (Lobato 1954). There was never time to record Portuguese adaptation to the environment, before the fortunate had returned to their lands of prior adaptation and the rest had contributed their genes, if they left any descendents at all, to the mestiços or ultimately the Negroes.

This stands in stark contrast to the other, later-coming Caucasoids. The settlements promoted by the Dutch East India Company and their successors have resulted in true colonization, with the occupation and utilization of lands by immigrants who, for a time at least, retained ties with their distant homelands. People from the first of these settlements moved into novel ecosystems, to which they had not undergone prior adaptation, long enough ago for us to be able already to identify, however tentatively, some selectively significant traits in their descendants. Furthermore, some of them contributed to the emergence of a variety of hybrid populations in several rather dissimilar environments, and the Caucasoid stock was sufficiently alike, the groups with which miscegenation took place sufficiently varied, for their descendents to be biologically of some interest. Very little advantage has been taken yet of these unprovoked experiments which could provide so much information. With the exceptions of the Basters and a few other groups, the Southern African hybrid peoples have been little studied.

Nevertheless, their affinities, if not their selection paths, are adequately documented. So, too, are those of their Caucasoid cousins and forebears. Any judgement of the internal affinities among the Khoisan-speaking peoples, on the other hand, needs to be made very warily. The surviving populations form too many isolates of small size, subjected to intensive drift, for differentiation among them to be meaningful. Though there are slight contrasts among the ecosystems of the San, differences among such miniscule population groups are unlikely to be ascribable to recognizable selection. Despite this, there are some pronounced frequency differences in mono-genic characters. The speakers of Southern and Central Bush languages have appreciably higher frequencies of the M-bearing alleles in the MNSs Henshaw blood group system than do Northern speakers, while in the ABO blood group their frequencies of the stronger A alleles $A^1$ and $A^{int}$ are lower. In the acid phosphatase system the frequency of $P^b$ shows a distinct geographical cline, being highest in the north. These are not sufficient to justify any adjustment to San taxonomies on biological grounds.

Where the question of affinities becomes especially approachable from the strictly biological standpoint, and without overmuch reliance on historical records, is among the Negro populations. There are enough of them for a study to be possible of their population structure as revealed by their biological traits, and such a study forms part of this chapter. For the rest, having dealt with each population segment principally in terms of the impact of the environment on its members, it remains for us to consider in more particular detail the effect which each group is having on the environment as it presently changes, and the possible consequences these will have on future ongoing human adaptive selection in Southern Africa.

## Hybridization and selective adaptation

Selection and adaptation are likely to follow different courses in the different modes of hybridization outlined in Chapter 9, and in conse-quence the individual hybrid groups are likely to vary in the degree to which they are adapted to their present-day environments. The various groups have principally one characteristic in common; they are all the products of miscegenation between stocks which have followed differing paths of microevolution in contrasting environ-ments. It is widely believed, though still an occasion for controversy, that hybridization in itself conveys a measure of increased fitness. The belief derives in the first instance from folk observation before much was known about the mechanisms of genetics. It has been supported by the very reasonable supposition that an extension of the array of genetic variation within the individual confers on that individual an enhanced capacity to cope with a variety of environments. This phenomenon has been described in several terms; as heterozygote advantage, as hybrid vigour, or as overdominance; all amount to the same thing.

It is advisable to be cautious about accepting this concept too wholeheartedly. Certainly it is relatively easy to conceive that in an environment totally new to both parent stocks the first-generation offspring of their miscegenation will be at an advantage, since they will possess among them the total array of the genetic material of both stocks. In the second and succeeding generations, however, there is likely to be a steady erosion of the genic variance, since selection will progressively have operated against the less advantageous genes. A hybrid people, therefore, which receives no further genetic contributions from either parent stock, but exposes itself endogamously to the stresses of its elective environment, will have the advantage of blending in itself all those factors which may have contributed to the prior adaptations of both parental lines, and tend to evolve much more quickly towards phenotypes most suited to the environment. On the other hand, if the selection is for individual genes or for the conservation of gene combinations which have already entered into linkage disequilibrium in one of other parent stock, rather than for reassortments and new gene combinations, the genic variance must necessarily decline as the population approaches its maximum fitness (Morton et al. 1967).

The only populations described above which appear to have followed this latter path of development are the Basters, to whom Fischer (1913) ascribed augmented fitness for just such reasons, and a few of the local Caucasoid/Negro hybrid groups, particularly those of the Buys and Dunne type. Yet when genetic marker studies were carried out on the Rehoboth Basters (Stellmacher et al. 1967; Nurse et al. 1982) it was somewhat surprisingly found that approximately equal quantities of 'typical' genes had survived from either parent stock. This might have been significant if the origins of the Basters had been any other than from approximately equal numbers of Caucasoids and Khoi. As it is, it implies that their Caucasoid forebears were no less well adapted to the semi-desert interior of the subcontinent than were their Khoi ancestors. A little reflection, however, will show that this can well have been the case. With the first penetration of the interior by relatively ill-prepared young Caucasoid men, selection is likely to have been intense. Mainly those of them equipped through pre-adaptation to accommodate themselves to semi-desert conditions would have been likely to survive and reproduce. The maladapted would be likely to have died without, or with less than the average number of, progeny. There is, moreover, a strong probability that those genes which have been taken as indicative of racial ancestry are totally unconnected, or only slightly relevant, to environmental adaptation.

Such pre-adaptation as was present in the Caucasoids would most probably involve genes and gene combinations totally different from those which had already become established in the Khoi. The mixture would consequently provide an enrichment to either stock, and, as Fischer (1913) says, lead to a degree of superior fitness in the Basters (and by extension in the Buyses and the Dunnes) which would equip them to survive in their environment rather better than either the Caucasoids or the Khoi can. That in practice this has not turned out to be the biological case, even though the Basters have flourished, can be ascribed on one hand to the economic advantages possessed by the Caucasoids, but on the other is, unexpectedly, inexplicable as anything but an indication of the ecological disadvantage of the persistence of the Caucasoid genetic material; a disadvantage which the Basters can only have overcome in much the same way as the Caucasoids have, by exploiting a command of resources and a method for their utilization superior to those available in a cash economy to the Khoi.

There is, of course, no reason to suppose that such abilities do not themselves proceed from inheritance, and that the heritable factors are not selected for. Ingenuity and technological adaptability do appear to be more characteristic of certain stocks than of others; both the Khoisan and the Caucasoids have initiated novelties of technique and apparatus which enable them more fitly to occupy their native niches, and both have demonstrated an ability to take over or improvise new techniques rapidly when these are required. That the Basters, the Buyses and the Dunnes have succeeded overall even while their susceptibility to environmental rigours may have been greater than that of their neighbours probably indicates that they have inherited more of such traits than have hybrid strains which have been less successful.

Hybrid communities which have been less deterministically established would, of course, possess a wider genic variance with a more extensive range of alleles and even polymorphisms, and could be expected to be especially ready for the operation of selection

in any of a number of directions. The main body of 'Coloureds' at
the Cape would fall pre-eminently into this category, but once again
there has been little indication of the development of superior biological fitness as a result of the mixture of genetic material. Here
economic forces may have acted adversely, and a lack of privilege
and opportunity inhibited what chances there were for the emergence of a Southern African people especially suited to the Southern
African environment; though the economy has not always been
adverse for the 'Coloureds', and yet they have remained especially
subject to poor health and early death. The tendency mentioned
in an earlier section, for social selection to proceed with the loss
of the more enterprising persons, if they are of an acceptable lightness of colour, into the Caucasoid community (Findlay 1936), may
be responsible for this, as it probably is for equivalent disadvantages
to the Basters; though the erosion of Baster unity is selectively different, and brought about especially by the marriage of Baster women
to Caucasoid men.

### Negro adaptation to the urban environment

Today there are numerous urban Negroes who disavow any connection with the rural areas at all. Many retain their language and some
traditional practices, but there are those who have gone so far as to
adopt either other dialects of their own language or other languages
entirely. Sometimes these are the languages of spouses, and it may be
claimed that one of the most considerable biological consequences
of the drift of Negroes to the towns has been the facilitation of
mating between members of groups which have evolved at a geographical remove from each other. This claim needs qualification,
however. At least as great a breakdown of traditional demarcations
between peoples happened at the time of the *Mfecane*, and similar
but unrecorded processes could earlier have produced similar results.
The impact of the *Mfecane* on individual Southern African Negro
peoples admittedly differed in degree, but not one outside the
south-western group was left entirely unaffected. The growth of
urbanization has likewise meant more to some peoples than to
others, but with the possible exception of the Kavango peoples it
has produced some effect on all of them.

In biological terms the effect is extremely difficult to define. The
three most notable factors are rather obvious ones, but their total
and durable influence on the Negro constitution cannot be measured
with any precision as yet. In the towns the agricultural cycle of food
availability is muted and to some extent replaced by rhythms

Fig. 11.1. Aerial view of Soweto, the large Black residential area near Johannesburg, South Africa. (Photograph by Mike McCann, Financial Mail.)

generated by industrial demands. There is always an excess of labour and a shortage of employment, but the distribution of food resources is relatively uniform. The unemployed and unemployable are linked by ties of kinship and obligation to those who have work; often it is such ties which have drawn them to the towns in the first place, and they are rarely without some source of food. When unemployment and starvation become overt, the urban dweller is conspicuous enough to be an easy object of the charity which hardly ever reaches deep into the rural areas. On the other hand dietary patterns change; the tedious preparation of the staple is supplanted by the purchase of more expensive foodstuffs which have to be consumed in smaller quantities, and commodities such as sugar which are modest luxuries in the villages become everyday matters in the towns. There is an increased use made of tinned foods, sometimes leading to a somewhat higher protein intake relative to the carbohydrate, and a much higher consumption of processed fats like butter, margarine and vegetable oils.

Changing dietary habits have had notable consequences on the pattern of disease. It is clear that the initial impact of urbanization on health would be in terms of the familiar patterns of urban infection, with crowding, sanitation and hygiene of primary importance in its regulation. In fact, with the reduction in infectious disease due to adequate public health measures, there has been an increased prominence of the metabolic diseases, such as diabetes mellitus and hypertension, which are often considered typical of the cities of the West. This has almost certainly been due to the adoption of a number of Western food practices, learned by Negro women either in domestic service or from books or the cinema or at adult education or women's association centres. The influence of the Afrikaner cuisine, in which large quantities of sugar and fat are used, has been particularly marked.

The accessibility of primary health care on the Western model has undoubtedly led to the survival in the towns of numerous infants and some adults who would not have survived in the rural areas. It would be inaccurate to claim that any town or city in Southern Africa provides an adequate curative service for all its Negro inhabitants, and there are naturally pronounced differences in standards among the urban centres, but in general health care is a great deal better than it is in a traditional society. Nor is it only in the context of Western medicine that this is the case. The traditional herbalist or diviner is far more exposed to competition and comparison with his peers in the cities than he is in the rural areas. Variability in the standard of Western-style attention, and the not always ill-based retention of faith in tribal medicine, mean that such practitioners have had to abandon part of their traditional roles and assume part of the role played by the more orthodox medical men in the West. This has been accomplished to a great extent by their taking advantage of the very factors which regulate Western urban medical practice: the presence of high concentrations of population, the use of public transport for the carriage of drugs and supplies, and the demand for a range of choice among medical options available, as well as enhanced opportunities for follow-up of patients.

In addition to these biological considerations, migration to an urban environment has produced a variable range of social adaptations. The formation of social groups in the cities has been regulated to a great extent by the particular provenances of their Negro inhabitants. In the smaller towns the Negroes are often all speakers of a single language or even dialect, with some variation in local allegiance within the linguistic uniformity. This applies even to some of the larger cities, such as East London with its overwhelming preponderance of Xhosa speakers. Conversely Grootfontein in Namibia, much smaller than East London, lies on a principal route of communication between the northern and central parts of the country, and has attracted Ambo, Kavango, Dama and Herero. The strength of traditional groupings in the urban environment is preserved in no very regular or predictable way. It depends largely on the degree of coherence felt in the original rural group, and the extent to which it is possible or desirable to maintain this in the towns. It naturally depends also on the temporal depth of urbanization; first generation immigrants may be expected to be more conscious of tribal ties and more given to tribal ways than their children or grandchildren. Particular groups, even after some generations, may feel more vulnerable than others, and as a result either strengthen ingroup ties or relax them in the hope of being less conspicuous; or perhaps set up alliances with stronger or equally threatened groups. Something of this sort seems to happen among the Tsonga of the Witwatersrand, many of whom are content to be labelled 'Shangaan' rather than try to conserve any more local identity, while others on the strength, apparently, of the Nguni hegemony in Gazaland, try to pass themselves off as 'Zulu' and often succeed owing to lack of any strong cohesion within the migrant Natal Nguni groups.

Hellman (1948) has shown how irrelevant the adoption of Western material culture is to social group formation among the urban Southern African Negroes. She took as criteria of 'detribalization' permanent migration, abandonment of allegiance to traditional chiefs and the complete severance of interdependence relationships with rural relatives. This somewhat superficial definition reflects the essential superficiality of the concept defined. Even though traditional ties may in substance be broken, the alternative ties subsequently developed often structurally resemble the old ones quite closely. A church may come to play for an individual the role fulfilled for his ancestors by the territorial community, or a residential sector which throws up a strong leader may impose its more immediate territoriality and loyalties on the weakening links to tribal land without changing their shape. Already the cities have been in existence long enough for entirely urban kinship groups to develop; as Mayer (1962) has suggested, this may be the most important single token of urbanization, and coincide with the essentially demographic concept of stabilization (Mitchell 1956a) and the cessation of circulation between town and country. Nevertheless, there are strong and irregular forces which may continue to elevate the importance of extra-urban links even for persons who have never been outside the town; these can best be assessed in individual cases only, by examination of the social networks to which each is central (Barnes 1954b; Bott 1957). In any case their quantitation, or even definition, for any particular society would be difficult (Dubb 1974).

Urban conditions have also already dictated certain variations in household type. Pauw (1963) distinguishes four main types: the nuclear family plus one or two dependants, the multi-generational male-headed household consisting of the nuclear family plus the children of unmarried daughters, and two types of female-headed household. One consists of the obvious combination of an unattached woman and her children; the other, often the consequence of this, the multi-generational female-headed household, is made up of a chain of daughters and mothers without any permanently associated male, and is becoming increasingly common. This represents a startling innovation in societies which are traditionally strongly patrilineal in ideology, and comes about as the consequence both of male migrant labour and of the relative weakness of ties within the urban family. The rural family demands fairly rigid role-allocation, since it operates as a productive unit and any failure of one member to perform appropriately will rebound on the productivity of the rest. In the towns and cities, however, production is unitary and not co-ordinated within the family, which as an entity functions only as a consumer. Males in the cash economy are relatively mobile, while women are constrained by the necessity of servicing dependant family members (Dubb 1974). Such women often rotate the offices of wage-earners, depending on whether they still have suckling infants or not. This tends to depress their wage-earning capacity, as they often do not remain in employment for continuous long periods.

Very little study has been made of the population genetic structure of urban communities anywhere in Africa. With the wide-spread assumption of new affiliations in the towns and cities, it is frequently almost impossible to secure an adequate urban sample to compare with a well-defined rural population. In any case it seems doubtful whether urbanization has had time to produce any specific changes, and whether even if an urban environment is acting selectively on certain loci the process could be identified yet. It has been shown (Ojikutu *et al.* 1977) that despite the very long period of urbanization of the Yoruba of Ibadan and Lagos in Nigeria, the patterns of enzyme gene markers in them reveal no significant difference from those in the Ambo peoples of Namibia, who are geographically as remote from them as any ostensibly unmixed Negro people could be. The studies of Jenkins (1977) and de Villiers (1977) on the rural and urban Pedi and Venda, carried out with the specific object of identifying biological changes due to urbanization, disclosed no significant genetic differences and only such morphological differences as could be ascribed to dietary changes. These workers carried out their investigations in full awareness that the choice of migration to the cities might not be completely random, and that the genetic constitutions of the migrants could possibly have played a part in their decisions to migrate. It was not explicitly envisaged that this would be, in fact, a type of pre-adaptation or perhaps even prior adaptation, and neither at the time of investigation nor even today is the range of available markers wide enough to provide more than a hint of any such process. Rather was it suggested that a more general type of hardiness, a conscious ability to withstand and flourish among the rigours of a new environment, was involved, and that this might show up in association with particular constellations of markers. This did not, however, prove to be the case.

### The rural Negro and adaptation to a changing rural environment

The biological loss to the rural Negro communities produced by the migration of certain of their members to the towns and cities could be a detrimental aspect of prior adaptation to specific rural environments. If there were any pre-adaptation to urban conditions, which is highly improbable, it is more likely that the degree of specificity contained in the process would rather produce biological enrichment, constituting as it would a fission removing those less suited to rural life. If, on the other hand, it is the hardy who migrate most readily, biological impoverishment of the rural populations might be expected to result; and there is some indication that this is the case.

Ogbu (1973) has claimed that labour migration is among the factors which exacerbate seasonal hunger among the Poka of Malaŵi, by removing labour at a time where it is most necessary locally. This has been contradicted by van Velsen (1964) for the Lakeshore Tonga and Nurse (1975b) for the Ngoni and Ntumba of the same country. Labour migration provides a very widespread means of relief of seasonal hunger by placing at the disposal of rural societies which traditionally suffer from it a means of alleviation through entry into a cash economy in which the main demand which can be satisfied by rural subsistence agriculturists is for the surplus of labour which they almost invariably possess. The cultivation of cash crops is not always a satisfactory substitute, as it sometimes leads to a relative neglect of the subsistence gardens and may even occasion a worsening of the nutritional status. Effective labour migration presupposes, however, the maintenance of strong links between the urban worker and his rural family. The growth of the cities of Southern Africa has led, as shown above, to the development in the cities of some Negro population elements who no longer have any ties with the rural areas. These people are cash-earners whose economic life produces only remote secondary effects on the rural populations. They do not constitute a majority yet, but their very existence does to some extent deprive the rural areas of resources. The greatest number of migrant workers still return some of their earnings to their areas of origin, however, though the degree to which this compensates or ever compensates for the loss of their rural productive potential cannot be estimated at all precisely. The amounts remitted almost certainly fall steadily as the worker's social ties within the city become stronger.

The rural environment has consequently undergone change insofar as there has been a removal of part of its population. We cannot define this loss in genetic terms, as we have insufficient information on which we might assess it. Economically, one would expect the expansion of the rural population to compensate steadily for the labour losses, though the men who continue to leave still tend to be the fittest, and the residuum consists of women, old men and children. All the largest cities and the most tempting employment possibilities are in the Republic of South Africa or Namibia, so that the human changes in the environment apply even to the four adjacent independent African states despite the relatively small degree of urbanization which has taken place in them. The result has been a change of the household authority structure in the rural areas, with, as in the cities, an increase in the power of women. The narrowing availability of land for cultivation and the congestion on what remains have led to some well-publicized press accounts of resultant food shortage in parts of South Africa, and there is a government dole for relief in a number of areas. The congestion is not yet particularly marked in Lesotho, Swaziland, Botswana, Namibia or southern Moçambique, where there are still some reserves of unexploited arable land and pasturage; but Lesotho in particular is reaching the limits of its unaided agricultural potential, and in Swaziland and southern Moçambique this cannot be far off.

The effect of steadily more concentrated occupation on the land itself has been dealt with in Chapters 1 and 6. Cultivation with reduced opportunity for fallowing depletes soil nutrients, and the confinement of herds to a restricted range of pastures accelerates erosion and leaching. This is not a new set of processes in Southern Africa, but it has been occurring at a steadily increasing rate.

The changed rural environment to which the rural Southern African has to adapt is thus a constricted and comparatively infertile land which needs to be cultivated by the relatively weakest members of the community. The genome which has been adapted to the rigours of the African agricultural cycle faces another set of adversely operating factors in addition to those. Hunger is more apparent due to the shrinking productivity of the land, which has also led to a shortage of fuel which can be life-threatening in a Highveld winter. Diminished fodder cuts down the condition and the productivity of the flocks and herds, with a consequent reduction both in available protein and in capital convertible to cash. The country becomes steadily more reliant on the town: adaptation takes the vicious path of the enforced migration of those best equipped physically to cope with the changes, so that they remit part of their urban earnings: but increasing generations break ties, the feebler remain at home to procreate, and in the longer run the rural population may increase in numbers but is likely to decline in vigour.

## The population structure of the modern Southern African Negro peoples

Even under traditional circumstances there was, and is, a considerable diversity of structure among the negro populations in Southern Africa. Settlement patterns varied from the concentrations preferred or enforced by their environments on the Sotho/Tswana peoples to the dispersed single household or rudimentary hamlets of the more northerly Nguni and the large patriarchal kraals of the Nguni of the

Cape. Shifting riverside cultivation along the Kavango kept the sparse peoples of the area to small separate villages until the rise of missionary or, in the case of the Mbukushu, mercantile activities promoted a few bigger aggregations, while the fertility of their land favoured the large adjacent multiple households of the Ambo. The pastoral Herero used their towns as depots and meeting-places, always inhabited though not always by the same persons. The Negro hunter–gatherers, whether Bantu-speaking like the Cimba of the Kaokoveld and the Kgalagadi, or Khoisan-speaking like the Dama, Kwengo, Masarwa and other inhabitants of the northern fringes of the Kalahari, either live in dispersed hamlets like the Dama or, like the Cimba, migrate in small groups along a cyclical route.

The large Tswana towns are divisible into wards, and a similar pattern sometimes emerges in the Negro townships associated with the more modern cities. Urban household types have already been described: over and above these lie the specifically urban varieties of association which are replacing the social groupings of the country-side. There has been an effort on the part of the authorities to preserve or bring about ethnic exclusivity of residence in the townships, when this is not already secured by the nature of inward migration from the hinterland. Cultural groups aimed at the preservation of individual values, whether or not these values can validly be translated to an urban setting, tend to be encouraged. Language is certainly a potent force towards the retention of ethnic identity, at least during the first generations of urbanization; yet even then marriage or cohabitation across ethnic or linguistic barriers occurs, and the offspring are sometimes presented with a hard choice of identities and loyalties. Forms of municipal government arise spontaneously or are imposed, erecting or recognizing boundaries within the townships, and these in their turn create new groupings and new potentiations of or limitations to directions of matings. Churches and social clubs may either reinforce or break down ethnic barriers, but in either case the probability of outbreeding is a good deal higher in the cities than it is in the rural areas.

Nevertheless, in the rural areas themselves the mating patterns traditionally varied considerably. Restrictions on kin mating were and are much less strict in the societies which formed large settlements, probably because exogamy will occur more readily at random where the choice of potential partners is wide. This does not imply in the Sotho/Tswana any intense communal preoccupation with exogamy such as characterized the Natal Nguni, but even the former, for all their preferences for cross-cousin marriage and relative tolerance (among the Ngwaketse and Rolong at least) of even more closely consanguineous unions, did preserve bans on marriage between certain categories of sometimes biologically fairly remote but classificatorily close kin. Clan exogamy has always been strict among the Nguni, and after the *Mfecane* and the consequent geographical dispersion of many clans the range of mating choice would have been wider than it had been before. Any reduction in population heterozygosity that had been occasioned by cumulation of remote common ancestors would have tended to be disrupted, and in the generations following the *Völkerwanderung* there would have been a lower but progressively increasing degree of inbreeding among the resettled groups until their semi-isolation was once more disturbed, this time by the extension of European influence.

For the Negroes of Namibia the Herero–Hottentot Wars took the place of the *Mfecane*, but their effect was not nearly so widespread. The Ambo were sometimes receptive and sometimes resentful of those who sought refuge or passage, but did not intervene; their principal nineteenth-century contacts were with the people to the north of them. The Mbukushu seem to have profited from the disturbances, and to have traded in slaves and ivory from a market at or near modern Andara. It is possible that the movements of the other Kavango peoples along and across the river were the last faint ripples of the *Mfecane* itself insofar as it extended into Barotseland. The Dama were shifted around, forming larger settlements under the protection of the Herero, later of the missionaries and then of the German government. The principal geographical changes in settlement pattern, though, were those of the Khoi, and the structure of the Negro populations was not likely to have changed much before the effective introduction of European government and the rapid allocation of land to Caucasoid settlers and European companies.

There has been only negligible gene flow from Caucasoids into the Southern African Negroes in modern times, and most of the undoubtedly large genetic contribution made to them by the San and Khoi cannot be dated accurately. Much of it probably happened early, though there are reliable indications that it still continues, particularly in Botswana (Silberbauer 1965; Nurse 1977*b*). The extent of the contributions can be assessed, mainly with the help of the Gm genetic marker system, in which the haplotype $Gm^{1,13,17}$ appears to be characteristically and exclusively Khoisan. Using this, Jenkins *et al.* (1970) determined a more than 50 per cent concentration of Khoisan genes in the Cape Nguni, with the other Negro peoples having lower but still appreciable proportions (see Figure 11.2). Even a sample of unsorted Malaŵians, who may consequently

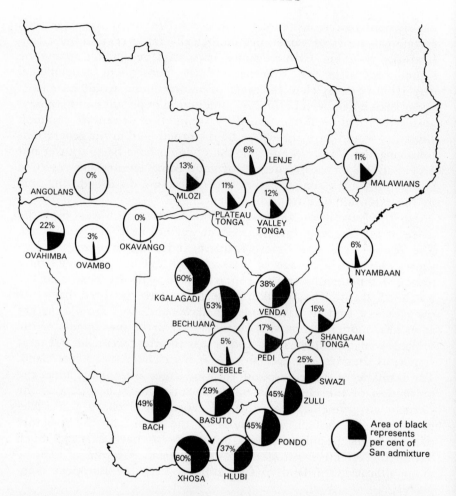

Fig. 11.2. Map of Southern Africa showing the estimated percentage of San admixture in each of several tribes as determined by the frequency of $Gm^{1,13,17}$. (From Jenkins *et al*. 1970.)

have included a disproportionate number of Ngoni, showed about 10 per cent Khoisan admixture. The proportions were high in most of the Sotho/Tswana and Nguni peoples, though only 14 per cent in the Pedi, and fell off fairly abruptly in the Tsonga to about 12 per cent. None of the Namibian Negro peoples shows nearly as much evidence of Khoisan gene flow, and the Kavango other than the Mbukushu, and the Kwambi/Ndonga moeity of the Ambo, show no evidence at all. It is, somewhat surprisingly, only in the Herero and Himba that proportions above 20 per cent appear to be present.

It has been claimed (Shapiro 1951*a*, *b*) that the Southern African

Negroes constitute a homogenous 'race', and that between the extant groups there is no significant difference. This hypothesis was based on the rather narrow range of gene markers which could be investigated at the time it was propounded, but was even then made on singularly unconvincing evidence. A large sample of Witwatersrand 'Bantu' was divided according to the ethnic affinities stated by its constituent members; the fact that for the Rhesus system the resultant sub-samples were not in anything approaching Hardy–Weinberg equilibrium was to all intents and purposes disregarded, and the sub-samples compared with one another by the $\chi^2$ test and found not to be significantly different. The method of substantiation of the stated ethnic allegiances is not given, and they appear to have been accepted on trust, without any scrutiny of clan-names, which we believe would have been advisable. It is consequently impossible to accept the conclusions of these papers, particularly as later more careful work in a number of additional systems has shown them to be invalid. We can only suppose that the sub-samples consisted in fact of random divisions of a random undifferentiated sample, which may or may not itself have been in apparent Hardy–Weinberg equilibrium.

The tables (pp. 299–346) reveal several striking differences among the Negro populations, though all, indeed, can be fitted into a single constellation of gene markers common to the rest of the continent south of the Sahara. The presence of Khoisan markers such as $Gm^{1,13}$, the acid phosphatase $P^r$ allele, $A^{bantu}$ and possibly $PGM_2^1$ is more pronounced than in other Negro populations, though these are virtually absent from the Dama and some other Namibian Negro peoples. The presence nonetheless in the Dama of a relatively elevated frequency of $AK^2$, which in all the other peoples appears approximately to parallel the degree of Khoisan admixture, can probably be ascribed in them to genetic drift, though the possibility of a selective advantage for this allele in arid environments cannot be excluded. The Dama, in fact, figure as the most 'typically' Negro of all these populations, with characteristic arrays of alleles in almost all the systems except, curiously, Rhesus, in which the frequency of $R^0$ is comparatively low and the frequencies of $r$ and $r''$ are unexpectedly high.

Individual examination of the marker systems gives good support for the migration hypotheses propounded in preceding chapters. In the ABO system the speakers of South-Western Bantu languages have on the whole lower frequencies of $A^2$ and $A^{bantu}$, while $B$ is lower in the South-Eastern group. Of the Khoisan-speakers, the 'Masarwa' affinities appear to be to the south-east and those of the

Dama and Kwengo to the west. The cultural similarities of the Herero to the Nguni are given biological support by the finding of the rare coupling of the Henshaw antigen with $Ns$ in them and in the Natal Nguni, while the Dama coupling with $NS$ is similar to that which Chalmers *et al.* (1953) found in the Negroes of West Africa. The antigen appears to be absent from the 'Masarwa', but then in the Pedi and Gciriku, whom they most closely resemble, it attains only low frequencies. In the former, however, coupling is with the $M$ alleles, while in the latter it is with $NS$. Frequencies of $NS$ tend on the whole to be higher in the South-Eastern group, though the highest frequency of all was found in the very small sample of Nkolonkadhi, who are aberrant in several other ways as well. The Rhesus system again distinguishes South-Eastern from South-Western by generally higher frequencies of $R^2$ in the latter, and the Dama from the rest by the presence at a polymorphic level of $r''$. It would be possible to regard the finding of either Duffy antigen as evidence of Khoisan or Caucasoid admixture, but the relative commonness of $Fy^a$ in the Ambo is at variance with the proportion of received Khoisan genes suggested by the Gm results. The Ambo, however, are characterized also by the presence of $K$ in the Kell system and $P^c$ in the acid phosphatase system, both alleles of probably Caucasoid origin. It may be hazarded, without adequate historical support, that these peoples have received indirect genetic contributions from the Portuguese of Angola.

Hardly any South-Eastern sample investigated lacks the transferrin $D^1$ variant, which is rather less often found in the peoples of the south-west except for the Kwangali, in whom its relatively high frequency could have come about due to drift. Once again, the 'Masarwa' possess more of it then do the other Khoisan-speaking Negroes. The haemoglobin qualitative variants are not very revealing in the context of migration: the high frequencies of $Hb\delta^{B_2}$ in the Herero-speakers are not reflected in the South-Eastern peoples, and the apparent absence of the allele from the Negroes of the Kalahari is of doubtful significance, while $Hb\beta^S$ is well known to be subject to selection and its absence from all the South-Eastern peoples except the Venda probably indicates (Jenkins 1972) that the mutation for it occurred in the Negroes after the ancestors of the South-Eastern peoples had already crossed the Limpopo into relatively malaria-free areas. In this they resemble the Herero and, less foreseeably, the Mbukushu. The mutations for the other polymorphic variants known to be protective against malaria, $Gd^A$ and $Gd^{A-}$, are apprently of rather greater antiquity. The much higher frequencies of $Gd^A$ than of $Gd^{A-}$ suggest that the former allele in fact either confers greater fitness than the latter, or is older and, unlike the latter, had attained relatively high frequencies before the earliest migrations of Negroes southward. Frequencies of the deficient variant are low to zero in the hunter–gatherer peoples, the Cimba and Kgalagadi as well as the Khoisan-speakers, which may indicate selection against it in hunting and gathering situations. This would imply that the mutation would be unlikely to have become established before the introduction of a culture of settlement among the Negroes and consequently that it dates only from the Iron Age.

It has been suggested (Nurse and Jenkins 1977a) that the $PGM_1^6$ variant at the first phosphoglucomutase locus, like $Gm^{1,5,14}$, may indicate Khoi descent. Both are present in the speakers of South-Eastern Bantu. The $PGM_2^2$ allele at the second locus certainly is more frequent in those populations which have received Khoisan gene flow, and among the Negroes reaches its highest frequency in the 'Masarwa', who present other signs of an unusually large Khoisan component in their ancestry. Acid phosphatase $P^r$, which is also held (Jenkins and Corfield 1972) to indicate Khoisan admixture, is less consistently distributed in the Southern African Negro peoples. Unexpectedly high frequencies, inconsistent with the degree of Khoisan admixture postulated for them, are found in the Kwangali and Sambyu, and the Kwengo frequency is higher than that in the 'Masarwa'. This could, however, suggest a Kavango focus of the variant, possibly derived from an antecedent Khoisan population there in which the $P^r$ frequency was unusually high, and which has been assimilated by the Negro later comers. The G!ang!ai, the extant San of the Kavango valley, do not possess frequencies of this allele at all excessive for San (Nurse et al. 1977).

The Richmond variant in the 6-phosphogluconate dehydrogenase system, $PGD^R$, is found far more often among the South-Western than the South-Eastern peoples, but its highest frequency (except for the regularly unusual Nkolonkadhi) is in the Dama. So striking is the Dama concentration of this allele that it has been suggested (Jenkins and Nurse 1974a) that it may have been from a focus in northern Namibia or southern Angola and by way of the slave trade that this allele has attained its present widespread distribution.

The remaining marker systems are either not notably informative or have not been investigated for a sufficient range of peoples for them to have much bearing on the study of overall genetic structure among the Southern African Negroes. Those which have been studied reveal that the geographical separation of the peoples is in the main reflected in their genetic structure, but that there are tokens of divisions and affinities which would not be suggested by geography though they reinforce historical traditions and reconstructions. Among the most notable are the persistent signs of a Nguni affinity for the Herero, and the dichotomy among the Ambo (Jenkins, Rootman, and Nurse, unpublished) which indicates a possible connection with the Nguni for the Ndonga, Kwambi and Kwaluudhi and a closer affinity with the Sotho/Tswana for the Kwanyama and Ngandjera, with the Mbalantu intermediate and the Nkolonkadhi fairly remote. These add weight both to Herero traditions of eastern origin and the equally persistent traditions claiming ancestral contact between them and the Ambo. Other linguistic and cultural evidence, however, combines with many of the biological findings to indicate that, substantial as such contacts and gene exchanges may have been, there is nevertheless a very fundamental division among the Negro peoples of Southern Africa, and that this is, as suggested in an earlier chapter, brought about by the presence of the Kalahari Desert.

The study of monogenic markers among the Southern African Negroes has come about just too late for the investigation of other customary parameters of population structure to be particularly meaningful. There is little point in investigating such factors as mean and relative heterozygosity in the midst of a period of flux such as has been proceeding since the Second World War. Few of the rural communities are any longer basically representative of traditional populations, and the urban communities are changing too rapidly for their biological characteristics not to change with them. The situation in the 1950s, 1960s and 1970s had already proceeded beyond the possibility of the provision even of a baseline valid for later studies, and we will have to wait for some stabilization before any such can be embarked on with any confidence. Meanwhile all we can do is to describe trends, such as those in urban marriage and movement patterns and in rural environmental impoverishment, and indicate their possible effects on genetic structure.

## The affinities of the Southern African populations with other populations of sub-Saharan Africa

Biological changes resulting from urbanization and associated cultural and technological innovations have not yet proceeded so far that the Southern African populations cannot be fitted any longer into the wider context of Africa. In fact, as the foregoing discussion of monogenic markers has indicated, there are occasional agglomerations of similarities which suggest an improbably close relationship between some Southern African Negroid populations presumed on archaeological and linguistic evidence to have inhabited the region for a very long time, and others at a great distance.

It is impossible to reconstruct the precise patterns of relationship among the African peoples. The best that can be done is to assume simple divergences among them, with no further contact. On this basis, a variant of principal components analysis, known as correspondence analysis, has been used to assess the relationship of some Southern African populations to a number of other sub-Saharan populations. The twenty-three populations included in the analysis, together with their country of origin, linguistic affiliation and source of the genetic data are shown in Table 11.1. The thirty-two alleles used in the analysis belonged to five red cell antigen, two serum protein and four red cell enzyme (including two haemoglobin) systems and are shown on the graphic representation (Figure 11.3). Their individual frequencies can be found in the original sources, for which references are provided in the Bibliography. A full account of the technique employed by Dr Michael Greenacre, who kindly carried out the analysis, was given originally by Greenacre and Degos (1977), who applied it to a study of the HLA-A and -B systems in 124 populations. Correspondence analysis has the advantage of representing simultaneously the populations and gene frequencies with respect to the same axes in multidimensional space. Visual representation of the differences among the twenty-three populations, as well as the relative participation of each gene in the dispersion, are presented in Figures 11.3 and 11.4.

The percentages of inertia explained by the first and second axes are 28 per cent and 16 per cent, respectively, so that nearly one half of the total inertia of the analysis is represented in this plane. This plane demonstrates the proximity of the two San populations, the Dobe !Kung (DOB) and Nharo (NAR) to each other, as well as their wide separation from all the other populations. The Nharo are speakers of a Tshu-Khwe or 'Central Bush' language, but follow a hunting/gathering way of life. Biologically they are San rather than Khoi, a fact attested by the results of the present analysis. The Khoi of Keetmanshoop (KEE) occupy a position on the first axis approximating the two San populations and their position on the second falls in the direction of the East African populations known to have Caucasoid admixture: the Sudanese (inhabitants of Khartoum (SUD) and the Beja (BEJ)), as well as the Amhara of Ethiopia (ETH). It is well known that the Khoi show genetic signs of Caucasoid admixture. There is, however, no biological evidence to support the view of Jeffreys (1968) that the genes originated among the Semitic populations of the Middle East. Not all the Caucasoids of East Africa

Table 11.1. The 23 sub-Saharan populations included in the correspondence analysis

| Abbreviation | Population | Country | Linguistic affiliation | Source of data |
|---|---|---|---|---|
| DOB | Dobe !Kung | Botswana/SWA | Khoisan | Jenkins 1972 |
| NAR | Nharo | Botswana | Khoisan | Jenkins 1972 |
| KEE | Khoi | SWA | Khoisan | Jenkins 1972 |
| DAM | Dama | SWA | Khoisan | Nurse et al. 1975 |
| HER | Herero | SWA | Bantu | Jenkins 1972 |
| TSW | Tswana | RSA | Bantu | Jenkins 1972 |
| SOT | Sotho | Lesotho | Bantu | Beaumont et al. 1979 |
| SWA | Swazi | Swaziland | Bantu | Jenkins 1972 |
| PED | Pedi | RSA | Bantu | Jenkins 1972 |
| MOZ | 'Negroes' | Moçambique | Bantu | Jenkins 1972 |
| NJI | Negro | Angola | Bantu | Nurse et al. 1979 |
| BAB | Babinga Pygmy | RCA | Bantu | Cavalli-Sforza et al. 1969 |
| SAR | Sara Majingay | Chad | Nilo-Saharan | Hiernaux 1976 |
| ETH | Amhara | Ethiopia | Afro-Asiatic | Harrison et al. 1969 |
| GAM | Gambians | Gambia | Sudanic | Facer and Brown 1979; Mourant et al. 1976 Welch et al. 1977, 1978a, 1978b, 1979 |
| KEN | Luo | Kenya | Nilo-Saharan | Luzzatto 1973; Mourant 1976; Herzog et al. 1970 |
| NIG | Yoruba | Nigeria | Sudanic | Tills et al. 1979; Otjikutu et al. 1977 Luzzatto 1973 |
| BEJ | Beja | Sudan | Afro-Asiatic | El Hassan et al. 1968 |
| SUD | Khartoum | Sudan | Afro-Asiatic | Luzzatto 1973; Saha et al. 1978 |
| SAN | Sandawe | Tanzania | Khoisan (?) | Godber et al. 1976 |
| NYA | Nyaturu | Tanzania | Bantu | Godber et al. 1976 |
| HAD | Hadza | Tanzania | Khoisan (?) | Mourant et al. 1976 |
| ZAI | Shi | Zaire | Bantu | Govaerts et al. 1972; Luzzatto 1973 |

Tables 1–28 appear at the end of the text on pages 299–346.

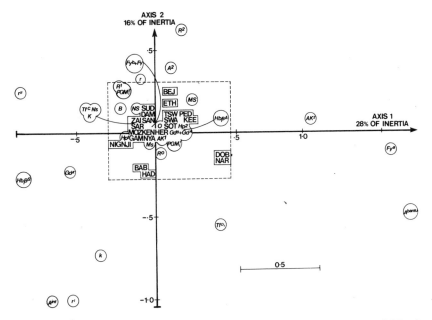

Fig. 11.3. Plane of the first two principal axes showing the positions of 23 sub-Sarahan populations and the 32 alleles belonging to 11 systems. (From Jenkins 1982.)

were necessarily entirely Semitic, and besides, the admixture is just as likely to have occurred in the last two or three centuries, following the settlement of European Caucasoids at the Cape and their gradual spread into the interior.

The position in space of the various populations conforms very strikingly with the geographical distances between them as well as with reconstructions of African pre-history which are becoming available from linguistic and archaeological data (Oliver and Fagan 1975); this is clearly shown in the summary population distribution (Figure 11.5). An outline map of Africa has been superimposed on this plot and, although a little manipulation of the lie of the outline was required, the final result is revealing. The north-east Africans (the two Sudanese populations (SUD and BEJ)) and the Ethiopians (ETH), all linguistically Afro-Asiatic, clearly distinguish themselves from the non-Caucasoid populations which seem to 'cluster' into at least four groups: (i) the far removed Khoisan; (ii) the Southern African Bantu-speaking groups comprising the Pedi, Tswana and Sotho (all belonging to one major linguistic group), the Swazi and the Herero; (iii) the eastern Bantu-speakers represented by the Nyaturu of Tanzania and a mixed Moçambique Negro population.

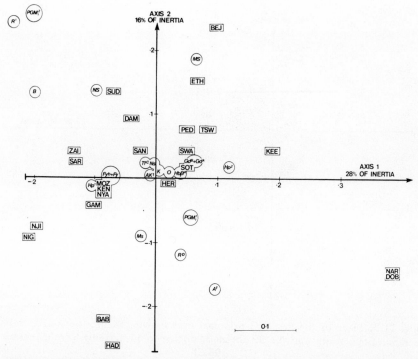

Fig. 11.4. Enlargement of the area indicated in Fig. 11.3. The 23 populations are shown but the alleles falling outside the area have been excluded. (From Jenkins 1982.)

One Kenyan population, the Luo, also falls in this group but speaks not a Bantu but a Nilo-Saharan language; (iv) the Nigerians and the Njinga of Angola who remain strikingly close even though widely separated geographically. The Gambian Negroes are intermediate between the latter group and group (iii). The Sandawe of Tanzania are close neighbours of the Nyaturu, with whom they have inter-married. They speak a click-language which is quite different from the Bantu languages of their neighbours and is classified as Khoisan by Greenberg (1963). There was even a claim by Trevor (1950) that morphologically the Sandawe resembled the Khoi. The present study does not confirm such a claim though the position of the Sandawe is somewhat removed from that of the Nyaturu. It can be explained by the striking differences in their frequencies of the malaria-protective genes $Hb\beta^S$ and $Gd^{A-}$. Otherwise their serogenetic resemblance to the Nyaturu is quite close (Godber et al. 1976).

When the projections of the alleles in this plane are examined it is evident that they cluster in accordance with their predominance in a particular group. For instance, $A^{bantu}$, $Fy^a$ $Tf^{D_1}$ and $AK^2$

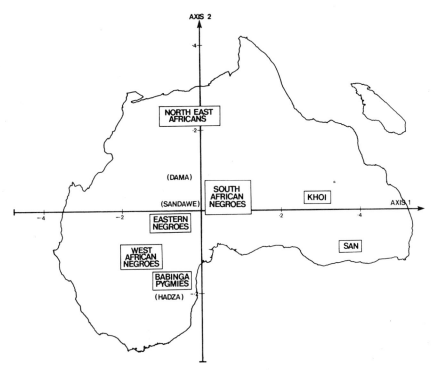

Fig. 11.5. Summary population distribution from the plane of the first two principal axes. The outline map of Africa has, after being subjected to a slight degree of 'continental drift', been superimposed on the plot. (From Jenkins 1982.)

contribute to the position of the San (DOB and NAR) populations, while $Hb\beta^S$, $Gd^{A-}$ and $r''$ ($cdE$) are largely responsible for the West African and Central African clusters. $R^2(cDE)$, $r(cde)$ and $A^2$ are the main alleles responsible for the East African cluster, consisting mostly of speakers of Afro-Asiatic languages.

### Affinities among some Southern African populations

Correspondence analysis has also been used to assess the relationship of forty-two Southern African populations. These are listed, together with the abbreviations and the sources of the data, in Table 11.2. The fifteen polymorphic systems employed, with the number of alleles presented in parenthesis, were: blood groups: ABO (6); MNSs (8), Rhesus (6); Duffy (2); serum proteins: haptoglobins (2), transferrins (3); red cell enzymes: acid phosphatase (4), phosphoglucomutase (3 alleles at first locus and 2 at second locus), adenylate kinase (2),

Table 11.2 The 42 Southern African populations included in the correspondence analysis

| Abbreviation | Population | Abbreviation | Population |
|---|---|---|---|
| KGA | Kgalagadi[1] | AM5 | Ambo-Mbalantu[7] |
| MAF | Subya/Fwe[1] | AM6 | Ambo-Kwaluudhi[7] |
| MBN | Mbanderu[1] | AM7 | Ambo-Nkolonkadhi[7] |
| PED | Pedi[1] | KA1 | Kavango-Kwangali[8] |
| ZUL | Zulu[1] | KA2 | Kavango-Sambyu[8] |
| TSW | Tswana (Fokeng)[2] | KA3 | Kavango-Gciriku[8] |
| MSE | Mseleni (Zulu/Tonga)[3] | KA4 | Kavango-Mbukushu[8] |
| NJI | Njinga[4] | KA5 | Kavango-Kwengo[8] |
| DAM | Dama[5] | GWI | G/wi[9] |
| BAS | Basters[6] | GAN | G//ana[9] |
| HER | Herero[1] | TSU | Tsumkwe !Kung[10] |
| HIM | Himba[1] | SAA | Saman!aika[10] |
| CIM | Cimba[1] | GAG | G!ang!ai[10] |
| BAM | Ngwato[1] | NAR | Nharo[10] |
| NKE | Nama-Keetmanshoop[1] | HEI | Hei//om[10] |
| NKE | Nama-Rehoboth[1] | G!A | G!aokx'ate |
| NKU | Nama-Kuboes[1] | MAS | Masarwa[11] |
| AM1 | Ambo-Kwanyama[7] | JCO | JHB Coloured[2] |
| AM2 | Ambo-Ndonga[7] | REO | Rehoboth[1] |
| AM3 | Ambo-Kwambi[7] | DBA | Afrikaners[1] |
| AM4 | Amgo-Ngandjera[7] | SWG | S.W.A. Germans[12] |

[1] present study;     [2] Jenkins (1972);
[3] Nurse et al. (1976);     [4] Nurse et al. (1979);
[5] Nurse et al. (1976);     [6] Nurse et al. (1982);
[7] Nurse et al. (1983, unpublished);     [8] Nurse and Jenkins (1977);
[9] Jenkins et al. (1975);     [10] Nurse et al. (1977);
[11] Chasko et al. (1979);     [12] Palmhert-Keller et al. (1983).

Tables 1–28 appear at the end of the text on pages 299–346.

6-phosphogluconate dehydrogenase (3), peptidase A (3); and haemoglobin loci $\beta$ (3) and $\delta$ (2). The individual alleles are shown on the graphic representation of the analysis (Figures 11.6 and 11.7).

The clustering of many of the populations is so striking that at the scale used for the plot there is, in fact, a great deal of actual overlap. As a result, the approximate positions of some of the populations have needed to be indicated by arrows. The plot is, nevertheless, presented because it indicates very clearly which alleles are contributing most to the spatial representation of the various populations. The 'enlargement' of the central area (Figure 11.7) clearly shows the positions of all the populations.

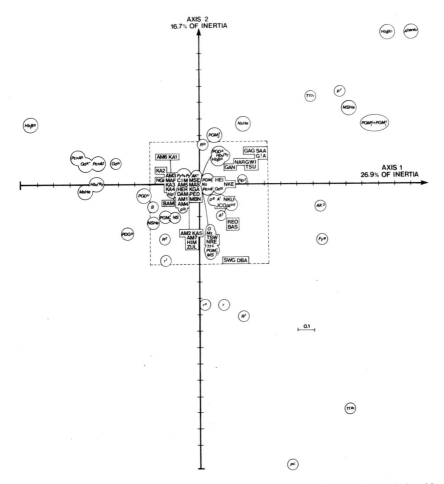

Fig. 11.6. Plane of the first two principal axes showing the position of the 42 Southern African populations and the 48 alleles of the 14 systems. (From Jenkins 1982.)

The Germans of Namibia and an Afrikaans population from Johannesburg cluster closely to each other and well away from the indigenous populations; the known hybrid samples (two Rehoboth populations and the Johannesburg 'Coloured') occupy a position intermediate between the Caucasoids and the Nama Khoi, whether they are those living in Rehoboth, Kuboes or Keetmanshoop. The last-named is the only Khoi population in a quadrant otherwise almost exclusively occupied by the eight San population samples. One Bantu-speaking people, the Tswana, just falls within its confines. This is a population which, on the basis of Gm studies not included in the present analysis, shows one of the highest proportions of San admixture.

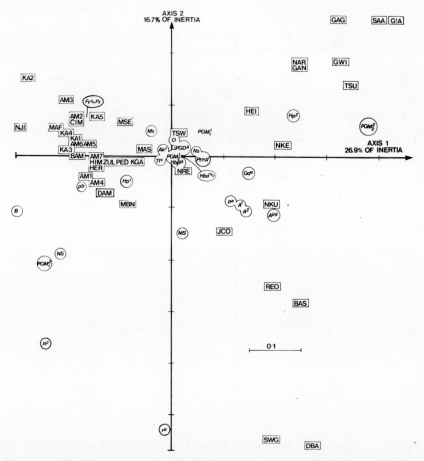

Fig. 11.7. Enlargement of the area indicated in Fig. 11.6. The 42 populations are shown but many alleles falling outside the area have been excluded. (From Jenkins 1982.)

All the Bantu-speaking populations (with the exception of the one Tswana sample referred to above) are distributed among the two left-hand quadrants. In general, those nearest to Axis 2 are, as might be expected, those with most Khoisan admixture. These include the Kgalagadi, the Masarwa (Chasko *et al.* 1979), the Pedi, the Zulu and the Mseleni Zulu/Tonga population (Nurse *et al.* 1974). The Njinga of Angola are the farthest removed while the Ambo and Kavango populations occupy an intermediate position. This distribution provides ample confirmation of the suggestion on the basis of the distribution of the Gm polymorphism in Southern Africa (Jenkins *et al.* 1970) that it was in the eastern half of the sub-continent that the Khoisan people were encountered. As has been described in an earlier chapter, they were there assimilated in large numbers by the southward-moving Bantu-speaking pastoralists/agriculturalists. In the western half of the sub-continent the migrations did not penetrate so far south and the Khoisan peoples were not encountered in such numbers. These findings are compatible with the claim by Tobias (1962c) on the basis of the archaeological and palaeoanthropological evidence then available that the area formerly populated by the San extended along the eastern half of the continent from the Mediterranean to the Cape.

The Herero-speaking Mbanderu for no obvious anthropological reasons are separated from the other Herero-speakers, simply called Herero and Himba on the plot, by the Dama. Genetic distance studies, simple cluster analysis and three-dimensional representations of relative genetic distances of Southern African populations have shown the Dama to be the most aberrant population in Southern Africa (Jenkins 1972; Jenkins *et al.* 1978; Nurse *et al.* 1976). The present analysis does not offer confirmation of this.

The contributions of the various alleles to the distribution of the populations are evident from Figure 11.6: $P^c$, $Tf^B$, $R^1$, $r''$ and $r$, as well as $AK^2$ and $Fy^a$ contribute most markedly to the position of the Afrikaners and the Germans (and less so to the hybrid populations, including the Khoikhoi with their apparently significant Caucasoid contribution); $P^r$, $Hb\beta^{D \text{ Bushman}}$, $A^{bantu}$, $Tf^{D_1}$, $MSHe$, $PGM_1^6$ and $PGM_1^7$ contribute to the position of the San, and $Hb\beta^S$, $Gd^{A-}$ and $Gd^A$ in addition to $Hb\delta^{B_2}$, $MsHe$ and the peptidase $A^8$ and $A^2$ alleles, are largely responsible for distinguishing the Bantu-speaking peoples of Northern Namibia and Angola from the South African Bantu-speakers.

Perhaps the most interesting point brought out by these analyses is the extreme dissimilarity of the San populations to their Negro fellow-Africans. Indeed, in Figure 11.7 the San are almost, though not quite, as far removed from the Negro core as are the Germans and the Afrikaners. This occurs in spite of the evidence not only of shared 'African' genes but also of occasional gene flow between San and Negro, predominantly from the former to the latter but also occasionally in the other direction. It is obvious that at the level of affinity studies the evolutionary relationships of Caucasoids, Khoisan and Negroes deserve extended study. Whether at the level of taxonomy these findings are enough to indicate that the Khoisan should be regarded as an individual race is more arguable. The common possession of a rare allele is a more certain indication of affinity than are the cumulative similarities in the frequencies of ordinary alleles, and the method of analysis does take this into consideration. The size of the sample, and the mode of its collection, may insinuate biases for which no kind of multivariate analysis can compensate. Nevertheless, the patterns which emerge do approximate roughly to the population affinities which we have already deduced from linguistics, traditional and recorded history, and the results of investigations of the monogenic marker systems.

The position of the Khoisan-speakers, however, shows just how labile language and culture can be. The evidence of biological affinities we have presented here is necessarily contemporary. Random processes are not easily dated, since they operate with fluctuating intensity. In this book we have attempted to describe the biological influences which are operating on the populations of Southern Africa today, while at the same time seeking to interpret as past structures with present relevance the comparatively small range of gene frequencies which has been determined for these peoples. To these we have added some salient facts about the health status, though the extent to which this is known differs so markedly from one people to another that it can distort any assessment depending too uncritically on it. Each additional variable impairs the stability, or reliability, of the unified structure, and we have consequently abstained from drawing too many firm conclusions.

## The impact of the Khoi and San on the environment

The people who retained their identity as Khoi during and after the social and biological traumata of the eighteenth century did so in part by conscious choice and in part because they dismissed or were not aware of the alternatives. Those hybrids who elected to join the Khoi rather than the 'Coloured' community, and those who coupled a 'Coloured' identity with a Khoi way of life, did so more deliberately. Because of their closer links with the politically dominant and technically more advanced Caucasoids, they were able in their first generations to use their fathers as the agents of procurement of some of these advantages, and to take a cultural leap which, despite any superficial similarities in their ways of life, was denied to the Khoi. Those of them whose genes remained in the Khoi were almost certainly those who managed least well in the communities of settled 'Coloureds' or pastoral Basters.

The remaining Khoi, then, politically shattered by epidemics and Caucasoid and Negro confrontation, deprived of the best pastures and harrowed by competition for waterholes, were confined increasingly to marginal situations. Many of those from the Cape migrated during the eighteenth and nineteenth centuries to inflate the numbers of the still politically effective Khoi of Namibia and ultimately to precipitate the wars against the Herero which led to German annexation. Others moved to the north-east, to constitute or augment the !Ora and supplement a pastoral life by brigandage. In the circumstances in which they now were, with available land decreasing in extent and quality, the pastoral life could not have been as profitable or agreeable as it had been in the past. The productivity of the available land had to be exploited to its maximum to ensure comfortable survival. From being commensal with the rest of the ecology, the Khoi were forced to operate in a way which would tend to disturb it and degrade the environment to which they were best adapted.

The Germans undoubtedly played some part in salvaging the ecology of Namibia, though that was hardly their primary object. Loss of life and stock during the wars of 1904–1905 did not impose on the Khoi, as it did on the Herero, a further population bottleneck; Khoi losses were nowhere near matching the depletion of the Herero by 80 per cent of their numbers. Herero land, however, proved more attractive to German settlers than Khoi land did, and the land alienated in southern Namibia went to Afrikaners who on the whole acquired it fairly. It is incidentally noteworthy that both Samuel Maharero and that remarkable mystic Hendrik Witbooi gave instructions that no Afrikaners were to be molested in the course of hostilities. Nevertheless, the German pacification led to further confinement of the Khoi into the least hospitable areas of the country, and consequent over-population, over-stocking and over-grazing. South African policies following the two World Wars accentuated this trend, and the Khoi probably are now as impoverished as they have ever been.

There have never been enough Khoi for their impact on the environment to have been more than just locally considerable. The over-grazing in Namibia, whose causes and consequences have been dealt with at greater length in an earlier chapter, and for which the Khoi are only partly responsible, represents the worst they can do to their surroundings. That worst is hardly advertent; with better opportunities they could have acted much better. There is no indication that before the coming of the Caucasoids and the Negroes the wandering Khoi could ever have brought any long-standing degradation of the land. They were somewhat more exploitative of it than the San were; but there was so much land and they were so mobile that at the earliest signs of diminishing productivity they could move on to somewhere better and leave the land to recover. Nowadays the ebbing utility of those areas they still inhabit helps to accelerate the withdrawal of the Khoi still further from the traditional Khoi way of life. Some of them are moving into towns; many continue the centuries-old drift on to the farms of Caucasoids and absorption into 'Coloured' communities. Those who remain either isolate themselves further or, as their forefathers did, act as repositories for the outcasts from the other peoples. The conflicting trends help to keep their numbers just about constant, and their genetic profile in a state of flux.

### The Caucasoids and the environment

It would appear that the selective effects of the Southern African environment on the Caucasoids are not reflected in the genetic systems most accessible to individual study, except perhaps insofar as some 'African' alleles acquired by admixture may have tended to be selectively retained. Any such process would, of course, bias quantitative estimates of admixture; and we cannot be sure that this has not been the case. The likelihood that any particular gene marker is individually ecosensitive is, however, small; some are known to be involved in resistance and susceptibility to certain diseases, but not always on their own. Epistatic selection, or selection of genes in linkage disequilibrium with those whose products are readily detectable, could have occurred or be occurring; but both epistasis and linkage disequilibrium may vary in strength, and it is perfectly possible that the Caucasoid presence in Southern Africa has not lasted long enough for such effects to be discernible yet.

The prior adaptation of the individual groups, and the more general and less easily quantifiable selective pressures which probably adapted the Afrikaner to the environment of the interior plateau (Fischer 1913), have been discussed earlier. The ability of newcomers to accommodate themselves to environments to which they are not selectively adapted has been greatly enhanced by advances in technology during the nineteenth and twentieth centuries, and will have reduced proportionately the importance of prior adaptation to the individual and the group. Artificial micro-environments approximating to those of the 'home' environments are now fairly easily created through improvements in building techniques and materials and the deployment of relatively inexpensive energy towards heating, cooling and humidification. Clothing, too, has become cheaper and its design more rational, while the range of foodstuffs has widened to the extent that even the most minor dietary changes are no longer necessary.

Much of this has been made possible by deliberate manipulation, even irreparable exploitation, of the original Southern African environment in order to produce the necessary wealth. Deforestation of the interior had started long before the Caucasoids arrived there, but their arrival accelerated it, leading to a declining rainfall. The introduction of large herds and flocks of domesticated cattle into areas with fragile ecosystems have promoted desertification more rapidly than would have been the case had they been exposed only to the originally modest numbers of beasts belonging to the former inhabitants. Large-scale cultivation, often using mechanical means producing more than just a surface disturbance of the topsoil, combined with the substitution of artificial fertilizers for the original fallowing and regeneration, has led to a steady rise in the amount of resource expenditure necessary to keep the land productive. The appetite of Caucasoid farmers for more and larger tracts of land has confined the growing non-Caucasoid farming population to areas where equal exploitation is the price of survival. Negro and hybrid farmers are charged with an improvidence into which they have been forced by political restraints placed on their expansion, but in the long run their improvidence is no greater, and rather more excusable, than that of farmers who enrich themselves while preventing others from gaining more than the barest subsistence.

There has been some belated effort towards restoration, and afforestation is actively pursued with some success though often through the introduction of fast-growing exotic species whose ultimate effects are not yet foreseeable. Caucasoid industry is geared to the rapid utilization both of the large mineral resources and the very much more slender sources of energy. The effects of mining are not confined to the small areas occupied by the mines themselves. They provoke the growth of townships, often in Southern Africa haphazard and unplanned, around them; they provide markets which lead to the ever more intensive exploitation of the agricultural and pastoral potential of the nearby lands. The development of secondary industry, often close to the mining towns, accentuates this trend. It has been the aim of the South African Government to make the country as self-sufficient as possible. Since the Second World War the pace of industrialization has been rapid. To meet its needs primary production has become steadily more directed towards the complete utilization of resources even if this leads to their ultimate exhaustion. South Africa's energy reserves are not great, and for energy the country relies heavily on imported fuel while at the same time endeavouring to secure the most profitable use of what there is within her borders. Considerable ingenuity has been shown in the use made of the latter, but it may be doubted whether, when the mineral wealth has declined to a point, not yet imminent but already foreseeable, where it will no longer be able to finance the importation of fuel, enough of it will remain to sustain the economy at anything like its present level.

Large concentrations of populations, large tracts of land, some of them good agricultural land, covered in factories, eject into the atmosphere and the country-side their inevitable detritus, much of it permanent. Ever-growing dependence on artifical fertilizers and a sinking water-table require a continuation of the economy at its present level or higher. If the means to maintain the present technology is lost, much of the land at present in production will become useless, and a much larger population than inhabited the region formerly will be left without even the slender resources which sustained its ancestors.

These considerations apply mainly to the Republic of South Africa, though they are relevant to some extent to the surrounding countries as well, caught up as they are in the South African economic system, a system devised and fostered by the Caucasoids of the region. There is still a possibility that Botswana and Namibia, if governed wisely, may escape by example the worst consequences of their neighbour's policies; but there is a danger that they, too, may be drawn into a too great or too rapid utilization of their mineral resources and end by following the South African pattern. Lesotho is already too overpopulated and too dependent on the South African economy to be able to escape. Swaziland, with more possibility for diversification, and Moçambique, with a sparser population and large tracts of arable land, are probably in a more favourable position.

Technology has relaxed adaptive selection as far as the Southern African Caucasoids are concerned. The failure of technology is likely to prove catastrophic for them, but not perhaps as catastrophic as it would be for people living in a less temperate climate. Though it would probably lead to the reversion of much of the Cape interior and most of the Highveld to desert, there would still be some arable land left in the Cape and Natal lowlands. The expansion of the populations of all races and mixtures which has already taken place would mean that competition for this land would become intense, and all the populations would be subjected to Darwinian pressures once more. The probable outcome can only be guessed.

## The environmental impact of the Southern African hybrids

The amorphousness of the hybrid peoples of Southern Africa makes it difficult to specify any effects they may have had on the Southern African environment. The Basters, as pastoralists, would have produced through the grazing of their stock effects more considerable than those brought about by the Khoi, who are poorer. These would, however, have been less than the effects produced by the rather richer Caucasoid cattlemen. The Griqua, when they have had cattle, have followed the Khoi rather than the Caucasoid pattern. All the hybrid peoples have necessarily been opportunists, and emulated as far as possible those of their neighbours who appeared to be most successful and whose technology was within their reach. It is consequently impossible to ascribe to them any particular environmental impact independently of those produced by their parent peoples.

## Futures for hybrid peoples: predominance or disappearance

Hitherto the biological function of the hybrid peoples of Southern Africa appears to have been mainly to act as a channel for the flow of genes between the various other communities. This would have been of least importance in the context of gene flow into the San, whose acquisition of Negro and Caucasoid genes has been slight and direct, with no recognition of an intermediate hybrid category. Where hybrids derived from the San do come into prominence is where they form a stage in the flow of San genes into other delimited communities or into the gene pool formed by the general and unstructured 'Cape Coloured' peoples. From this pool gene flow has occurred in all directions, despite the legislative imposition of barriers against its flow into Caucasoids (and, more recently, into other peoples) and despite a partly spontaneous social stratification which militates against it. The degree of gene flow tends to vary geographically, and so cannot easily be quantified. Similarly, the genetic reservoir formed by the hybrid peoples fluctuates in its constitution both spatially and temporally. A few, generally socially and politically marginal, hybrid communities have succeeded in making themselves genetically distinctive, but at the cost of sacrificing some of their more enterprising members to upward social mobility often without compensatory recruitment. On the whole, however, the barriers to easy mixing among hybrid communities have been geographical, and have tended to break down as communications improve.

With the removal of legislated social barriers, one can expect some increase in gene flow, though this will not necessarily be very considerable. As the number of hybrid persons grows, however, the social significance of hybridity may be expected to diminish. The currently dominant Caucasoid societies of Southern Africa are themselves internally stratified, mainly on economic criteria, though a number of other factors, religious, political and pre-eminently once more ethnic, do also to a great extent condition the directions of mating. Economic parity is likely to contribute almost as much towards easing the marital mixing of the races as it has to the slow breakdown of barriers between Boer and Briton. Any change will be gradual, and there will be increasing stress placed on non-racial considerations, but the likelihood of the rapid development of a predominantly hybrid population for the subcontinent is remote. There have been no specifically racial barriers to miscegenation in Hawaii, and the tendency there has been not towards the formation of a large explicitly hybrid population but towards the choosing by individuals of the category in which they wish to be placed, of the society which they wish to join. This does not always correspond very closely with their stated ancestry (Morton et al. 1967).

Morton et al. (1967) have also confirmed that in Hawaii any heterozygote advantage which may occur operates most considerably in the $F_1$ generation and becomes steadily less important as the generations advance; but this may have been due to the large proportion of back-crosses in the populations they studied. Furthermore, the climate and resources of Hawaii do not require such specialized biological adaptation as do the harsh deserts, with their temperature extremes, of the interior of Southern Africa. The situation of the hybrid Hawaiians might be more comparable to that in Moçambique and Swaziland; but then there is no malaria in Hawaii, and conversely little leprosy in Southern Africa; the selective requirements would be quite different between the two regions, and the stage of hybridization likely to be best adapted to the environment would hardly be the same.

Were adaptive selection to continue to operate in Southern Africa as it did during the eighteenth and nineteenth centuries, we could anticipate that with the breakdown of mating barriers the most successful stock to emerge would probably be one rather resembling the Basters. The present century, however, has seen a proliferation of technological aids to adaptation, and these have diminished the immediate, though possibly not the ultimate future, requirements for actual biological adaptation. These technological aids have not been equally accessible to all categories of Southern African hybrid, so that there has recently been considerable variation in the degree of selective pressures to which their different divisions have been exposed. Nevertheless, the period during which they have been available has been too short for them to have much selective impact except under circumstances of catastrophe. Such circumstances could be represented by the geographical and economic variability in access to medical attention, particularly to chemotherapy for acute infectious diseases. It is, therefore, possible that the two quite different types of selection, one purely biological and the other modified by, indeed dependent on, technology, have been acting lately on what we have already seen to be a biologically amorphous set of peoples.

The outcome of these processes is no more foreseeable for the Southern African hybrid peoples than it is for any others. If the Hawaiian social pattern is eventually attained, there could be the progressive extension of miscegenation coupled, as its surreptitious present degree is, with largely optative social membership of racially described groups whose genetic boundaries have become blurred. If there is ever any drive to preserve 'Coloured' awareness, however, the 'Coloureds' could end up as the predominant ethnic element of the region. If racial stratification is supplanted by economic stratification without consideration of race, the 'Coloureds' and Griqua, though possibly not the Basters, are as surely likely to disappear as categories, as the mixed folk of Botswana and Swaziland, and the *mestiços* of Moçambique, are doing at present. On the other hand, there would then be a strong probability that a very large proportion of the population would ultimately be racially mixed, and a situation result resembling that in Brazil rather than that in Hawaii.

# GENETIC POLYMORPHISMS IN
# THE PEOPLES OF
# SOUTHERN AFRICA

Table 1–28, pages 299–346

# Table 1. ABO System

| Population | n | Phenotypes | | | | | | | | Gene Frequencies | | | | | | | Reference |
|---|---|---|---|---|---|---|---|---|---|---|---|---|---|---|---|---|---|
| | | A1 | A2 | Abantu | A1B | A2B | AbantuB | B | O | A1 | A2 | Abantu | B | O | x2 | df | |
| **Bantu-speaking Negroes:** | | | | | | | | | | | | | | | | | |
| **Nguni:** | | | | | | | | | | | | | | | | | |
| Swati | 92 | 15 | 9 | 2 | 2 | 3 | 0 | 9 | 52 | .092 | .075 | .013 | .079 | .737 | 4.747 | 3 | Jenkins (1972) |
| Zulu/Ronga | 123 | 26 | 9 | 1 | 1 | 0 | 0 | 16 | 70 | .122 | .047 | .005 | .072 | .754 | 1.550 | 3 | Nurse et al (1974) |
| Zulu | 91 | 17 | 10 | - | - | - | - | 13 | 49 | .106 | .069 | - | .086 | .739 | .272 | 2 | Present study |
| Ndebele | 252 | 43 | 18 | - | 6 | 12 | - | 57 | 116 | .103 | .069 | - | .161 | .667 | 10.163 | 2 | Lowe (1969) |
| **Sotho/Tswana:** | | | | | | | | | | | | | | | | | |
| S. Sotho (Highland) | 90 | 27 | 6 | 0 | 4 | 1 | 0 | 12 | 40 | .190 | .049 | .000 | .099 | .662 | .186 | 2 | Beaumont et al (1979) |
| S. Sotho (Lowland) | 92 | 15 | 10 | 2 | 1 | 9 | 0 | 19 | 36 | .092 | .121 | .014 | .170 | .604 | 11.421 | 3 | " |
| S. Sotho | 292 | 93 | | | 2 | 22 | | 56 | 121 | .092 | .220 | | .143 | .637 | 1.175 | 3 | Moullec et al (1966) |
| Pedi (Urban) | 238 | 35 | 26 | 4 | 2 | 5 | 1 | 42 | 123 | .081 | .074 | .013 | .111 | .721 | 1.942 | 3 | Jenkins (1972) |
| Pedi (Rural) | 210 | 80 | 23 | 3 | 3 | 7 | 0 | 40 | 103 | .085 | .082 | .009 | .127 | .698 | 3.059 | 3 | Jenkins (1972) |
| Fokeng | 428 | 64 | 64 | 9 | 5 | 19 | 0 | 74 | 177 | .105 | .116 | .014 | .122 | .643 | 9.983 | 3 | Jenkins (1972) |
| Ngwato | 83 | 15 | 5 | 1 | 0 | 2 | 1 | 20 | 40 | .102 | .049 | - | .150 | .699 | 1.760 | 2 | Present study |
| Kgalagadi | 124 | 27 | 7 | - | 0 | 2 | 0 | 23 | 66 | .116 | .038 | .000 | .103 | .743 | 3.751 | 2 | Present study |
| Pedi (Rural) | 107 | 18 | 15 | 0 | 2 | 0 | 0 | 25 | 47 | .098 | .083 | .000 | .137 | .682 | 3.046 | 2 | " |
| **Venda:** | | | | | | | | | | | | | | | | | |
| Urban | 248 | 33 | 26 | 1 | 0 | 6 | 1 | 43 | 138 | .069 | .073 | .005 | .107 | .747 | 8.091 | 3 | Jenkins (1972) |
| Rural | 257 | 31 | 26 | 0 | 1 | 1 | 0 | 54 | 144 | .065 | .059 | .000 | .116 | .761 | 5.046 | 2 | Jenkins (1972) |
| Rural | 105 | 9 | 3 | 0 | 0 | 0 | 0 | 29 | 64 | .044 | .015 | .000 | .150 | .791 | 2.333 | 2 | Present study |
| **Tsonga/Ronga:** | | | | | | | | | | | | | | | | | |
| Ronga | 45 | | 9 | | | | | 5 | 30 | .097 | .118 | | .069 | .813 | .121 | 3 | Matznetter & Spielmann (1969) |
| "Shangaan" | 122 | | 30 | | | 3 | | 20 | 69 | | .146 | .000 | .099 | .755 | .107 | 3 | Matznetter & Spielmann " |
| Tsonga | 1109 | 188 | 72 | 0 | 16 | 12 | 0 | 198 | 623 | .097 | .043 | | .108 | .752 | 2.964 | 2 | Cunha (1968) |
| **Kavango:** | | | | | | | | | | | | | | | | | |
| Mbukushu | 57 | 8 | 1 | 2 | 3 | 2 | 0 | 14 | 29 | .091 | .012 | .000 | .184 | .713 | 4.030 | 2 | Nurse & Jenkins (1977b) |
| Gciriku | 111 | 20 | 10 | 0 | 5 | 3 | 0 | 24 | 49 | .143 | .065 | .000 | .156 | .665 | .65 | 1 | Nurse & Jenkins (1977b) |
| Sambyu | 93 | 21 | 2 | 0 | 2 | 3 | 0 | 26 | 39 | .153 | .016 | .000 | .183 | .648 | .140 | 1 | Nurse & Jenkins (1977b) |
| Kwangali | 107 | 22 | 4 | 0 | 2 | 1 | 0 | 33 | 45 | .138 | .028 | .000 | .185 | .649 | 2.770 | 3 | Nurse & Jenkins (1977b) |
| **Ambo:** | | | | | | | | | | | | | | | | | |
| Kwambi | 107 | 20 | 7 | 2 | 3 | 1 | 0 | 16 | 58 | .114 | .043 | .011 | .098 | .734 | .52 | 2 | Nurse et al (1983) |
| Kwaluudhi | 69 | 16 | 3 | 0 | 2 | 3 | 0 | 24 | 21 | .143 | .053 | .000 | .241 | .563 | 4.97 | 3 | " |
| Ndonga | 113 | 28 | 2 | 0 | 3 | 1 | 1 | 29 | 52 | .131 | .032 | .009 | .159 | .678 | 2.05 | 3 | " |
| Mbalantu | 72 | 18 | 1 | 0 | 5 | 0 | 0 | 19 | 28 | .174 | .008 | .009 | .192 | .617 | .248 | 2 | " |
| Kwanyama | 128 | 25 | 3 | 0 | 5 | 2 | 0 | 36 | 57 | .130 | .017 | .000 | .185 | .667 | 2.56 | 3 | " |
| Ngandjera | 68 | 12 | 1 | 0 | 1 | 1 | 0 | 22 | 31 | .135 | .011 | .000 | .189 | .665 | 3.87 | 2 | " |
| Nkolonkadhi | 32 | 6 | 0 | 0 | 0 | 1 | 0 | 9 | 16 | .100 | .018 | .000 | .171 | .711 | 1.107 | 1 | " |
| **Herero:** | | | | | | | | | | | | | | | | | |
| Botswana | 62 | 11 | 4 | 0 | 1 | 0 | 0 | 14 | 32 | .107 | .033 | .000 | .130 | .730 | 1.089 | 2 | Jenkins (1972) |
| SWA (Namibia) | 89 | 16 | 1 | 0 | 4 | 1 | - | 19 | 48 | .105 | .007 | .008 | .146 | .734 | 1.380 | 1 | Jenkins et al (1983) |
| Mbanderu | 97 | 21 | 4 | 1 | 3 | | | 21 | 49 | .125 | .026 | - | .127 | .721 | 1.094 | 2 | " |
| Himba | 86 | 22 | 0 | 0 | | | | 10 | 51 | .151 | .000 | .000 | .079 | .770 | .590 | 1 | " |
| Chimba | 33 | 3 | 1 | 0 | | | | 0 | 29 | .047 | .016 | .000 | .000 | .937 | .650 | 1 | " |

299

# Table 1 (cont.)

| Population | n | Phenotypes | | | | | | | | Gene Frequencies | | | | | $\chi^2$ | df | Reference |
|---|---|---|---|---|---|---|---|---|---|---|---|---|---|---|---|---|---|
| | | $A_1$ | $A_2$ | $A_{bantu}$ | $A_1B$ | $A_2B$ | $A_{bantu}B$ | B | O | $A^1$ | $A^2$ | $A_{bantu}$ | B | O | | | |
| **Khoisan-speaking Negroes:** | | | | | | | | | | | | | | | | | |
| Sarwa | 79 | 11 | 5 | 1 | 2 | 1 | 2 | 16 | 41 | .042 | .085 | .009 | .143 | .721 | 1.760 | 1 | Chasko et al (1979) |
| Kwengo | 35 | 12 | 3 | 0 | 0 | | 0 | 10 | 10 | .169 | .075 | .000 | .221 | .535 | 5.540 | 2 | Nurse & Jenkins (1977b) |
| Dama | 448 | 38 | 5 | 0 | 14 | 4 | 0 | 177 | 210 | .061 | .006 | .000 | .249 | .685 | .350 | 1 | Krussmann (1968) |
| Dama | 119 | 16 | 0 | 0 | 9 | 0 | 0 | 47 | 47 | .099 | .000 | .000 | .272 | .629 | 1.120 | 1 | Nurse et al (1976) |
| **Khoi:** | | | | | | | | | | | | | | | | | |
| Keetmanshoop Nama | 153 | 36 | 14 | 5 | 2 | 17 | 0 | 28 | 51 | .120 | .126 | .022 | .171 | .561 | 28.02 | 4 | Jenkins (1972) |
| Keetmanshoop Nama | 398 | 94 | 34 | 9 | 6 | 32 | 0 | 98 | 125 | .137 | .103 | .015 | .189 | .557 | 38.95 | 3 | " |
| Topnaars (Kuiseb) | 58 | | 15 | | | 9 | | 20 | 14 | | .229 | | .218 | .553 | 2.93 | 1 | Jenkins & Brain (1967) |
| Sesfontein Nama | 42 | 9 | | 0 | 1 | 2 | | 14 | 16 | .129 | .028 | .000 | .228 | .615 | 1.118 | 1 | Jenkins (1972) |
| **San:** | | | | | | | | | | | | | | | | | |
| **Northern:** | | | | | | | | | | | | | | | | | |
| Glanglai | 36 | 11 | 1 | 6 | 1 | | | 2 | 15 | .176 | .018 | .118 | .042 | .646 | .000 | 1 | Nurse & Jenkins (1977a) |
| Tsumkwe !Kung | 122 | 41 | 4 | 8 | | | | 5 | 64 | .190 | .021 | .044 | .021 | .724 | 1.660 | 1 | Nurse et al (1977) |
| Samanlai !kaiKung | 34 | 11 | 0 | 2 | | | | 5 | 16 | .196 | .000 | .042 | .077 | .686 | 1.610 | 1 | " |
| Dobe | 436 | 181 | 10 | 25 | 2 | 2 | | 13 | 203 | .239 | .018 | .040 | .020 | .684 | 2.461 | 2 | Jenkins (1972) |
| /Ai/ai | 65 | 28 | 4 | 0 | | | | 2 | 31 | .247 | .042 | .000 | .016 | .696 | .833 | 1 | " |
| /Du/da | 102 | 33 | 1 | 4 | | | | 2 | 62 | .178 | .006 | .024 | .009 | .782 | .199 | 1 | " |
| //Au//en | 264 | 79 | *2 | 5 | 0 | 1 | | 16 | 161 | .163 | *.007 | .011 | .033 | .785 | 5.000 | 2 | Nurse et al (1977) |
| **Central:** | | | | | | | | | | | | | | | | | |
| Hei//om | 96 | 17 | 4 | 3 | 1 | | | 10 | 61 | .099 | .025 | .019 | .059 | .797 | .030 | 1 | Nurse et al (1977) |
| Nharo | 149 | 39 | *4 | 4 | 1 | | | 3 | 98 | .141 | .020 | .016 | .013 | .809 | .230 | 1 | " |
| G/wi | 94 | 16 | 16 | 6 | 1 | | | 2 | 52 | .091 | .102 | .042 | .021 | .744 | .000 | 1 | Jenkins et al (1975) |
| G//ana | 50 | 2 | 1 | 7 | 1 | | | 2 | 37 | .021 | .011 | .078 | .030 | .860 | -1.81 | 1 | Jenkins et al (1975) |
| **Southern:** | | | | | | | | | | | | | | | | | |
| !Xõ | 51 | 3 | 7 | 2 | 0 | | 2 | 1 | 36 | .040 | .099 | .036 | .037 | .788 | 5.24 | 1 | Nurse & Jenkins (1977a) |
| G!aokx'ate | 33 | 12 | 3 | 5 | 0 | | | 1 | 12 | .206 | .061 | .115 | .015 | .603 | .660 | 1 | " |
| **Uncertain:** | | | | | | | | | | | | | | | | | |
| Eastern ǂHuã | 36 | 0 | 2 | 3 | 0 | 0 | 0 | 3 | 28 | .000 | .029 | .046 | .043 | .882 | .021 | 1 | Nurse & Jenkins (1977a) |
| **Caucasoid:** | | | | | | | | | | | | | | | | | |
| **Afrikaans-speaking:** | | | | | | | | | | | | | | | | | |
| Cape Town | 761 | 289 | | 0 | 18 | | | 92 | 362 | .180 | .229 | .000 | .075 | .696 | 52.290 | 1 | Botha (1972) |
| Johannesburg | 77 | 24 | 10 | 0 | 3 | | | 13 | 27 | | .103 | .000 | .111 | .606 | 1.911 | 2 | Present study |
| Johannesburg | 160 | 52 | 9 | 0 | 8 | | | 18 | 73 | .204 | .041 | | .085 | .670 | .407 | 1 | " |
| SWA/Namibia | 56 | 17 | 6 | 0 | 2 | | | 4 | 27 | .177 | .073 | | .055 | .694 | -.550 | 2 | Palmhert-Keller et al (1983) |
| **English-speaking:** | | | | | | | | | | | | | | | | | |
| Cape Town | 467 | 183 | | 0 | 18 | | | 54 | 212 | .192 | .245 | | .080 | .674 | .017 | 1 | Botha (1972) |
| **German-speaking:** | | | | | | | | | | | | | | | | | |
| SWA/Namibia | 118 | 37 | 10 | 0 | 7 | | | 13 | 51 | .192 | .062 | .000 | .089 | .657 | 1.720 | 3 | Palmhert-Keller et al (1983) |

| | n | | | | | | | | | | | | | | | | Reference |
|---|---|---|---|---|---|---|---|---|---|---|---|---|---|---|---|---|---|
| **Indian:** | | | | | | | | | | | | | | | | | |
| _Transvaal:_ | | | | | | | | | | | | | | | | | |
| Hindu | 53 | 6 | 3 | – | 2 | 1 | – | 24 | 17 | .086 | .048 | – | .300 | .566 | .420 | 2 | Present study |
| Moslem | 56 | 14 | 2 | – | 3 | 4 | – | 17 | 16 | .189 | .033 | – | .244 | .534 | .180 | 1 | " |
| **Natal:** | | | | | | | | | | | | | | | | | |
| Tamil-speaking | 197 | 34 | 9 | – | 7 | 3 | – | 71 | 73 | .111 | .035 | | .234 | .620 | 1.060 | 1 | Moores (1980) |
| Telegu-speaking | 101 | 14 | 4 | – | 5 | 2 | – | 32 | 44 | .099 | .033 | | .215 | .653 | .450 | 1 | " |
| Hindi-speaking | 126 | 27 | 2 | – | 6 | 1 | – | 45 | 45 | .142 | .014 | | .235 | .609 | .570 | 1 | " |
| Moslem | 242 | 48 | 7 | – | 19 | 6 | – | 80 | 82 | .149 | .032 | | .246 | .573 | 1.190 | 1 | " |
| **Hybrid Peoples:** | | | | | | | | | | | | | | | | | |
| _Baster-like:_ | | | | | | | | | | | | | | | | | |
| Rehoboth | 120 | 41 | 8 | 1 | 7 | 3 | 0 | 16 | 44 | .221 | .052 | .007 | .115 | .605 | 11.330 | 3 | Nurse et al (1982) |
| !Kuboes | 137 | 28 | 6 | – | 0 | – | – | 22 | 73 | .109 | .059 | – | .116 | .717 | 32.603 | 3 | Jenkins (1972) |
| Riemvaasmaak | 89 | 26 | 6 | – | {4 | 8 | – | 14 | 39 | .182 | .049 | – | .107 | .662 | .005 | 1 | Nurse & Jenkins (1978) |
| Griqua | 267 | 66 | 17** | 16 | {21 | – | | 48 | 99 | .158 | .047† | .047 | .139 | .609 | .39 | 1 | Nurse & Jenkins (1975) |
| _'Cape Coloureds'_ | | | | | | | | | | | | | | | | | |
| Cape Town | 714 | {223 | – | – | {59 | – | – | 149 | 283 | {.182 | {.221 | – | .157 | .622 | 2.671 | 1 | Botha (1972) |
| Johannesburg | 109 | 31 | 4 | 2 | 5 | 3 | 0 | 23 | 41 | | .040 | .012 | .154 | .612 | 3.233 | 3 | Jenkins (1972) |
| Orange River Valley | 179 | 34 | 16 | 4 | 4 | 11 | 0 | 39 | 71 | .113 | .089 | .014 | .164 | .620 | 10.336 | 3 | Jenkins " |

*(A₂+A$_{int}$); **includes 1 individual A$_{int}$; †includes A$^{int}$ with frequency of .003

301

# Table 2. MNSs System

| Population | n | MS | MSHe | MSs | MSsHe | Ms | MsHe | MNS | MNSHe | MNSs | MNSsHe | MNs | MNsHe | NS | NSHe | NSs | NSsHe | Ns | NsHe |
|---|---|---|---|---|---|---|---|---|---|---|---|---|---|---|---|---|---|---|---|
| **Bantu-speaking Negroes:** | | | | | | | | | | | | | | | | | | | |
| *Nguni:* | | | | | | | | | | | | | | | | | | | |
| Swati | 92 | 39 | | 5 | | 27 | | | | 13 | | 26 | | | | 5 | | 12 | |
| Zulu | 91 | | | | 6 | 15 | | | 4 | | 13 | | | | 2 | | 3 | | 9 |
| Zulu/Tonga | 123 | 4 | 1 | 16 | 0 | 13 | 0 | 8 | 0 | 22 | 2 | 30 | 0 | 6 | 0 | 6 | 1 | 13 | 1 |
| *Sotho/Tswana:* | | | | | | | | | | | | | | | | | | | |
| S.Sotho (Highland) | 89 | 7 | 3 | | | 14 | 0 | 10 | 4 | | | 34 | 0 | 5 | 1 | | | 11 | 0 |
| S.Sotho (Lowland) | 91 | 9 | 5 | | | 16 | 0 | 16 | 7 | | | 29 | 0 | 2 | 0 | | | 7 | 0 |
| S.Sotho | 292 | | | 23 | | 62 | | | | 52 | | 93 | | | | 16 | | 46 | |
| Pedi (rural) | 116 | 15 | 3 | | | 14 | 0 | 26 | 1 | | | 27 | 1 | 10 | 0 | | | 19 | 0 |
| Pedi (urban) | 130 | 16 | 2 | | | 11 | 0 | 23 | 4 | | | 48 | 0 | 7 | 0 | | | 19 | 0 |
| Fokeng | 119 | 20 | 4 | | | 17 | 0 | 25 | 2 | | | 31 | 0 | 5 | 0 | | | 15 | 0 |
| Ngwato | 95 | 3 | 0 | 8 | 3 | 21 | 0 | 2 | 1 | 9 | 3 | 26 | 0 | 1 | 0 | 3 | 0 | 15 | 0 |
| Kgalagadi | 124 | 8 | 5 | | | 27 | | 27 | 5 | | | 33 | | 6 | 2 | | | 11 | |
| *Venda:* | | | | | | | | | | | | | | | | | | | |
| Rural | 75 | | | 5 | | 17 | | 18 | | | | 18 | | | | 6 | | 11 | |
| Urban | 97 | | 8 | | | 31 | | | 15 | | | | 28 | | | | 5 | | 10 |
| Rural | 105 | 0 | 0 | 5 | 0 | 28 | 0 | 1 | 0 | 10 | 0 | 35 | 2 | 1 | 1 | 5 | 0 | 16 | 1 |
| *Tsonga/Ronga:* | | | | | | | | | | | | | | | | | | | |
| Ronga | 45 | | 1 | | 4 | 4 | | | 2 | | 9 | 16 | | | | 2 | 0 | 7 | |
| "Shangaan"/Tsonga | 122 | 4 | | 9 | | 23 | | 5 | | 14 | | 44 | | | | 2 | 11 | 10 | |
| *Kavango:* | | | | | | | | | | | | | | | | | | | |
| Mbukushu | 57 | 2 | 0 | 9 | 0 | 14 | 0 | 0 | 1 | 4 | 0 | 16 | 1 | 0 | 2 | 0 | | 8 | 0 |
| Gciriku | 111 | 2 | 0 | 6 | 0 | 21 | 0 | 1 | 1 | 12 | 2 | 47 | 0 | 0 | 2 | 4 | 1 | 12 | 0 |
| Sambyu | 94 | 2 | 1 | 10 | 1 | 24 | 0 | 0 | 0 | 12 | 0 | 33 | 0 | 0 | 2 | 2 | 0 | 9 | 0 |
| Kwangali | 87 | 3 | 0 | 11 | 0 | 16 | 3 | 1 | 0 | 15 | 3 | 19 | 2 | 1 | 0 | 5 | 0 | 8 | 0 |
| *Ambo:* | | | | | | | | | | | | | | | | | | | |
| Kwambi | 107 | 1 | 0 | 2 | 0 | 29 | 0 | 3 | 0 | 12 | 2 | 33 | 0 | 0 | 2 | 7 | 0 | 16 | 0 |
| Kwaluudhi | 69 | 2 | 0 | 6 | 0 | 14 | 0 | 1 | 1 | 11 | 0 | 19 | 1 | 0 | 0 | 5 | 0 | 9 | 0 |
| Ndonga | 113 | 4* | 1 | 5 | 0 | 20 | 1 | 4 | 2 | 16 | 4 | 34 | 0 | 1 | 0 | 4 | 0 | 17 | 0 |
| Mbalantu | 73 | 2 | 0 | 8 | 0 | 15 | 0 | 2 | 2 | 8 | 3 | 19 | 0 | 0 | 0 | 5 | 2 | 7 | 0 |
| Kwanyama | 128 | 0 | 0 | 8 | 1 | 18 | 1 | 4 | 1 | 21 | 5 | 40 | 2 | 2 | 1 | 5 | 4 | 7 | 0 |
| Ngandjera | 69 | 0 | 0 | 7 | 1 | 11 | 1 | 1 | 1 | 17 | 3 | 19 | 0 | 0 | 1 | 2 | 1 | 15 | 0 |
| Nkolonkathi | 32 | 0 | 0 | 2 | 0 | 5 | 0 | 3 | 1 | 4 | 0 | 6 | 0 | 0 | 1 | 5 | 0 | 5 | 0 |
| *Herero:* | | | | | | | | | | | | | | | | | | | |
| Botswana | 46 | 2 | 0 | | | 15 | 0 | 5 | 0 | 13 | 2 | | | 2 | 0 | 7 | 0 | | |
| S.W.A.(Namibia) | 89 | 0 | 0 | 5 | 0 | 17 | 0 | 1 | 0 | 7 | 2 | 41 | 1 | 3 | 0 | 2 | 0 | 10 | 0 |
| Mbanderu | 97 | 5 | | 4 | | 36 | | 3 | | 4 | | 29 | | 0 | | 2 | | 14 | |
| Himba | 86 | 0 | 0 | 3 | 0 | 17 | 1 | 2 | 0 | 9 | 2 | 33 | 0 | 0 | 0 | 3 | 1 | 15 | 0 |
| Chimba | 33 | 0 | 0 | 1 | 0 | 6 | 0 | 0 | 0 | 0 | 0 | 9 | 0 | 0 | 0 | 3 | 0 | 14 | 0 |
| **Khoisan-speaking Negroes:** | | | | | | | | | | | | | | | | | | | |
| Sarwa | 79 | 0 | 0 | 1 | 0 | 24 | 0 | 1 | 0 | 2 | 0 | 33 | 0 | 0 | 0 | 3 | 0 | 15 | 0 |
| Kwengo | 36 | 2 | 0 | 4 | 1 | 14 | 0 | 1 | 1 | 3 | 0 | 8 | 0 | 0 | 0 | 0 | 0 | 2 | 0 |
| Dama | 119 | 1 | 0 | 4 | 0 | 17 | 0 | 1 | 0 | 8 | 1 | 51 | 0 | 2 | 2 | 3 | 2 | 27 | 0 |
| **Khoisan:** | | | | | | | | | | | | | | | | | | | |
| *Khoi:* | | | | | | | | | | | | | | | | | | | |
| Keetmanshoop Nama | 391 | | | 89 | | 76 | | | | 73 | | 113 | | | | 12 | | 28 | |
| Keetmanshoop Nama | 153 | | | 40 | | 40 | | | | 19 | | 39 | | | | 5 | | 10 | |
| Sesfontein Nama | 42 | | | 3 | | 3 | | | | 8 | | 15 | | | | 5 | | 8 | |
| **San:** | | | | | | | | | | | | | | | | | | | |
| *Northern:* | | | | | | | | | | | | | | | | | | | |
| Glanglai | 36 | 1 | 0 | | | 5 | | 0 | 3 | | | 12 | 0 | 0 | 0 | | | 15 | 0 |
| Tsumkwe !Kung | 122 | 1 | 1 | | | 18 | | 4 | 9 | | | 49 | 0 | 1 | 1 | | | 38 | 0 |
| Saman!aika " | 34 | *1 | 1 | | | 11 | | 0 | 1 | | | 13 | 0 | 0 | 0 | | | 7 | 0 |
| Dobe " | 258 | | | 25 | | 58 | | | | 42 | | 75 | | | | 4 | | 54 | |
| /Ai/ai " | 65 | | | 5 | | 17 | | | | 4 | | 27 | | | | 1 | | 11 | |
| /Du/du " | 90 | | | 2 | | 10 | | | | 6 | | 32 | | | | 2 | | 38 | |
| //Au//en " | 111 | 3 | 10 | | | 17 | | 7 | 11 | | | 32 | 1 | 2 | 6 | | | 22 | 0 |
| *Central:* | | | | | | | | | | | | | | | | | | | |
| Hei//om | 96 | 9 | 5 | | | 19 | | 7 | 7 | | | 37 | 0 | 0 | 0 | | | 11 | 1 |
| Nharo | 125 | 10 | 10 | | | 32 | | 14 | 2 | | | 45 | 0 | 0 | 0 | | | 12 | 0 |
| G/wi | 94 | 0 | 0 | 6 | 2 | 21 | | 1 | 1 | 11 | 1 | 32 | | | | 3 | | 16 | 0 |
| //ana | 50 | 2 | 0 | 8 | 0 | 13 | | 0 | 0 | 10 | 3 | 9 | | | | 2 | | 3 | 0 |
| *Southern:* | | | | | | | | | | | | | | | | | | | |
| !Xõ | 51 | 2 | 4 | | | 14 | | 1 | 1 | | | 23 | 0 | 0 | 0 | | | 6 | 0 |
| G!aokx'ate | 33 | 1 | 8 | | | 2 | | 5 | 3 | | | 12 | 0 | 1 | 0 | | | 1 | 0 |
| *Uncertain:* | | | | | | | | | | | | | | | | | | | |
| Eastern ╪Huã | 35 | 2 | 1 | | | 8 | | 2 | 0 | | | 11 | 0 | 1 | 1 | | 0 | 9 | 0 |
| **Caucasoid:** | | | | | | | | | | | | | | | | | | | |
| *Afrikaners:* | | | | | | | | | | | | | | | | | | | |
| Johannesburg | 77 | 1 | | 6 | | 7 | | 4 | | 19 | | 21 | | 0 | | 7 | | 12 | |
| Johannesburg | 160 | 7 | | 22 | | 15 | | 3 | | 34 | | 45 | | 3 | | 6 | | 25 | |
| SWA/Namibia | 57 | 2 | | 10 | | 5 | | 3 | | 14 | | 12 | | 0 | | 0 | | 11 | |

| | | Gene Frequencies | | | | | | | | |
|---|---|---|---|---|---|---|---|---|---|---|
| MS | MSHe | Ms | MsHe | NS | NSHe | Ns | NsHe | $\chi^2$ | df | Reference |

| MS | MSHe | Ms | MsHe | NS | NSHe | Ns | NsHe | $\chi^2$ | df | Reference |
|---|---|---|---|---|---|---|---|---|---|---|
| .125 | | .478 | | .038 | | .359 | | 7.364 | 2 | Jenkins (1972) |
| .545 | | .207 | | .070 | | .178 | | 76.48 | 5 | Present study |
| .171 | .010 | .348 | - | .137 | - | .321 | .013 | 11.430 | 7 | Nurse *et al* (1974) |
| | | | | | | | | | | |
| .077 | .034 | .428 | - | .063 | .012 | .386 | - | 1.515 | 2 | Jenkins (1972) |
| .111 | .071 | .434 | - | .061 | - | .323 | - | 4.503 | 2 | " " |
| .093 | | .446 | | .077 | | .384 | | 3.014 | 2 | Moullec *et al* (1966) |
| .182 | .016 | .309 | .006 | .082 | - | .405 | - | .382 | 2 | Jenkins (1972) |
| .142 | .023 | .346 | - | .060 | - | .428 | - | 4.671 | 2 | " " |
| .214 | .026 | .349 | - | .025 | - | .387 | - | 4.551 | 2 | " " |
| .113 | .037 | .435 | - | .061 | - | .355 | - | 9.206 | 8 | Present study |
| .079 | .033 | .473 | - | .114 | .017 | .284 | - | 3.681 | | " " |
| | | | | | | | | | | |
| .084 | | .450 | | .133 | | .334 | | 3.358 | 2 | Jenkins (1972) |
| .072 | | .551 | | .084 | | .292 | | .855 | 2 | " " |
| .042 | .000 | .501 | .000 | .077 | .005 | .361 | .014 | 1.239 | 3 | Present study |
| .119 | | .342 | | .136 | | .403 | | .717 | 4 | Matznetter & Spielmann (1969) |
| .124 | | .433 | | .116 | | .327 | | 9.838 | 7 | " " " " |
| | | | | | | | | | | |
| .132 | - | .491 | .009 | .026 | .009 | .333 | - | 4.18 | 2 | Nurse & Jenkins (1977b) |
| .077 | - | .468 | - | .063 | .027 | .365 | - | 8.89 | 5 | " " " " |
| .106 | .011 | .527 | - | .048 | - | .308 | - | 2.55 | 5 | " " " " |
| .134 | - | .435 | .051 | .089 | - | .291 | - | 8.38 | 7 | " " " " |
| | | | | | | | | | | |
| .056 | - | .476 | - | .089 | .019 | .360 | - | 9.83 | 5 | Nurse *et al* (1983) |
| .145 | - | .413 | .007 | .062 | .007 | .362 | - | 2.42 | 3 | " " " " |
| .137 | .027 | .392 | .009 | .071 | - | .364 | - | 10.47 | 5 | " " " " |
| .144 | - | .432 | - | .068 | .048 | .308 | - | 4.41 | 7 | " " " " |
| .109 | - | .371 | .023 | .102 | .035 | .359 | - | 7.60 | 10 | " " " " |
| .145 | - | .413 | .022 | .086 | .036 | .298 | - | 7.40 | 6 | " " " " |
| .109 | - | .328 | - | .188 | .031 | .344 | - | 5.80 | 4 | " " " " |
| | | | | | | | | | | |
| .040 | - | .547 | - | .063 | - | .333 | .017 | .873 | 2 | Jenkins (1972) |
| .056 | - | .460 | .014 | .060 | .013 | .397 | - | 7.760 | 4 | Jenkins *et al* (1983) |
| .101 | - | .549 | - | .033 | - | .317 | - | 36.669 | | " " " " |
| .065 | - | .447 | .014 | .054 | .014 | .406 | - | 2.560 | 5 | " " " " |
| .013 | - | .346 | - | .054 | - | .587 | - | 6.020 | 2 | " " " " |
| | | | | | | | | | | |
| .019 | - | .525 | - | .032 | - | .424 | - | 5.14 | 3 | Chasko *et al* (1979) |
| .141 | .029 | .592 | - | - | - | .238 | - | 3.39 | 2 | Nurse & Jenkins (1977b) |
| .049 | - | .404 | - | .069 | .022 | .456 | - | 7.31 | 5 | Nurse *et al* (1976) |
| | | | | | | | | | | |
| .205 | | .455 | | .050 | | .291 | | 1.974 | 2 | Jenkins (1972) |
| .197 | | .516 | | .041 | | .247 | | 2.089 | 2 | " " " " |
| .099 | | .318 | | .114 | | .469 | | 1.024 | 2 | " " " " |
| | | | | | | | | | | |
| .014 | .044 | .323 | .000 | .000 | .000 | .619 | .000 | 2.38 | 2 | Nurse *et al* (1977) |
| .107 | .028 | .304 | - | .019 | .008 | .534 | .000 | 3.97 | 2 | " " " " |
| .168 | .037 | .405 | | .000 | .000 | .390 | .000 | 1.59 | 3 | " " " " |
| .123 | | .425 | | .025 | | .427 | | 10.935 | 2 | Jenkins (1972) |
| .065 | | .512 | | .016 | | .407 | | .150 | 2 | " " " " |
| .039 | | .305 | | .018 | | .638 | | .696 | 2 | " " " " |
| .042 | .081 | .377 | | .024 | .000 | .431 | .045 | 3.90 | 4 | Nurse *et al* (1977) |
| | | | | | | | | | | |
| .172 | .073 | .379 | | .019 | .000 | .338 | .019 | 7.58 | 6 | Nurse *et al* (1977) |
| .095 | .060 | .505 | | .000 | .000 | .340 | .000 | 7.48 | 3 | " " " " |
| .069 | .023 | .489 | | .048 | | .373 | | 1.540 | 4 | Jenkins *et al* (1975) |
| .164 | .028 | .507 | | .037 | | .264 | | 7.030 | 4 | " " " " |
| | | | | | | | | | | |
| .033 | .049 | .555 | | .000 | .000 | .363 | .000 | 1.18 | 3 | Nurse & Jenkins (1977a) |
| .056 | .149 | .417 | | .057 | .000 | .321 | .000 | 10.32 | 5 | Nurse *et al* (1977) |
| | | | | | | | | | | |
| .046 | .014 | .426 | | .058 | .014 | .442 | .000 | 4.56 | 3 | Nurse & Jenkins (1977a) |
| | | | | | | | | | | |
| .158 | | .309 | | .114 | | .418 | | 3.27 | 5 | Present study |
| .205 | | .326 | | .070 | | .399 | | 8.31 | 5 | " " " " |
| .252 | | .299 | | .045 | | .404 | | 5.740 | 4 | Palmhert-Keller *et al* (1983) |

# Table 2 (*cont.*)

| Population | n | Phenotypes | | | | | | | | | | | | | | | | | |
|---|---|---|---|---|---|---|---|---|---|---|---|---|---|---|---|---|---|---|---|
| | | MS | MSHe | MSs | MSsHe | Ms | MsHe | MNS | MNSHe | MNSs | MNSsHe | MNs | MNsHe | NS | NSHe | NSs | NSHe | Ns | NsHe |
| **Caucasoid (cont):** | | | | | | | | | | | | | | | | | | | |
| *German-speaking:* | | | | | | | | | | | | | | | | | | | |
| SWA/Namibia | 118 | 5 | | 16 | | 11 | 1 | | | 31 | | 27 | | 0 | | 6 | | 21 | |
| *Indian:* | | | | | | | | | | | | | | | | | | | |
| Transvaal: | | | | | | | | | | | | | | | | | | | |
| Hindu | 53 | 1 | | 14 | | 10 | 6 | | | 6 | | 10 | | 0 | | 0 | | 6 | |
| Moslem | 56 | 2 | | 9 | | 15 | 2 | | | 4 | | 13 | | 0 | | 0 | | 11 | |
| Natal: | | | | | | | | | | | | | | | | | | | |
| Tamil-speaking | 199 | 11 | | 33 | | 29 | 8 | | | 40 | | 48 | | 3 | | 12 | | 15 | |
| Telegu-speaking | 102 | 10 | | 19 | | 12 | 5 | | | 19 | | 23 | | 1 | | 4 | | 9 | |
| Hindi-speaking | 113 | 8 | | 19 | | 20 | 3 | | | 19 | | 29 | | 1 | | 7 | | 7 | |
| Moslem | 204 | 13 | | 35 | | 27 | 12 | | | 47 | | 41 | | 3 | | 7 | | 19 | |
| **Hybrid Peoples** | | | | | | | | | | | | | | | | | | | |
| *Baster-Like:* | | | | | | | | | | | | | | | | | | | |
| Rehoboth | 120 | 1 | 2 | 10 | 3 | 15 | | 2 | 2 | 12 | 5 | 48 | 3 | 1 | 1 | 1 | 2 | 13 | |
| !Kuboes | 132 | | 20 | | | 33 | | | 24 | | | 33 | | | 10 | | | 12 | |
| Riemvaasmaak | 89 | 3 | 2 | 7 | 2 | 7 | 0 | 1 | 1 | 7 | 4 | 33 | 0 | 1 | 1 | 0 | 1 | 19 | 0 |
| **Griqua:** | 264 | 3 | 5 | 26 | 13 | 52 | 0 | 5 | 1 | 31 | 8 | 87 | 0 | 1 | 0 | 5 | 0 | 27 | 0 |
| *"Cape Coloureds":* | | | | | | | | | | | | | | | | | | | |
| Johannesburg | 109 | | 20 | | | 25 | | | 22 | | | 26 | | | 3 | | | 13 | |

*includes one MS[u] individual

+ *NS*[u] gene frequency of .080 excluded

304

| | | Gene Frequencies | | | | | | $\chi^2$ | df | Reference |
|---|---|---|---|---|---|---|---|---|---|---|
| MS | MSHe | Ms | MsHe | NS | NSHe | Ns | NsHe | | | |
| .220 | | .301 | | .055 | | .424 | | 3.36 | 5 | Palmhert-Keller *et al* (1983) |
| .321 | | .358 | | .000 | | .321 | | 5.960 | 3 | Present study |
| .187 | | .447 | | .000 | | .366 | | 4.570 | 3 | " " |
| .214 | | .394 | | .110 | | .282 | | 1.50 | 2 | Moores (1980) |
| .212 | | .421 | | .151 | | .217 | | 13.50 | 2 | " " |
| .210 | | .431 | | .095 | | .264 | | 2.80 | 2 | " " |
| .264 | | .349 | | .092 | | .295 | | 3.40 | 2 | " " |
| .080 | .037 | .416 | | .057 | .024 | .305† | .000 | 11.330 | 6 | Nurse *et al* (1982) |
| | .134 | .461 | | .089 | | .316 | | 2.876 | 2 | Jenkins (1972) |
| .118 | .039 | .337 | .000 | .028 | .028 | .450 | .000 | 14.11 | 7 | Nurse & Jenkins (1978) |
| .117 | .055 | .452 | .000 | .039 | .000 | .338 | .000 | .000 | 8 | Nurse & Jenkins (1975) |
| .180 | | .454 | | .059 | | .308 | | 2.337 | 2 | Jenkins (1972) |

# Table 3. Rhesus System

| Population | n | ccDee | CcDee | ccDEe | CcDEe | CCDee | ccDEE | ccdee | Ccdee | ccdEe |
|---|---|---|---|---|---|---|---|---|---|---|
| **Bantu-speaking Negroes:** | | | | | | | | | | |
| *Nguni:* | | | | | | | | | | |
| Swati | 92 | 63 | 7 | 11 | 3 | 1 | 0 | 6 | 1 | |
| Zulu/Tonga | 123 | 81 | 19 | 19 | 3 | 0 | 1 | | | |
| Zulu | 403 | 256 | 55 | 45 | 22 | 0 | 0 | 16 | 9 | 0 |
| Zulu | 91 | 60 | 13 | 13 | 0 | 1 | 1 | 3 | | |
| *Sotho/Tswana:* | | | | | | | | | | |
| S.Sotho | 292 | 215 | 23 | 45 | 2 | 0 | | 7 | 0 | 0 |
| S.Sotho(Highlands) | 87 | 62 | 10 | 11 | 3 | 0 | 0 | 1 | 0 | 0 |
| S.Sotho(Lowlands) | 91 | 73 | 6 | 12 | 0 | 0 | 0 | 0 | 0 | 0 |
| Pedi(urban) | 237 | 155 | 39 | 33 | 2 | 0 | 2 | 6 | 0 | |
| Pedi(rural) | 115 | 85 | 11 | 16 | 2 | 0 | 0 | 1 | 0 | |
| Pedi | 106 | 76 | 9 | 12 | 4 | 0 | 1 | 3 | 1 | 0 |
| Fokeng | 119 | 87 | 15 | 13 | 1 | 0 | 1 | 2 | 0 | 0 |
| Ngwato | 95 | 70 | 8 | 9 | 1 | 0 | 3 | 4 | | |
| Kgalagadi | 124 | 83 | 5 | 20 | 0 | 0 | 9 | 6 | 1 | |
| *Venda:* | | | | | | | | | | |
| Rural | 105 | 72 | 7 | 12 | 4 | 0 | 4 | 2 | 2 | |
| *Tsonga/Ronga:* | | | | | | | | | | |
| Ronga | 45 | 30 | 4 | 6 | 3 | 0 | 0 | 2 | 0 | 0 |
| "Shangaan"/Tsonga | 121 | 84 | 15 | 15 | 2 | 0 | 1 | 2 | 2 | 0 |
| *Kavango:* | | | | | | | | | | |
| Mbukushu | 56 | 35 | 6 | 10 | 1 | | 4 | 0 | 0 | |
| Gciriku | 110 | 75 | 11 | 17 | 2 | | 2 | 3 | 0 | |
| Sambyu | 94 | 64 | 6 | 19 | 2 | | 2 | 0 | 1 | |
| Kwangali | 106 | 69 | 9 | 23 | 1 | | 2 | 2 | 0 | |
| *Ambo:* | | | | | | | | | | |
| Kwambi | 107 | 79 | 10 | 12 | 1 | 0 | 5 | 0 | 0 | |
| Kwaluudhi | 69 | 45 | 11 | 11 | 0 | 1 | 1 | 0 | 0 | |
| Ndonga | 113 | 74 | 11 | 23 | 1 | 0 | 4 | 0 | 0 | |
| Mbalantu | 72 | 48 | 8 | 14 | 1 | 0 | 1 | 0 | 0 | |
| Kwanyama | 128 | 69 | 19 | 25 | 5 | 1 | 7 | 1 | 1 | |
| Ngandjera | 69 | 47 | 9 | 7 | 2 | 0 | 3 | 1 | 0 | |
| Nkolonkadhi | 32 | 22 | 6 | 4 | 0 | 0 | 0 | 0 | 0 | |
| *Herero:* | | | | | | | | | | |
| Botswana | 61 | 40 | 10 | 11 | 0 | 0 | 0 | | | |
| SWA/Namibia | 89 | 46 | 19 | 13 | 5 | 2 | 4 | 0 | 0 | 0 |
| Mbanderu | 97 | 49 | 22 | 9 | 5 | 2 | 7 | 2 | | |
| Himba | 86 | 41 | 16 | 20 | 5 | 0 | 4 | 0 | 0 | 0 |
| Chimba | 33 | 25 | 8 | 0 | 0 | 0 | 0 | 0 | 0 | 0 |
| **Khoisan-speaking Negroes:** | | | | | | | | | | |
| Sarwa | 79 | 60 | 4 | 8 | 0 | 0 | 4 | 3 | | |
| Kwengo | 35 | 24 | 5 | 4 | 0 | 0 | 2 | 0 | | |
| Dama | 52 | 44 | 1 | 1 | 0 | 0 | 0 | 6 | | |
| Dama | 92 | 55 | 8 | 21 | 0 | 0 | 3 | 2 | | 3 |
| Dama | 442 | 257 | 35 | 81 | 10 | 0 | 7 | 36 | | 15 |
| **Khoisan:** | | | | | | | | | | |
| *Khoi:* | | | | | | | | | | |
| Keetmanshoop Nama | 152 | 104 | 28 | 13 | 2 | 3 | 1 | 1 | | |
| Topnaars | 58 | | | | | | | | | |
| Sesfontein Nama | 42 | 28 | 7 | 3 | 2 | 2 | 0 | | | |
| *San:* | | | | | | | | | | |
| *Northern:* | | | | | | | | | | |
| G!ang!ai | 36 | 36 | 0 | 0 | 0 | 0 | 0 | 0 | | |
| Tsumkwe !Kung | 122 | 113 | 5 | 1 | 0 | 2 | 0 | 1 | | |
| Saman!aika " | 34 | 32 | 2 | 0 | 0 | 0 | 0 | 0 | | |
| Dobe " | 383 | 337 | 35 | 2 | 0 | 0 | 0 | 9 | | |
| /Ai/ai " | 63 | 52 | 8 | 1 | 0 | 0 | 0 | 2 | | |
| /Du/da " | 102 | 95 | 7 | 0 | 0 | 0 | 0 | 0 | | |
| //Au//en | 210 | 194 | 9 | 6 | 0 | 0 | 0 | 1 | | |

| | | | Gene Frequencies | | | | | | |
|---|---|---|---|---|---|---|---|---|---|
| cDe | CDe | cDE | cde | Cde | cdE | CDE ±\nCdE | χ² | df | Reference |
| .610 | .035 | .076 | .244 | .036 | | | 4.711 | 1 | Jenkins (1972) |
| .821 | .081 | .098 | | | | | 1.810 | 3 | Nurse et al (1974) |
| .614 | .045 | .083 | .197 | .062 | | | 40.485 | 4 | Jenkins (1972) |
| .653 | .082 | .082 | .182 | | | | 1.750 | 3 | Present study |
| .718 | .043 | .084 | .155 | | | | .600 | 2 | Moullec et al (1966) |
| .740 | .075 | .081 | .105 | | | | 12.032 | 3 | Jenkins (1972) |
| .901 | .033 | .066 | .000 | | | | 1.076 | 3 | Jenkins (1972) |
| .671 | .087 | .082 | .161 | | | | 3.197 | 3 | Jenkins (1972) |
| .772 | .057 | .078 | .093 | | | | 2.038 | 3 | Jenkins (1972) |
| .684 | .032 | .085 | .165 | .033 | | | 7.767 | 3 | Present study |
| .736 | .067 | .067 | .130 | | | | .160 | 2 | Jenkins (1972) |
| .667 | .047 | .084 | .202 | | | | 5.176 | 1 | Present study |
| .630 | .009 | .125 | .221 | .015 | | | 1.000 | 2 | Present study |
| .677 | .000 | .109 | .143 | .047 | .000 | .025 | 10.106 | 2 | Present Study |
| .619 | .070 | .100 | .211 | | | | 2.831 | 1 | Matznetter & Spielmann (1969) |
| .715 | .025 | .079 | .128 | .053 | | | .405 | 3 | Matznetter & Spielmann (1969) |
| .733 | .063 | .176 | .000 | .000 | .028 | | 5.180 | 2 | Nurse & Jenkins (1977b) |
| .671 | .059 | .105 | .165 | .000 | .000 | | 1.580 | 3 | Nurse & Jenkins (1977b) |
| .807 | .033 | .125 | .015 | .020 | .000 | | 2.150 | 1 | Nurse & Jenkins (1977b) |
| .682 | .047 | .133 | .138 | .000 | .000 | | .370 | 3 | Nurse & Jenkins (1977b) |
| .862 | .052 | .086 | .000 | .000 | | | .380 | 3 | Nurse et al (1983) |
| .812 | .094 | .094 | .000 | .000 | | | 1.780 | 2 | Nurse et al (1983) |
| .805 | .053 | .142 | .000 | .000 | | | 2.330 | 3 | Nurse et al (1983) |
| .819 | .063 | .118 | .000 | .000 | | | .360 | 3 | Nurse et al (1983) |
| .655 | .044 | .172 | .088 | .040 | | | 4.740 | 3 | Nurse et al (1983) |
| .715 | .081 | .105 | .099 | .000 | | | 1.360 | 2 | Nurse et al (1983) |
| .844 | .094 | .062 | .000 | .000 | | | .690 | 1 | Nurse et al (1983) |
| .828 | .082 | .090 | | | | | 4.054 | 5 | Jenkins (1972) |
| .697 | .157 | .146 | .000 | | | | 4.170 | 3 | Jenkins et al (1983) |
| .556 | .158 | .137 | .137 | | | .012 | 19.801 | 3 | Jenkins et al (1983) |
| .690 | .138 | .172 | .000 | | | | 3.830 | 3 | Jenkins et al (1983) |
| .879 | .121 | .000 | .000 | | | | .650 | 1 | Jenkins et al (1983) |
| .689 | .029 | .093 | .189 | | | | .940 | 1 | Chasko et al (1979) |
| .828 | .082 | .090 | .000 | .000 | .000 | .000 | 1.290 | 1 | Jenkins and Brain (1967) |
| .640 | .010 | .010 | .341 | | | | .232 | 5 | Jenkins (1972) |
| .638 | .044 | .084 | .147 | .086 | | | 3.630 | 3 | Nurse et al (1976) |
| .518 | .039 | .083 | .289 | .018 | .053 | | 5.663 | 6 | Knussmann (1969) |
| .744 | .118 | .056 | .081 | | | | 2.388 | 5 | Jenkins (1972) |
| .810 | .129 | .060 | .000 | .000 | | | 6.550 | 3 | Jenkins & Brain (1967) |
| .786 | .155 | .060 | | | | | 4.758 | 2 | Jenkins (1972) |
| 1.000 | .000 | .000 | .000 | .000 | | | | | Nurse et al (1977) |
| .876 | .031 | .004 | .089 | | | | | | Nurse et al (1977) |
| .971 | .029 | .000 | .000 | | | | | | Nurse et al (1977) |
| .798 | .046 | .003 | .154 | | | | .987 | 1 | Jenkins (1972) |
| .929 | .064 | .008 | .000 | | | | .335 | 2 | Jenkins (1972) |
| .966 | .034 | .000 | .000 | | | | .129 | 1 | Jenkins (1972) |
| .895 | .021 | .014 | .069 | | | | .280 | 1 | Jenkins (1972) |

# Table 3 (*cont.*)

| Population | n | | | | Phenotypes | | | | | |
|---|---|---|---|---|---|---|---|---|---|---|
| | | ccDee | CcDee | ccDEe | CcDEe | CCDee | ccDEE | ccdee | Ccdee | ccdEe |
| **Central:** | | | | | | | | | | |
| Hei//om | 97 | 72 | 16 | 4 | 2 | 2 | 0 | 1 | | |
| Nharo | 147 | 139 | 5 | 2 | 0 | 0 | 1 | 0 | | |
| G/wi | 94 | 89 | 2 | 1 | 0 | 1 | 1 | 0 | | |
| G//ana | 50 | 41 | 1 | 8 | 0 | 0 | 0 | 0 | | |
| **Southern:** | | | | | | | | | | |
| !Xo | 72 | 68 | 2 | 1 | 0 | 0 | 0 | 0 | | |
| G!aokx'ate | 33 | 27 | 6 | 0 | 0 | 0 | 0 | 0 | | |
| **Uncertain:** | | | | | | | | | | |
| Eastern ǂHuã | 36 | 26 | 2 | 7 | 1 | 0 | 0 | 0 | | |
| **Caucasoid:** | | | | | | | | | | |
| **Afrikaans-speaking:** | | | | | | | | | | |
| Cape Town | 641 | 27 | 231 | 73 | 73 | 116 | 12 | 89 | 20* | |
| Johannesburg | 77[4] | 2 | 29 | 9 | 11 | 7 | 5 | 13 | | 1 |
| Johannesburg | 160[v] | 5 | 62 | 10 | 26 | 34 | 9 | 12 | 1 | |
| SWA/Namibia | 57 | 1 | 19 | 9 | 8 | 8 | 2 | 10 | | |
| **English-speaking:** | | | | | | | | | | |
| Cape Town | 474 | 34 | 180 | 52 | 49 | 85 | 7 | 63 | 4* | |
| **German-speaking:** | | | | | | | | | | |
| SWA/Namibia | 118 | 4 | 47 | 19 | 11 | 19 | 3 | 15 | | |
| **Indian:** | | | | | | | | | | |
| **Transvaal:** | | | | | | | | | | |
| Hindu | 53 | 0 | 14 | 2 | 5 | 29 | 0 | 3 | | |
| Moslem | 56 | 2 | 22 | 4 | 7 | 15 | 0 | 6 | | |
| **Natal:** | | | | | | | | | | |
| Tamil-speaking | 423[2] | 4 | 138 | 13 | 44 | 201 | 4 | 15 | 3 | 0 |
| Telegu-speaking | 171[3] | 3 | 66 | 8 | 14 | 70 | 4 | 5 | 0 | 0 |
| Hindi-speaking | 260[5] | 4 | 78 | 8 | 14 | 138 | 1 | 14 | 2 | 0 |
| Moslem | 314[6] | 7 | 107 | 21 | 39 | 114 | 7 | 15 | 2 | 1 |
| **Hybrid Peoples:** | | | | | | | | | | |
| **Baster-like:** | | | | | | | | | | |
| Rehoboth | 120 | 40 | 44 | 7 | 10 | 9 | 3 | 5 | 1 | 1 |
| !Kuboes | 136[6] | 79 | 40 | 9 | 3 | 0 | 0 | 5 | | |
| Riemvaasmaak | 89[v] | 43 | 24 | 10 | 4 | 7 | 0 | | | |
| **Griqua** | 264 | 178 | 45 | 27 | 4 | 2 | 1 | 4 | 3 | |
| **"Cape Coloureds":** | | | | | | | | | | |
| Cape Town | 486 | 116 | 205 | 27 | 20 | 78 | 1 | 19 | 20* | |
| Johannesburg | 109 | 47 | 40 | 13 | 2 | 3 | 3 | 1 | 0 | |
| Orange River Valley | 202 | 99 | 71 | 17 | 6 | 6 | 0 | 3 | 0 | |

[1] these individuals not distinguished from those of phenotype ccDDEE

[2] includes two individuals with phenotype CcDEE

[3] include one individual with phenotype CCDEE

[4] includes one individual with phenotype ccdEE

[5] includes one individual with phenotype CCDEe

[6] includes one individual with phenotype CcDEE

* includes individuals of other rare phenotypes

| | Gene Frequencies | | | | | | | | |
| $cDe$ | $CDe$ | $cDE$ | $cde$ | $Cde$ | $cdE$ | $CDE + CdE$ | $\chi^2$ | df | References |
|---|---|---|---|---|---|---|---|---|---|
| .786 | .155 | .060 | .095 | | | | 3.940 | 2 | Nurse *et al* (1977) |
| .901 | .017 | .007 | .075 | | | | 3.060 | 1 | Nurse *et al* (1977) |
| .963 | .021 | .016 | | | | | .580 | 1 | Jenkins *et al* (1975) |
| .865 | .010 | .125 | | | | | 1.980 | 1 | Jenkins *et al* (1975) |
| .861 | .014 | .007 | .118 | | | | .033 | 1 | Jenkins (1972) |
| .909 | .091 | .000 | .000 | | | | .370 | 1 | Nurse *et al* (1977) |
| .847 | .042 | .111 | .000 | | | | 2.240 | 2 | Nurse & Jenkins (1977a) |
| .051 | .425 | .135 | .373 | .004 | .012 | | 20.060 | 2 | Botha (1972) |
| .031 | .350 | .181 | .417 | | .002 | | 3.919 | 4 | Present study |
| .054 | .489 | .171 | .282 | .004 | | | 10.450 | 3 | Present study |
| .020 | .377 | .184 | .419 | | | | 0.010 | 3 | Palmhert-Keller *et al* (1983) |
| .085 | .423 | .122 | .364 | .003 | .000 | .003 | .859 | 2 | Botha (1972) |
| .048 | .407 | .152 | .393 | | | | 2.550 | 3 | Palmhert-Keller *et al* (1983) |
| .000 | .727 | .066 | .207 | .000 | | | .750 | 3 | Present study |
| .050 | .527 | .098 | .325 | .000 | | | .850 | 3 | Present study |
| .025 | .675 | .076 | .202 | .019 | .000 | .002 | 1.600 | 3 | Moores (1981) |
| .055 | .645 | .086 | .209 | .000 | .000 | .005 | 4.900 | 3 | Moores (1981) |
| .028 | .697 | .045 | .211 | .016 | .000 | .003 | 3.300 | 3 | Moores (1981) |
| .048 | .584 | .113 | .231 | .017 | .006 | .002 | 1.900 | 3 | Moores (1981) |
| .390 | .278 | .075 | .195 | .030 | .032 | | 7.550 | 4 | Nurse *et al* (1982) |
| .604 | .158 | .044 | .194 | | | | 5.269 | 4 | Jenkins (1972) |
| .675 | .242 | .074 | | | | .009 | 1.820 | 2 | Nurse & Jenkins (1978) |
| .708 | .050 | .063 | .123 | | .056 | | 1.14 | 4 | Nurse & Jenkins (1975) |
| .329 | .401 | .052 | .198 | .010 | .005 | .005 | 43.668 | 1 | Botha (1972) |
| .534 | .220 | .076 | .150 | | | | 11.760 | 3 | Jenkins (1972) |
| .638 | .163 | .057 | .085 | | | | 10.385 | 4 | Jenkins (1972) |

# Table 4. Duffy, Kell and P Systems

| Population | n | Fy(a+b-) | Fy(a+b+) | Fy(a-b+) | Fy(a-b-) | $Fy^a$ | $Fy^b$ | $Fy$ | $\chi^2_{(1)}$ |
|---|---|---|---|---|---|---|---|---|---|
| **Bantu-speaking Negroes:** | | | | | | | | | |
| **Nguni:** | | | | | | | | | |
| Swati | 92 | | 10 | 82 | | .056 | .944 | | |
| Zulu/Ronga | 123 | | 21 | 102 | | .089 | .911 | | |
| Zulu | 91 | 8 | 0 | 11 | 72 | .045 | .062 | .893 | .572 |
| Xhosa | 73 | | 12 | 61 | | .086 | .914 | | |
| **Sotho/Tswana:** | | | | | | | | | |
| S. Sotho (Highland) | 89 | | 7 | 82 | | .040 | .960 | | |
| S. Sotho (Lowland) | 92/30 | | 12 | 80 | | .067 | .933 | | |
| Pedi (Urban) | 130 | | 11 | 119 | | .043 | .957 | | |
| Pedi (Rural) | 116 | | 12 | 104 | | .055 | .945 | | |
| Fokeng | 204/119 | | 17 | 187 | | .042 | .958 | | |
| Ngwato | 95 | 5 | 0 | 13 | 77 | .027 | .071 | .902 | .398 |
| Kgalagadi | 124 | 17 | 1 | 20 | 86 | .075 | .089 | .836 | .313 |
| Venda: Rural | 105 | 4 | 0 | 1 | 100 | .019 | .005 | .976 | .022 |
| **Tsonga/Ronga** | | | | | | | | | |
| Ronga | 45 | 0 | 0 | 0 | 45 | .000 | .000 | 1.000 | |
| Shangaan/Tsonga | 122 | 0 | 0 | 0 | 122 | .000 | .000 | 1.000 | |
| **Kavango:** | | | | | | | | | |
| Mbukushu | 37/26 | 1 | 1 | 0 | 35 | .026 | .014 | .960 | |
| Gciriku | 86/57 | 3 | 1 | 8 | 74 | .023 | .054 | .923 | 3.44 |
| Sambyu | 72/92 | 4 | 0 | 6 | 62 | .028 | .042 | .930 | .22 |
| Kwangali | 37/107 | 2 | 1 | 3 | 31 | .041 | .054 | .905 | 3.53 |
| **Ambo:** | | | | | | | | | |
| Kwambi | 106/38/18 | | 8 | 98 | | .039 | .961 | | |
| Kwaluudhi | 69/50 | | 13 | 56 | | .099 | .901 | | |
| Ndonga | 57/56/22 | | 6 | 51 | | .053 | .947 | | |
| Mbalantu | 67/66/30 | | 17 | 50 | | .136 | .864 | | |
| Kwanyama | 116/65 | | 33 | 83 | | .154 | .846 | | |
| Ngandjera | 69/61 | | 19 | 50 | | .149 | .851 | | |
| Nkolonkadhi | 31/27 | | 4 | 27 | | .067 | .933 | | |
| **Herero:** | | | | | | | | | |
| Botswana | 62/61 | | 3 | 59 | | .045 | .955 | | |
| SWA/Namibia | 75/89 | 7 | 2 | 10 | 56 | .058 | .078 | .864 | 2.520 |
| Mbanderu | 97 | 7 | 0 | 6 | 84 | .037 | .031 | .932 | .240 |
| Himba | 86 | 4 | 2 | 2 | 78 | .064 | .041 | .895 | 8.750 |
| Chimba | 33/16 | | 1 | 32 | | .015 | .985 | | |
| **Khoisan-speaking Negroes:** | | | | | | | | | |
| Sarwa | 79 | 14 | 0 | 5 | 60 | .093 | .032 | .875 | .580 |
| Kwengo | 35 | 2 | 0 | 6 | 27 | .029 | .086 | .885 | .230 |
| Dama | 90/92 | 3 | 2 | 7 | 78 | .046 | .065 | .889 | 8.680 |
| **Khoisan:** | | | | | | | | | |
| **Khoi:** | | | | | | | | | |
| Keetmanshoop Nama | 153 | | 78 | 75 | | .300 | .700 | | |
| Topnaars Kuiseb | 13 | | 11 | 2 | | .608 | .392 | | |
| Sesfontein Nama | 42 | | 12 | 30 | | .155 | .845 | | |
| **San:** | | | | | | | | | |
| **Northern:** | | | | | | | | | |
| G!anglai | 36 | 8 | 12 | 14 | 2 | .304 | .450 | .246 | .650 |
| Tsumkwe !Kung | 279 | | 151 | 128 | | .323 | .677 | | |
| Saman!aika " | 34 | | 18 | 16 | | .314 | .686 | | |
| Dobe " | 385 | | 174 | 211 | | .260 | .740 | | |
| /Ai/ai " | 64 | | 32 | 32 | | .293 | .707 | | |
| /Du/da " | 54 | | 28 | 26 | | .307 | .693 | | |
| //Au//en | 185 | | 106 | 79 | | .347 | .654 | | |
| **Central:** | | | | | | | | | |
| Hei//om | 96 | | 22 | 74 | | .122 | .878 | | |
| Nharo | 147 | | 70 | 77 | | .276 | .724 | | |
| G/wi | 94 | | 58 | 36 | | .381 | .619 | | |
| G//ana | 50 | | 24 | 26 | | .279 | .721 | | |

| Kell system | | | | P system | | | | Reference |
|---|---|---|---|---|---|---|---|---|
| Phenotypes | | Gene frequencie | | Phenotypes | | Gene frequencies | | |
| K(+) | K(-) | $K$ | $k$ | $P_1(+)$ | $P_1(-)$ | $p^1$ | $P$ | |
| 0 | 92 | .000 | 1.000 | | | | | Jenkins (1972) |
| 2 | 121 | .008 | .992 | | | | | Nurse *et al.* (1974) |
| 0 | 91 | .000 | 1.000 | 89 | 2 | .852 | .148 | Present study |
| | | | | | | | | Jenkins (1972) |
| 0 | 89 | .000 | 1.000 | | | | | Beaumont *et al* (1979) |
| 0 | 30 | .000 | 1.000 | | | | | " " |
| 0 | 130 | .000 | 1.000 | | | | | Jenkins (1972) |
| 0 | 116 | .000 | 1.000 | | | | | Jenkins (1972) |
| 0 | 119 | .000 | 1.000 | | | | | Jenkins (1972) |
| 0 | 95 | .000 | 1.000 | 92 | 3 | .822 | .178 | Present study |
| | | | | | | | | Present study |
| 4 | 101 | .019 | .981 | | | | | Present study |
| 0 | 45 | .000 | 1.000 | 44 | 1 | .851 | .149 | Matznetter & Spielmann (1969) |
| 0 | 122 | .000 | 1.000 | 119 | 3 | .843 | .157 | Matznetter & Spielmann (1969) |
| 0 | 37 | .000 | 1.000 | 25 | 1 | .804 | .196 | Nurse & Jenkins (1977b) |
| 0 | 86 | .000 | 1.000 | 57 | 0 | 1.000 | .000 | Nurse & Jenkins (1977b) |
| 0 | 92 | .000 | 1.000 | 90 | 2 | .853 | .147 | Nurse & Jenkins (1977b) |
| 0 | 107 | .000 | 1.000 | 106 | 1 | .903 | .097 | Nurse & Jenkins (1977b) |
| 0 | 38 | .000 | 1.000 | 18 | 0 | 1.000 | .000 | Nurse *et al* (1983) |
| 3 | 47 | .031 | .969 | | | | | Nurse *et al* (1983) |
| 3 | 53 | .027 | .973 | 22 | 0 | 1.000 | .000 | Nurse *et al* (1983) |
| 2 | 64 | .015 | .985 | 30 | 0 | 1.000 | .000 | Nurse *et al* (1983) |
| 3 | 113 | .013 | .987 | 63 | 2 | .825 | .175 | Nurse *et al* (1983) |
| 2 | 59 | .016 | .984 | | | | | Nurse *et al* (1983) |
| 2 | 25 | .038 | .962 | | | | | Nurse *et al* (1983) |
| 0 | 61 | .000 | 1.000 | | | | | Jenkins (1972) |
| 4 | 85 | .023 | .977 | 89 | 0 | 1.000 | .000 | Jenkins *et al.* (1983) |
| | | | | | | | | Jenkins *et al.* (1983) |
| 0 | 86 | .000 | 1.000 | 84 | 2 | .847 | .153 | Jenkins *et al.* (1983) |
| 0 | 33 | .000 | 1.000 | 16 | 0 | 1.000 | .000 | Jenkins *et al.* (1983) |
| 0 | 79 | .000 | 1.000 | | | | | Chasko *et al.* (1979) |
| 0 | 35 | .000 | 1.000 | 31 | 4 | .662 | .338 | Nurse & Jenkins (1977b) |
| 0 | 92 | .000 | 1.000 | 88 | 4 | .792 | .208 | Nurse *et al.* (1976) |
| 2 | 151 | .007 | .993 | | | | | Jenkins (1972) |
| | | | | | | | | Jenkins (1972) |
| 0 | 42 | .000 | 1.000 | | | | | Jenkins (1972) |
| 0 | 279 | .000 | 1.000 | | | | | Nurse *et al.* (1977 ) |
| | | | | | | | | Jenkins (1972) |
| | | | | 26 | 8 | .515 | .485 | Nurse *et al* (1977) |
| 0 | 385 | .000 | 1.000 | | | | | Nurse & Jenkins (1977a) |
| 0 | 64 | .000 | 1.000 | | | | | " " " |
| | | | | | | | | " " " |
| | | | | | | | | " " " |
| | | | | | | | | Nurse *et al* (1977) |
| 2 | 145 | .007 | .993 | | | | | Nurse *et al.* (1977 ) |
| 0 | 94 | .000 | 1.000 | | | | | Nurse *et al.* (1977 ) |
| 0 | 50 | .000 | 1.000 | | | | | Jenkins *et al.* (1975) |
| | | | | | | | | Jenkins *et al.* (1975) |

# Table 4 (*cont.*)

| Population | n | Phenotypes | | | | Gene frequencies | | | |
|---|---|---|---|---|---|---|---|---|---|
| | | Fy(a+b-) | Fy(a+b+) | Fy(a-b+) | Fy(a-b-) | $Fy^a$ | $Fy^b$ | $Fy$ | $\chi^2_{(1)}$ |
| **Southern:** | | | | | | | | | |
| !Xõ | 51 | 9 | 22 | 17 | 3 | .348 | .466 | .186 | 4.100 |
| G!aokx'ate | 33 | 5 | 16 | 12 | 0 | .394 | .606 | .000 | .000 |
| **Uncertain:** | | | | | | | | | |
| Eastern ǂHuã | 36 | 7 | 9 | 14 | 6 | .240 | .370 | .390 | 1.580 |
| **Caucasoid:** | | | | | | | | | |
| Afrikaans-speaking: | | | | | | | | | |
| Johannesburg | 77 | 14 | 38 | 24 | 1 | .406 | .527 | .067 | 2.790 |
| Johannesburg | 160 | 49 | 77 | 34 | 0 | .547 | .453 | .000 | .134 |
| SWA/Namibia | 57 | 13 | 30 | 14 | | .491 | .509 | | 0.160 |
| German-speaking: | | | | | | | | | |
| SWA/Namibia | 118 | 23 | 58 | 37 | | .441 | .559 | | 0.000 |
| **Indian:** | | | | | | | | | |
| Transvaal: | | | | | | | | | |
| Hindu | 53 | 45 | | 8 | | .612 | .388 | | |
| Moslem | 56 | 47 | | 9 | | .599 | .401 | | |
| Natal: | | | | | | | | | |
| Tamil-speaking | 144/399/211 | 130 | | 14 | | .688 | .312 | | |
| Telegu-speaking | 118/181/115 | 102 | | 16 | | .632 | .368 | | |
| Hindi-speaking | 147/289/142 | 130 | | 17 | | .660 | .340 | | |
| Moslem | 100/296/248 | 72 | | 28 | | .471 | .529 | | |
| **Hybrid Peoples:** | | | | | | | | | |
| Baster-like: | | | | | | | | | |
| Rehoboth | 116/120/119 | 36 | 54 | 25 | 1 | .521 | .431 | .048 | 1.810 |
| !Kuboes | 137 | 85 | | 52 | | .384 | .616 | | |
| Riemvasmaak | 89 | 27 | 11 | 28 | 23 | .243 | .250 | .507 | .010 |
| Griqua | 262/262/161 | 111 | | 151 | | .241 | .759 | | |
| 'Cape Coloureds': | | | | | | | | | |
| Johannesburg | 109 | 52 | | 57 | | .276 | .724 | | |

312

| Kell system | | | | P system | | | | Reference |
|---|---|---|---|---|---|---|---|---|
| Phenotypes | | Gene frequencie | | Phenotypes | | Gene frequencies | | |
| K(+) | K(-) | $K$ | $k$ | $P_1(+)$ | $P_1(-)$ | $P^1$ | $P$ | |
| | | | | | | | | Nurse & Jenkins (1977a) |
| | | | | | | | | Nurse *et al* (1977) |
| | | | | | | | | |
| | | | | | | | | Nurse & Jenkins (1977 a) |
| | | | | | | | | |
| | | | | | | | | Present study |
| 14 | 146 | .045 | .955 | 99 | 61 | .383 | .617 | Present study |
| 5 | 52 | .045 | .955 | 43 | 14 | .504 | .496 | Palmhert-Keller *et al* (1983) |
| | | | | | | | | " " " |
| 4 | 114 | .017 | .983 | | | | | Palmhert-Keller *et al* (1983) |
| | | | | | | | | |
| 14 | 39 | .142 | .858 | | | | | Present study |
| 9 | 47 | .084 | .916 | | | | | Present study |
| | | | | | | | | |
| 8 | 391 | .010 | .990 | 153 | 58 | .476 | .524 | Moores (1980) |
| 2 | 179 | .006 | .994 | 80 | 35 | .448 | .552 | Moores (1980) |
| 2 | 287 | .004 | .996 | 117 | 25 | .580 | .420 | Moores (1980) |
| 16 | 280 | .027 | .973 | 173 | 75 | .450 | .550 | Moores (1980) |
| | | | | | | | | |
| 28 | 92 | .117 | .883 | 93 | 26 | .533 | .467 | Nurse *et al.* (1982) |
| 3 | 134 | .011 | .989 | | | | | Jenkins (1972) |
| 3 | 86 | .017 | .983 | 84 | 5 | .763 | .237 | Nurse & Jenkins (1978) |
| 18 | 244 | .035 | .865 | 149 | 12 | .729 | .271 | Nurse & Jenkins (1975) |
| | | | | | | | | |
| 10 | 99 | .046 | .954 | | | | | Jenkins (1972) |

## Table 5. Kidd and Lutheran Systems

| Population | n | Kidd Phenotypes | | | Kidd Gene frequency | | Lutheran Phenotypes | | | Lutheran Gene frequency | | Reference |
|---|---|---|---|---|---|---|---|---|---|---|---|---|
| | | Jk(a+b-) | Jk(a+b+) | Jk(a-b+) | $Jk^a$ | $Jk^b$ | Lu(a+b-) | Lu(a+b+) | Lu(a-b+) | $Lu^a$ | $Lu^b$ | |
| **Bantu-speaking Negroes** | | | | | | | | | | | | |
| _Sotho/Tswana:_ | | | | | | | | | | | | |
| S. Sotho (Highland) | 89 | 86 | 3 | | .983 | .017 | | | | | | Beaumont et al. (1978) |
| S. Sotho (Lowland) | 30 | 28 | 2 | | .966 | .034 | | | | | | Beaumont et al. (1978) |
| Pedi (Urban) | 128 | 79 | 49 | | .786 | .214 | | | | | | Jenkins (1972) |
| Pedi (Rural) | 116 | 90 | 26 | | .881 | .119 | | | | | | Jenkins (1972) |
| _Tsonga/Ronga:_ | | | | | | | | | | | | |
| Ronga | 45 | 25 | 17 | 3 | .744 | .256 | 0 | 1 | 44 | .011 | .989 | Matznetter & Spielmann (1969) |
| Shangaan/Tsonga | 122 | 79 | 39 | 4 | .807 | .193 | 0 | 21 | 101 | .090 | .910 | Matznetter & Spielmann (1969) |
| _Herero:_ | | | | | | | | | | | | |
| Botswana | 45 | 40 | 5 | | .943 | .057 | | | | | | Jenkins (1972) |
| SWA/Namibia | 89ᵛ | 61 | 28 | | .828 | .172 | | | | | | Jenkins et al. (1983) |
| Mbanderu | 97ᵛ | 80 | 1 | 3 | .618 | .020 | | | | | | Jenkins et al. (1983) |
| **Khoisan-speaking Negroes:** | | | | | | | | | | | | |
| Dana | 86 | | | | | | 5 | | 81 | .030 | .970 | Nurse et al. (1976) |
| **Khoisan:** | | | | | | | | | | | | |
| _Khoi:_ | | | | | | | | | | | | |
| Keetmanshoop Nama | 153 | 137 | 16 | | .946 | .054 | | | | | | Jenkins (1972) |

314

| | | | | | | | | | | |
|---|---|---|---|---|---|---|---|---|---|---|
| **San:** | | | | | | | | | | |
| **Northern:** | | | | | | | | | | |
| Dobe !Kung | 340 | 268 | 72 | .888 | .112 | | | | | Jenkins (1972) |
| /Ai/ai " | 64 | 48 | 16 | .866 | .134 | | | | | Jenkins (1972) |
| /Du/da " | 47 | 41 | 6 | .934 | .066 | | | | | Jenkins (1972) |
| //Au//en " | 54 | 39 | 15 | .850 | .150 | | | | | Nurse et al (1977) |
| **Central:** | | | | | | | | | | |
| Hei//om | 96 | 79 | 17 | .907 | .093 | | | | | Nurse et al. (1977) |
| Nharo | 127 | 101 | 26 | .892 | .108 | | | | | " |
| **Caucasoid** | | | | | | | | | | |
| **Indian:** | | | | | | | | | | |
| **Natal:** | | | | | | | | | | |
| Tamil-speaking | 124/255 | 55 | 69 | .666 | .334 | 0 | .000 | 1.000 | 255 | Moores (1980) |
| Telugu-speaking | 108 | 50 | 58 | .680 | .320 | 0 | .000 | 1.000 | 108 | Moores (1980) |
| Hindi-speaking | 125/143 | 56 | 69 | .669 | .331 | 0 | .000 | 1.000 | 143 | Moores (1980) |
| Moslem | 178/219 | 73 | 105 | .640 | .360 | 0 | .000 | 1.000 | 219 | Moores (1980) |
| **Hybrid Peoples:** | | | | | | | | | | |
| **Baster-like:** | | | | | | | | | | |
| !Kuboes | 137 | 89 | 48 | .806 | .194 | | | | | Jenkins (1972) |
| **'Cape Coloureds':** | | | | | | | | | | |
| Johannesburg | 109 | 73 | 36 | .819 | .181 | | | | | Jenkins (1972) |

[v] includes 13 individuals with Jk(a-b-) phenotypes with frequency $Jk$ .362

# Table 6. ABH Secretor and Lewis Systems

| Population | Lewis system | | | | | | Secretor system | | | | | Reference |
|---|---|---|---|---|---|---|---|---|---|---|---|---|
| | n | Phenotypes | | | Gene frequencies | | n | Phenotypes | | Gene frequencies | | |
| | | Le(a+b-) | Le(a-b+) | Le(a-b-) | Le | le | | Secretors | Non-secretors | Se | se | |
| **Bantu-speaking Negroes:** | | | | | | | | | | | | |
| *Nguni:* | | | | | | | | | | | | |
| Zulu | 171 | 38 | 93 | 40 | .516 | .484 | 181 | 132 | 49 | .479 | .521 | Moores & Brain (1968) |
| *Sotho/Tswana:* | | | | | | | | | | | | |
| S. Sotho (Highland) | 85 | 18 | 50 | 17 | .553 | .447 | 85 | 66 | 19 | .538 | .462 | Beaumont et al (1979) |
| S. Sotho (Lowland) | 30 | 5 | 13 | 12 | .368 | .632 | 30 | 19 | 11 | .395 | .605 | " |
| Pedi (Rural) | 114 | 21 | 57 | 36 | .438 | .562 | 201 | 147 | 54 | .482 | .518 | Jenkins (1972) |
| *Kavango:* | | | | | | | | | | | | |
| Mbukushu | 51 | 3 | 27 | 21 | .350 | .650 | 56 | 51 | 5 | .701 | .299 | Nurse & Jenkins (1977b) |
| Gciriku | 72 | 7 | 31 | 34 | .313 | .687 | 99 | 83 | 16 | .598 | .402 | Nurse & Jenkins (1977b) |
| Sambyu | 81 | 9 | 38 | 34 | .172 | .828 | 82 | 62 | 20 | .380 | .620 | Nurse & Jenkins (1977b) |
| Kwangali | 68 | 8 | 31 | 29 | .347 | .653 | 98 | 75 | 23 | .446 | .554 | Nurse & Jenkins (1977b) |
| *Herero:* | | | | | | | | | | | | |
| Botswana | 55 | 6 | 25 | 24 | .349 | .651 | 49 | 41 | 8 | .596 | .404 | Jenkins (1972) |
| SWA/Namibia | 87 | 16 | 44 | 27 | .443 | .557 | 75 | 55 | 20 | .484 | .516 | Jenkins et al. (1983) |
| Mbanderu | 97 | 12 | 41 | 44 | .307 | .693 | | | | | | Jenkins et al. (1983) |
| Himba | 61 | 11 | 21 | 29 | .311 | .689 | 61 | 43 | 18 | .457 | .543 | Jenkins et al. (1983) |
| **Khoisan-speaking Negroes:** | | | | | | | | | | | | |
| Kwengo | 35 | 6 | 22 | 7 | .553 | .447 | 30 | 18 | 12 | .367 | .633 | Nurse & Jenkins (1977b) |
| Dama | | | | | | | 82 | 69 | 13 | .602 | .398 | Nurse et al. (1976) |
| Dama | 52 | 6 | 26 | 20 | .380 | .620 | 52 | 42 | 10 | .562 | .438 | Jenkins (1972) |

Khoisan:

| | | | | | | | | | | | | |
|---|---|---|---|---|---|---|---|---|---|---|---|---|
| **Khoi:** | | | | | | | | | | | | |
| Keetmanshoop Nama | 115 | 17 | 43 | 55 | .309 | .691 | 129 | 106 | 23 | .578 | .422 | Jenkins (1972) |
| Topnaars (Kuiseb) | 29 | 0 | 26 | 3 | .679 | .321 | 53 | 45 | 8 | .611 | .389 | Jenkins & Brain (1967) |
| Topnaars (Sesfontein) | 42 | 7 | 29 | 6 | .622 | .378 | 42 | 34 | 8 | .558 | .442 | Jenkins (1972) |
| **San:** | | | | | | | | | | | | |
| **Northern:** | | | | | | | | | | | | |
| Dobe !Kung | 218[v] | 31 | 118 | 69 | .437 | .563 | 308 | 254 | 54 | .582 | .418 | Jenkins (1972) |
| /Ai/ai " | 65 | 6 | 37 | 22 | .419 | .581 | 60 | 53 | 7 | .658 | .342 | Jenkins (1972) |
| /Du/da " | 32 | 3 | 25 | 4 | .646 | .354 | 87 | 77 | 10 | .661 | .339 | Jenkins (1972) |
| **Central:** | | | | | | | | | | | | |
| Nharo | 104 | 12 | 69 | 23 | .530 | .470 | 43 | 40 | 3 | .916 | .084 | Jenkins (1972) |
| **Southern:** | | | | | | | | | | | | |
| !Xõ | 60 | 6 | 46 | 8 | .635 | .365 | 66 | 61 | 5 | .724 | .276 | Jenkins (1972) |
| **Caucasoid:** | | | | | | | | | | | | |
| **Indian:** | | | | | | | | | | | | |
| **Natal:** | | | | | | | | | | | | |
| Tamil-speaking | 426[†] | 125 | 228 | 69 | .598 | .402 | | | | | | Moores (1980) |
| Telegu-speaking | 198[*] | 43 | 125 | 28 | .624 | .376 | | | | | | Moores (1980) |
| Hindi-speaking | 269 | 63 | 167 | 39 | .619 | .381 | | | | | | Moores (1980) |
| Moslem | 282 | 50 | 199 | 33 | .658 | .342 | | | | | | Moores (1980) |
| **Hybrid Peoples:** | | | | | | | | | | | | |
| **Baster-like:** | | | | | | | | | | | | |
| Rehoboth | 115 | 25 | 76 | 14 | .651 | .349 | 115 | 87 | 28 | .507 | .493 | Nurse et al. (1982) |
| !Kuboes | 88 | 26 | 22 | 40 | .325 | .675 | 112 | 79 | 33 | .458 | .542 | Jenkins (1972) |
| Riemvasmaak | 89[*] | 19 | 49 | 19 | .538 | .462 | | | | .479 | .521 | Nurse & Jenkins (1978) |
| **'Cape Coloureds':** | | | | | | | | | | | | |
| Johannesburg | 108 | 19 | 64 | 25 | .518 | .482 | 108 | 85 | 23 | .538 | .462 | Jenkins (1972) |

[†]Four individuals Le(a+b+)  [*]Two individuals Le(a+b+)
[v]excluding seven individuals with phenotype Le(a+b+)

## Table 7. Xg System

| Population | n | Phenotypes | | | | Gene Frequencies | | Reference |
|---|---|---|---|---|---|---|---|---|
| | | Males | | Females | | $Xg^a$ | $Xg$ | |
| | | Xg(a+) | Xg(a-) | Xg(a+) | Xg(a-) | | | |
| **Bantu-speaking Negroes:** | | | | | | | | |
| Herero: | | | | | | | | |
| Botswana | 69 | 12 | 24 | 21 | 12 | .370 | .630 | Jenkins (1972) |
| Himba | 50 | 12 | 22 | 7 | 9 | .307 | .693 | Present study |
| Chimba | 15 | 2 | 6 | 4 | 3 | .305 | .695 | " |
| **Khoisan-speaking Negroes:** | | | | | | | | |
| Dama | 69 | 12 | 23 | 18 | 16 | .325 | .675 | Nurse et al (1976) |
| **Khoisan:** | | | | | | | | |
| San: | | | | | | | | |
| Central: | | | | | | | | |
| Hei//om | 94 | 21 | 31 | 28 | 14 | .414 | .586 | Nurse & Jenkins (1977a) |
| Southern: | | | | | | | | |
| !Xõ | 37 | 10 | 11 | 5 | 11 | .300 | .700 | Nurse & Jenkins (1977a) |
| Uncertain: | | | | | | | | |
| Eastern ǂHuã | 32 | 6 | 4 | 14 | 8 | .444 | .556 | Nurse & Jenkins (1977a) |
| **Caucasoids:** | | | | | | | | |
| Indian: | | | | | | | | |
| Natal: | | | | | | | | |
| Tamil-speaking | 105 | 81 | 24 | | | .771 | .229 | Moores (1980) |
| Telegu-speaking | 106 | 77 | 29 | | | .726 | .274 | " |
| Hindi-speaking | 101 | 72 | 29 | | | .713 | .287 | " |
| Moslem | 100 | 64 | 36 | | | .640 | .360 | " |

# Table 8. Haptoglobin and Transferrin Systems

| Population | \(n\) | \multicolumn{5}{c}{Haptoglobins — Phenotypes} | | | | | Gene Frequencies | | \(x^2[1]\) | \(n\) | \multicolumn{3}{c}{Transferrins — Phenotypes} | | | Gene Frequencies | | | \(x^2[1]\) | References |
|---|---|---|---|---|---|---|---|---|---|---|---|---|---|---|---|---|---|---|---|
| | | 1 | 2-1 | 2-1m | 2 | 0 | \(Hp^1\) | \(Hp^2\) | \(x^2[1]\) | \(n\) | C | CD1 | D1 | \(Tf^C\) | \(Tf^{D1}\) | \(Tf^{D1}\) | \(x^2[1]\) | References |
| **Bantu-speaking Negroes:** | | | | | | | | | | | | | | | | | | |
| **Nguni:** | | | | | | | | | | | | | | | | | | |
| Swati | 104 | 22 | 38 | 12 | 27 | 5 | .475 | .525 | .014 | 104 | 101 | 3 | 0 | .986 | .014 | | | Jenkins (1972) |
| Swati | 92 | 18 | 38 | 7 | 21 | 8 | .482 | .518 | .458 | 92 | 89 | 5 | 0 | .984 | .016 | | | Hitzeroth & Hummel (1978) |
| Swati | 68 | 21 | 24 | 6 | 15 | 2 | .545 | .455 | .504 | 68 | 63 | 5 | 0 | .963 | .037 | | | Jenkins (1972) |
| Zulu | 100 | 24 | 44 | 6 | 23 | 3 | .505 | .495 | .269 | 100 | 100 | 0 | 0 | 1.000 | .000 | | | McDermid & Vos (1971a) |
| Zulu | 304 | 71 | 137 | 24 | 63 | 9 | .511 | .489 | 2.512 | 304 | 292 | 12 | 0 | .980 | .020 | | | Hitzeroth & Hummel (1978) |
| Zulu | 118 | 36 | 49 | 8 | 25 | 4 | .548 | .452 | .551 | 118 | 109 | 8 | 1 | .958 | .042 | | 3.251 | Nurse et al (1974) |
| Zulu/Tonga | 75 | 20 | 29 | 8 | 10 | 8 | .575 | .425 | 1.100 | 75 | 75 | 0 | 0 | 1.000 | .000 | | | Jenkins (1972) |
| Xhosa | 59 | 15 | 25 | | 16 | 2 | .491 | .509 | .276 | 59 | 58 | 1 | 0 | .991 | .009 | | | Weissmann et al (1982) |
| Xhosa | 136 | 37 | 63 | 28 | 34 | | .511 | .489 | .182 | | | | | | | | | Giblett et al (1966) |
| Xhosa | 265 | 75 | 96 | 6 | 57 | 6 | .535 | .465 | .493 | 265 | 259 | 6 | 0 | .989 | .011 | | | Jenkins (1972) |
| Baca | 194 | 64 | 74 | | 44 | | .553 | .447 | 3.653 | 198 | 193 | 5 | 0 | .986 | .014 | | | " " |
| Hlubi | 117 | 33 | 55 | 1 | 22 | 5 | .549 | .451 | .081 | 114 | 113 | 1 | 0 | .996 | .004 | | | " " |
| Pondo | 56 | 11 | 26 | | 16 | 8 | .454 | .546 | .002 | 56 | 55 | 1 | 0 | .991 | .009 | | | " " |
| Ndebele | 167 | 35 | 61 | 9 | 54 | 8 | .440 | .560 | 1.812 | | | | | | | | | Hitzeroth & Hummel (1978) |
| Ndebele | 228 | 58 | 81 | 20 | 57 | 12 | .480 | .520 | 2.178 | 168 | 163 | 4 | 1 | .982 | .018 | | 15.530 | Jenkins (1972) |
| **Sotho/Tswana:** | | | | | | | | | | | | | | | | | | |
| S.Sotho | 218 | 63 | 83 | 15 | 46 | 11 | .541 | .459 | .451 | 218 | 198 | 20 | 0 | .954 | .046 | | | Giblett et al (1966) |
| S.Sotho | 284 | 87 | 95 | 24 | 63 | 15 | .545 | .455 | 3.146 | 284 | 263 | 21 | 0 | .963 | .037 | | | Moullec et al (1966) |
| S.Sotho (Highlands) | 90 | 28 | 32 | 3 | 23 | 4 | .529 | .471 | 2.849 | 89 | 86 | 3 | 0 | .983 | .017 | | | Beaumont et al (1979) |
| S.Sotho (Lowlands) | 92 | 21 | 46 | 4 | 19 | 2 | .511 | .489 | 1.112 | 91 | 89 | 2 | 0 | .998 | .011 | | | " " |
| Pedi (Rural) | 76 | 22 | 25 | 3 | 23 | 2 | .493 | .507 | 3.461 | | | | | | | | | Jenkins (1972) |
| Pedi (Rural) | 263 | 60 | 102 | 19 | 60 | 22 | .500 | .500 | .004 | 262 | 253 | 9 | 0 | .983 | .017 | | | Hitzeroth & Hummel (1978) |
| Pedi (Urban) | 110 | 33 | 39 | 5 | 17 | 16 | .585 | .415 | .115 | | | | | | | | | Jenkins (1972) |
| Pedi (Urban) | 97 | 33 | 36 | 5 | 25 | 0 | .541 | .459 | 3.514 | 96 | 91 | 5 | 0 | .974 | .026 | | | Present study |
| Fokeng | 197 | 56 | 64 | 14 | 52 | 11 | .511 | .489 | 4.777 | 197 | 182 | 15 | 0 | .962 | .038 | | | Jenkins (1972) |
| Ngwato | 56 | 19 | 20 | 8 | 12 | 4 | .565 | .435 | .965 | 56 | 55 | 1 | 0 | .991 | .009 | | | Jenkins (1972) |
| Tswana | 148 | 41 | 52 | 8 | 43 | 4 | .493 | .507 | 3.992 | 154 | 145 | 8 | 1 | .968 | .032 | | 4.468 | Hitzeroth & Hummel (1978) |
| Kgalagadi | 55 | 17 | 18 | 4 | 14 | 2 | .528 | .472 | 1.470 | 54 | 49 | 4 | 1 | .945 | .055 | | 3.724 | Jenkins & Steinberg (1966) |
| Kgalagadi | 111 | 27 | 42 | 7 | 26 | 9 | .505 | .495 | .156 | 111 | 102 | 9 | 0 | .959 | .041 | | | Present study |
| **Venda:** | | | | | | | | | | | | | | | | | | |
| Rural | 256 | 63 | 98 | 41 | 43 | 11 | .541 | .459 | .496 | 256 | 245 | 11 | 0 | .979 | .021 | | | Jenkins (1972) |
| Rural | 104 | 31 | 43 | 10 | 20 | 0 | .553 | .447 | .009 | 103 | 95 | 8 | 0 | .961 | .039 | | | Present study |
| Urban | 239 | 73 | 84 | 22 | 48 | 12 | .555 | .445 | .669 | 239 | 217 | 21 | 1 | .952 | .048 | | .027 | Jenkins (1972) |
| Urban | 41 | 15 | 12 | 6 | 8 | 0 | .585 | .415 | .374 | 42 | 41 | 1 | 0 | .988 | .012 | | | Hitzeroth & Hummel (1978) |
| Urban (Pretoria) | 76 | 24 | 30 | 4 | 15 | 3 | .562 | .438 | .215 | 76 | 74 | 2 | 0 | .987 | .013 | | | Jenkins (1972) |
| **Tsonga/Ronga:** | | | | | | | | | | | | | | | | | | |
| Ronga | 41 | 14 | 15 | 0 | 11 | 1 | .538 | .462 | 2.412 | | | | | | | | | Matznetter & Spielmann (1969) |
| "Shangaan"/Tsonga | 116 | 35 | 50 | 2 | 25 | 4 | .546 | .454 | .462 | | | | | | | | | " " |
| Shangaan-Tsonga | 104 | 30 | 34 | 7 | 21 | 12 | .549 | .451 | .938 | 104 | 103 | 1 | 0 | .995 | | .050 | | Jenkins (1972) |
| Shangaan-Tsonga | 155 | 46 | 57 | 13 | 30 | 9 | .555 | .445 | .127 | 155 | 150 | 5 | 0 | .984 | | .016 | | Hitzeroth & Hummel (1978) |

319

# Table 8 (*cont.*)

| Population | Haptoglobins |  |  |  |  |  |  |  |  | Transferrins |  |  |  |  |  |  | References |
|---|---|---|---|---|---|---|---|---|---|---|---|---|---|---|---|---|---|
|  | n | 1 | 2-1 | 2-1m | 2 | 0 | $Hp^1$ | $Hp^2$ | $x^2$[1] | n | C | $CD_1$ | $D_1$ | $Tf^C$ | $Tf^{D1}$ | $x^2$[1] |  |
| **Kavango:** |  |  |  |  |  |  |  |  |  |  |  |  |  |  |  |  |  |
| Mbukushu | 54 | 23 | 22 | 4 | 3 | 2 | .692 | .308 | 1.53 | 54 | 53 | 1 | 0 | .991 | .009 |  | Nurse & Jenkins (1977b) |
| Gciriku | 111 | 34 | 35 | 8 | 12 | 22 | .624 | .376 | .07 | 111 | 104 | 7 | 0 | .968 | .032 |  | " |
| Sambyu | 94 | 32 | 32 | 5 | 12 | 13 | .624 | .376 | .06 | 94 | 93 | 1 | 0 | .995 | .005 |  | " |
| Kwangali | 107 | 49 | 33 | 5 | 12 | 8 | .687 | .313 | 1.16 | 107 | 95 | 12 | 0 | .944 | .056 |  | " |
| **Ambo:** |  |  |  |  |  |  |  |  |  |  |  |  |  |  |  |  |  |
| Kwambi | 92 | 44 | 31 | 6 | 5 | 6 | .727 | .273 | .58 | 92 | 92 | 0 | 0 | 1.000 | .000 |  | Nurse *et al* (1983) |
| Kwaluudhi | 61 | 32 | 21 | 1 | 5 | 2 | .729 | .271 | .20 | 61 | 61 | 0 | 0 | 1.000 | .000 |  | Nurse *et al* (1983) |
| Ndonga | 100 | 33 | 43 | 2 | 17 | 5 | .584 | .416 | .06 | 100 | 100 | 0 | 0 | 1.000 | .000 |  | " |
| Mbalantu | 53 | 28 | 15 | 2 | 2 | 6 | .777 | .223 | .07 | 53 | 51 | 2 | 0 | .981 | .019 |  | " |
| Kwanyama | 97 | 49 | 28 | 5 | 8 | 7 | .728 | .272 | .50 | 97 | 95 | 2 | 0 | .990 | .010 |  | " |
| Ngandjera | 63 | 29 | 25 |  | 9 |  | .659 | .341 | .89 | 63 | 62 | 1 | 0 | .992 | .008 |  | " |
| Nkolonkadhi | 30 | 14 | 14 | 1 |  | 1 | .741 | .259 | 2.83 | 30 | 30 | 0 | 0 | 1.000 | .000 |  | " |
| **Herero:** |  |  |  |  |  |  |  |  |  |  |  |  |  |  |  |  |  |
| Botswana | 62 | 15 | 24 | 3 | 12 | 8 | .528 | .472 | .001 | 60 | 58 | 2 | 0 | .983 | .017 |  | Jenkins (1972) |
| SWA/Namibia | 83 | 22 | 34 | 1 | 15 | 11 | .549 | .451 | .013 | 83 | 82 | 1 | 0 | .994 | .006 |  | Present study |
| Mbanderu | 116 | 29 | 61 |  | 26 |  | .513 | .487 | .318 | 118 | 114 | 4 | 0 | .983 | .017 |  | " |
| Himba | 76 | 16 | 39 | 6 | 9 | 6 | .550 | .450 | 6.251 | 76 | 74 | 2 | 0 | .983 | .017 |  | " |
| Chimba | 33 | 8 | 18 |  | 4 | 3 | .567 | .443 | 1.480 | 33 | 33 | 0 | 0 | 1.000 | .000 |  | " |
| **Khoisan-speaking Negroes:** |  |  |  |  |  |  |  |  |  |  |  |  |  |  |  |  |  |
| Sarwa | 79 | 17 | 26 | 7 | 12 | 17 | .540 | .460 | .320 | 79 | 75 | 4 | 0 | .949 | .051 |  | Chasko *et al* (1979) |
| Kwengo | 35 | 10 | 11 | 3 | 4 | 7 | .607 | .393 | .060 | 35 | 34 | 1 | 0 | .986 | .014 |  | Nurse & Jenkins (1977b) |
| Dama | 117 | 49 | 36 | 9 | 17 | 6 | .664 | .356 | 1.480 | 117 | 117 | 0 | 0 | 1.000 | .000 |  | Nurse *et al* (1976) |
| Dama |  |  |  |  |  |  |  |  |  | 238 | 231 | 6 | 1 | .985 | .013 |  | Schumacher *et al* (1979) |
| **Khoi:** |  |  |  |  |  |  |  |  |  |  |  |  |  |  |  |  |  |
| Keetmanshoop Nama | 153 | 55 | 69 | 1 | 27 | 1 | .594 | .406 | .343 | 153 | 150 | 3 | 0 | .990 | .010 |  | Jenkins (1972) |
| Sesfontein Topnaars | 58 | 17 | 34 | 0 | 7 | 0 | .590 | .410 | 2.537 | 58 | 58 | 0 | 0 | 1.000 | .000 |  | Jenkins & Brain (1967) |
| Sesftonein Nama | 42 | 20 | 20 | 1 | 0 | 0 | .726 | .274 | 2.838 | 42 | 41 | 1 | 0 | .988 | .012 |  | Jenkins (1972) |
| **San:** |  |  |  |  |  |  |  |  |  |  |  |  |  |  |  |  |  |
| **Northern:** |  |  |  |  |  |  |  |  |  |  |  |  |  |  |  |  |  |
| Glanglai | 36 | 0 | 7 | 3 | 17 | 9 | .185 | .815 | 1.40 | 36 | 31 | 5 | 0 | .931 | .069 |  | Nurse *et al* (1977) |
| Tsumkwe !Kung | 272 | 21 | 111 | 4 | 118 | 18 | .309 | .691 | .60 | 272 | 192 | 70 | 10 | .834 | .166 | 1.248 | Jenkins (1972) |
| Tsumkwe !Kung | 121 | 9 | 51 | 3 | 53 | 5 | .310 | .690 | .86 | 121 | 86 | 34 | 1 | .851 | .149 | 1.45 | Nurse *et al* (1977) |
| Samanlaika " | 34 | 2 | 10 | 0 | 19 | 3 | .226 | .774 | .17 | 34 | 22 | 10 | 2 | .794 | .206 | .38 | " |
| Dobe " | 423 | 38 | 155 | 5 | 173 | 52 | .318 | .682 | .01 | 418 | 301 | 107 | 10 | .848 | .152 | .014 | Jenkins (1972) |
| /Ai/ai " | 60 | 8 | 20 | 0 | 24 | 8 | .346 | .654 | 1.12 | 60 | 36 | 22 | 2 | .783 | .217 | .371 | " |
| /Du/da " | 101 | 5 | 43 | 0 | 48 | 6 | .276 | .724 | 1.38 | 101 | 57 | 38 | 6 | .752 | .248 | .011 | " |
| //Au//en | 265 | 19 | 101 | 2 | 137 | 6 | .272 | .728 | .00 | 265 | 220 | 44 | 1 | .913 | .087 | .60 | Nurse *et al* (1977) |

Table of Tf (transferrin) type distributions and gene frequencies

| Population | n | | | | | | gene freq | | χ² | n | | | | gene freq | | | Reference |
|---|---|---|---|---|---|---|---|---|---|---|---|---|---|---|---|---|---|
| **Central:** | | | | | | | | | | | | | | | | | |
| Nharo | 147 | 23 | 65 | 3 | 56 | 0 | .388 | .612 | .10 | 147 | 124 | 22 | 1 | .918 | .082 | .000 | Nurse *et al* (1977) |
| G/wi | 94 | 10 | 36 | 2 | 44 | 2 | .315 | .685 | .18 | 93 | 87 | 6 | 0 | .968 | .032 | | Jenkins *et al* (1975) |
| G//ana | 50 | 3 | 23 | 0 | 22 | 2 | .302 | .698 | .92 | 50 | 49 | 1 | 0 | .990 | .010 | | " |
| **Southern:** | | | | | | | | | | | | | | | | | |
| Mainly !Xõ | 125 | 13 | 48 | 2 | 59 | 3 | .311 | .689 | 0.24 | 125 | 114 | 11 | 0 | .956 | .044 | | Jenkins & Steinberg (1966) |
| !Xõ | 50 | 2 | 21 | 0 | 27 | 0 | .250 | .750 | 0.69 | 50 | 44 | 6 | 0 | .940 | .060 | | Jenkins & Nurse (1976) |
| G!aokx'ate | 33 | 11 | 19 | 0 | 3 | 0 | .621 | .379 | 1.58 | 33 | 33 | 0 | 0 | 1.000 | .000 | | Nurse *et al* (1977) |
| **Uncertain:** | | | | | | | | | | | | | | | | | |
| Eastern ǂHuã | 33 | 5 | 14 | 0 | 14 | 0 | .364 | .636 | .25 | 35 | 35 | 0 | 0 | 1.000 | .000 | | Nurse & Jenkins (1977a) |
| **Caucasoids:** | | | | | | | | | | | | | | | | | |
| *Afrikaans-speaking:* | | | | | | | | | | | | | | | | | |
| Johannesburg | 77 | 11 | 44 | 0 | 21 | 1 | .434 | .566 | 2.417 | 77 | 75 | 2 | 0 | .987 | .013 | | Present study |
| Johannesburg | 146 | 26 | 66 | 0 | 51 | 3 | .413 | .587 | 0.327 | 146 | 145 | 1 | 0 | .997 | .003 | | " |
| SWA/Namibia | 57 | 9 | 36 | 0 | 11 | 1 | .482 | .518 | 4.58 | 57 | 57 | 0 | 0 | 1.000 | .000 | | Palmhert-Keller *et al* (1983) |
| *German-speaking:* | | | | | | | | | | | | | | | | | |
| SWA/Namibia | 117 | 38 | 43 | 0 | 19 | 17 | .595 | .405 | 1.16 | 117 | 117 | 0 | 0 | 1.000 | .000 | | Palmhert-Keller *et al* (1983) |
| **Indian:** | | | | | | | | | | | | | | | | | |
| *Transvaal:* | | | | | | | | | | | | | | | | | |
| Hindu | 46 | 1 | 17 | 0 | 24 | 4 | .226 | .774 | 1.060 | 46 | 45 | 1 | 0 | .987 | .011 | | Present study |
| Moslem | 42 | 1 | 19 | 0 | 20 | 2 | .262 | .738 | 2.080 | 42 | 41 | 1 | 0 | .988 | .012 | | " |
| *Natal:* | | | | | | | | | | | | | | | | | |
| Mixed | 193 | 1 | 42 | 0 | 147 | 3 | .116 | .884 | 1.202 | 193 | 193 | 0 | 0 | 1.000 | .000 | | McDermid & Vos (1971c) |
| **Hybrid peoples:** | | | | | | | | | | | | | | | | | |
| *Baster-like:* | | | | | | | | | | | | | | | | | |
| Rehoboth | 119 | 22 | 67 | 1 | 17 | 12 | .523 | .477 | 7.530 | 119[b] | 116 | 2 | 0 | .988 | .008 | | Nurse *et al* (1982) |
| !Kuboes | 136 | 50 | 59 | 0 | 26 | 1 | .589 | .411 | 1.290 | 133 | 132 | 1 | 0 | .996 | .004 | | Jenkins (1972) |
| Riemvaasmaak | 88 | 28 | 37 | 6 | 11 | 6 | .608 | .392 | .770 | 88 | 88 | 0 | 0 | 1.000 | .000 | | Nurse & Jenkins (1978) |
| Griqua: | 229 | 56 | 95 | 13 | 47 | 18 | .521 | .479 | .14 | 229[d] | 219 | 9 | 1 | .958 | .042 | 0.57 | Nurse & Jenkins (1975) |
| **"Cape-Coloureds":** | | | | | | | | | | | | | | | | | |
| Johannesburg | 336 | 61 | 124 | 3 | 138 | 10 | .382 | .618 | 9.98 | | | | | | | | Jenkins (1972) |
| Johannesburg | 108 | 20 | 47 | 3 | 35 | 3 | .429 | .571 | .079 | 108 | 102 | 6 | 0 | .972 | .028 | | " |
| Orange River Valley | 175 | 42 | 78 | 0 | 53 | 2 | .468 | .532 | 1.527 | 177 | 171 | 6 | 0 | .983 | .017 | | " |
| Sesfontein | 81 | 41 | 25 | 6 | 4 | 5 | .753 | .257 | .376 | 81 | 75 | 6 | 0 | .963 | .037 | | " |

[b] one individual TfCB giving TfB allele frequency of .004

[d] one individual TfCB giving TfB allele frequency of .002

321

## Table 9. Gm System in Negro, Khoi, and San Peoples of Southern Africa

| Population | n | Phenotypes | | | | | | | | | | Haplotype Frequencies | | | | | | | | $\chi^2$ | df | Ref. |
|---|---|---|---|---|---|---|---|---|---|---|---|---|---|---|---|---|---|---|---|---|---|---|
| | | 1,5,13,14 | 1,5,13,14 | 1,5,6,14 | 1,5,6,13 | 1,5,6,14 | 1,13 | 1,5,13 | 1 | 1,3,5,13,14 | 1,3,5,6,13,14 | 1,5,13,14 | 1,5,6,14 | 1,5,6 | 1,5,13 | 1 | 1,5,14 | 1,14 | 3,5,13,14 | | | |
| **Bantu-speaking Negroes:** | | | | | | | | | | | | | | | | | | | | | | |
| *Nguni:* | | | | | | | | | | | | | | | | | | | | | | |
| Swati | 126 | 62 | 49 | 6 | 5 | 2 | 1 | 1 | | 0 | 0 | .608 | .131 | .150 | .106 | | | | .004 | 3.475 | 4 | (1) |
| Zulu | 130 | 61 | 43 | 10 | 7 | 4 | 5 | 5 | | 0 | 0 | .496 | .130 | .163 | .188 | | .023 | | | .044 | 1 | (1) |
| Ndebele | 103 | 50 | 43 | 5 | 1 | 4 | | | | 2 | 1 | .676 | .099 | .200 | .021 | .005 | | | | | | (1) |
| Xhosa | 214 | 110 | 48 | 11 | 8 | 6 | 26 | | 2 | | | .442 | .108 | .083 | .254 | .105 | | | .007 | 1.747 | 2 | (1) |
| Baca | 137 | 67 | 41 | 10 | 8 | 4 | 6 | 9 | | | | .508 | .129 | .152 | .208 | | | | .004 | 2.110 | 2 | (1) |
| Hlubi | 145 | 78 | 43 | 8 | 8 | 2 | 9 | 4 | | 3 | | .521 | .155 | .050 | .157 | .117 | | | | | | (1) |
| Pondo | 112 | 53 | 32 | 16 | 5 | 2 | 4 | | 3 | | | .433 | .170 | .127 | .188 | | | .082 | | 1.788 | 1 | (1) |
| *Sotho/Tswana:* | | | | | | | | | | | | | | | | | | | | | | |
| Tswana | 155 | 93 | 36 | 9 | 4 | 4 | 9 | | | 0 | 2 | .565 | .106 | .107 | .222 | .102 | | | .003 | 9.146 | 2 | (1) |
| S.Sotho | 149 | 81 | 39 | 13 | 6 | 3 | 4 | | | 2 | 1 | .538 | .138 | .097 | .121 | .085 | | | .003 | 3.957 | 3 | (1) |
| Pedi | 146 | 65 | 54 | 18 | 5 | 0 | 2 | 2 | | 1 | 0 | .533 | .231 | .075 | .072 | | | | | 7.150 | 3 | (1) |
| Kgalagadi | 48 | 26 | 7 | 2 | 8 | 3 | 2 | 2 | | 0 | 0 | .413 | .145 | .240 | .268 | | | | | 1.936 | 2 | (1) |
| Venda | 80 | 38 | 31 | 5 | 1 | 2 | 3 | 3 | | 0 | 0 | .554 | .171 | .116 | .158 | | .079 | | | 2.671 | 2 | (1) |
| *Tsonga/Ronga:* | | | | | | | | | | | | | | | | | | | | | | |
| "Shangaan" | 152 | 73 | 67 | 7 | 2 | 2 | 1 | 1 | | 0 | 0 | .648 | .162 | .124 | .065 | | | | | 2.380 | 2 | (1) |
| *Kavango:* | | | | | | | | | | | | | | | | | | | | | | |
| Mbukushu | 115 | 52 | 53 | 7 | 0 | 2 | 0 | 0 | | 0 | 1 | .683 | .165 | .148 | .101 | .097 | .004 | .008 | | .806 | 1 | (1) |
| Gciriku | 59 | 25 | 24 | 1 | 0 | 7 | 0 | 0 | | 1 | 0 | .581 | .025 | .288 | .055 | | | | | .768 | 1 | (1) |
| Sambyu | 98 | 40 | 42 | 10 | 1 | 4 | 0 | 0 | | 0 | 0 | .569 | .185 | .145 | .027 | | | | | 2.540 | 1 | (1) |
| Kwangali | 46 | 24 | 17 | 3 | | 1 | 0 | | | 2 | 0 | .662 | .134 | .148 | | | | | | .319 | 1 | (1) |
| Kwangali | 105 | 54 | 40 | 7 | 2 | 2 | | 2 | | 2 | 0 | .664 | .144 | .141 | .051 | | | | | .131 | 1 | (2) |
| Mbunja | 111 | 45 | 50 | 12 | 1 | 2 | | 1 | | | 1 | .612 | .219 | .137 | .027 | | | .005 | | .048 | 1 | (1) |
| **Ambo:** | | | | | | | | | | | | | | | | | | | | | | |
| Kwambi | 119 | 70 | 44 | 3 | 0 | 2 | 0 | 0 | | 0 | 0 | .773 | .083 | .143 | | | | | | .346 | 1 | (1) |
| Kwanyama | 118 | 57 | 54 | 4 | 1 | 2 | 0 | 0 | | 0 | 0 | .696 | .119 | .165 | .020 | | | | | 2.581 | 2 | (1) |

Table of Gm haplotype data for Herero, Khoisan-speaking Negroes, and Khoisan (Khoi and San) populations. Column headers (phenotype and haplotype labels) are not present on this page; columns are given positionally.

| Population | n | c1 | c2 | c3 | c4 | c5 | c6 | c7 | c8 | h1 | h2 | h3 | h4 | h5 | h6 | h7 | χ² | d.f. | Ref |
|---|---|---|---|---|---|---|---|---|---|---|---|---|---|---|---|---|---|---|---|
| **Herero:** | | | | | | | | | | | | | | | | | | | |
| Botswana | 60[V] | 31 | 20 | 2 | 3 | 3 | 1 | 0 | 0 | .594 | .063 | .203 | .122 | .017 | | | .542 | 2 | (3) |
| SWA/Namibia | 108[V] | 54 | 36 | 5 | 3 | 5 | 1 | 0 | 4 | .619 | .082 | .192 | .085 | .005 | | .019 | 2.925 | 2 | (2) |
| Himba | 54 | 25 | 23 | 3 | 2 | 0 | 0 | 0 | 1 | .612 | .185 | .102 | .092 | | | .009 | 2.887 | 1 | (1) |
| **Khoisan-speaking Negroes:** | | | | | | | | | | | | | | | | | | | |
| Sarwa | 116 | 48[V] | 23 | 5[V] | 13 | 15 | 2[V] | 8 | 2 | .347 | .292 | .236 | | | .084 | .042 | 1.470 | 3 | (2) |
| Kwengo | 52 | 30 | 14 | 2 | 1 | 1 | 3[V] | 2 | | .463 | | .196 | .138 | | .145 | | 4.971 | 1 | (2) |
| Dama(Sesfontein) | 25 | 9 | 13 | 2 | 0 | | 1 | | | .601 | .233 | .000 | .166 | | .029 | | .150 | 1 | (4) |
| Dama(Okombahe) | 101 | 49 | 38 | 7 | 1 | 5 | | | | .592 | .272 | .014 | | | .102 | .020 | 3.357 | 3 | (2) |
| **Khoisan:** | | | | | | | | | | | | | | | | | | | |
| **Khoi:** | | | | | | | | | | | | | | | | | | | |
| Topnaars (Kuiseb) | 57[V] | 19 | 23 | 7 | 1 | 1 | 5 | | 0 | .330 | .280 | .280 | .040 | | .040 | .000 | 2.730 | 5 | (1) |
| Topnaars (Sesfontein) | 42 | 21 | 12 | 2 | 1 | 0 | 6 | | 0 | .429 | .163 | .333 | .048 | | .000 | .000 | .940 | 2 | (1) |
| **San:** | | | | | | | | | | | | | | | | | | | |
| **Northern:** | | | | | | | | | | | | | | | | | | | |
| G!ag!ai | 33[V] | 9 | 5 | 0 | 3 | 3 | 1 | 3 | | .179 | .076 | | .558 | .169 | .057 | .019 | 2.803 | 1 | (2) |
| Tsumkwe !Kung | 103 | 41 | 1 | 0 | 0 | 1 | 0 | 0 | | .195 | .000 | .021 | .628 | .099 | .029 | | 5.200 | 2 | (5) |
| Samaniaika " | 32 | 12 | 0 | 1 | 2 | 0 | 0 | 0 | | .213 | .000 | .017 | .593 | .148 | .029 | | 7.760 | 3 | (5) |
| Dobe " | 394 | 188 | 19 | 0 | 0 | 0 | 0 | 0 | | .288 | .024 | .004 | .603 | .081 | .000 | | 1.340 | 2 | (6) |
| /Ai/ai " | 62 | 37 | 2 | 0 | 0 | 0 | 0 | 0 | | .378 | .016 | | .501 | .105 | .000 | | 1.220 | 1 | (6) |
| /Du/da " | 100 | 49 | 6 | 0 | 0 | 0 | 0 | 7 | | .295 | .030 | .000 | .555 | .120 | .000 | .022 | 3.870 | 2 | (6) |
| //Au//en | 258 | 115 | 21 | 1 | 0 | 2 | 2 | 6 | | .276 | .041 | .006 | .499 | .155 | .000 | .022 | 9.370 | 5 | (6) |
| **Central:** | | | | | | | | | | | | | | | | | | | |
| Nharo | 138 | 60 | 13 | 2 | 2 | 1 | 1 | | | .270 | .050 | .015 | .526 | .134 | .000 | .005 | 3.060 | 3 | (6) |
| **Southern:** | | | | | | | | | | | | | | | | | | | |
| Mainly !Xõ | 72 | 30 | 14 | 2 | 1 | 0 | 2 | | | .290 | .111 | .012 | .452 | .135 | | .000 | 1.700 | 1 | (7) |
| !Xõ | 49[V] | 23 | 6 | 1 | 0 | 0 | 1 | | 0 | .302 | .071 | | .555 | .071 | | | 1.221 | 1 | (2) |
| G!aokx'ate | 27[V] | 8 | | 0 | 0 | 0 | 1 | | 0 | .163 | | | .630 | .185 | | .022 | .314 | 1 | (2) |
| **Uncertain:** | | | | | | | | | | | | | | | | | | | |
| Eastern ‡Hua | 33[V] | 16 | | 5 | | 3 | | 3 | | .173 | | | .399 | .121 | .205 | .102 | 6.371 | 3 | (2) |

[V] excludes one individual Gm(5,6,13,14) and one Gm(1,2,5,13,14) individual omitted; [V] includes one individual who was Gm(21+); [S] includes one individual Gm(3,5,13,14); omitting one Gm(1,5,13,21) individual; [V] includes the frequency of .032 for an unusual haplotype Gm[1,5,13,14,21]; one individual Gm(1,5). [V] one individual Gm(1,5); [V] two individuals Gm(1,5). [∇] one Gm(1,2,5,6,14) individual omitted; [∇] Gm[1,2] frequency .005; [∇] Gm[1,2] frequency .005.

References: (1) Jenkins et al (1970); (2) Present study; (3) Jenkins (1972); (4) Nurse and Jenkins (1977a); (5) Nurse et al (1977); (6) Steinberg et al (1975); (7) Jenkins and Steinberg (1966).

# Table 10. Gm System in Khoi, Caucasoid and 'Hybrid' Peoples of Southern Africa

| Population | n | Phenotypes | | | | | | | | | | | | | | | | | | |
|---|---|---|---|---|---|---|---|---|---|---|---|---|---|---|---|---|---|---|---|---|
| | | 1·5·13·14 | 1·5·13·14·21 | 1·5·6·13·14 | 1·5·6·13·14·21 | 1·5·6·14 | 1·5·6·14·21 | 1·13 | 1·5·13·21 | 1·5·6·13 | 1·5·6·21 | 1·6 | 1·2·13·(21) | 1·2·5·13·14·(21) | 1·2·5·6·(21) | 1·2·5·6·14·(21) | 1·2·3·5·13·14·(21) | 1·3·5·13·14 | 1·3·5·13·14·21 | 1·3·5·6·13·14 |
| **Khoisan:** | | | | | | | | | | | | | | | | | | | | |
| *Khoi:* | | | | | | | | | | | | | | | | | | | | |
| Topnaars ˇ₁ | 57 | 16 | | 23 | | 5 | | 5 | | 1 | | 0 | | 0 | 1 | 1 | 2 | 0 | 2 | 0 |
| Topnaars ˇ₃ | 42 | 20 | 0 | 12 | 0 | 1 | 0 | 4 | 0 | 1 | 0 | 0 | 2 | 1 | 0 | 1 | 0 | 0 | 0 | 0 |
| Keetmanshoop Nama | 149 | 86 | 19 | 5 | | 0 | 1 | 18 | 4 | 1 | 0 | 0 | 1 | 1 | 0 | 0 | 0 | 0 | 1 | 0 | 0 |
| **Caucasoid:** | | | | | | | | | | | | | | | | | | | | |
| *Indian:* | | | | | | | | | | | | | | | | | | | | |
| Natal (Hindu and Moslem) | 398 | 2 | | 0 | | 0 | | 26 | | 0 | | 0 | 63 | 7 | 1 | 0 | | 55 | | 137 | 0 |
| **Hybrid Peoples:** | | | | | | | | | | | | | | | | | | | | |
| *Baster-like:* | | | | | | | | | | | | | | | | | | | | |
| !Kuboes | 132 | 34 | 13 | 0 | 0 | 0 | 0 | 9 | 14 | 0 | 0 | 0 | 10 | 18 | 9 | 0 | 0 | 2 | 5 | 4 | 0 |
| *"Cape Coloureds"* | | | | | | | | | | | | | | | | | | | | |
| Johannesburg | 112 | 18 | | 10 | | 1 | | 6 | 2 | 2 | | 4 | 2 | 2 | 0 | 1 | 4 | | 43 | | 9 |

₁ˇ (Kuiseb Valley)

₃ˇ (Sesfontein)

| | | | | | Haplotypes | | | | | | | | | | | |
|---|---|---|---|---|---|---|---|---|---|---|---|---|---|---|---|---|
| 3/5/13/14 | 1/5/14 | 1/5/14/21 | 1/2/14/(21) | 1/(21) | 1/5/13/14 | 1/13 | 1/(21) | 1/5/6 | 1/5/6/14 | 1/5/14 | 1/2/21 | 3/5/13/14 | 1/3/5/13/14 | $x^2$ | df | References |
| 1 | 0 | | 0 | 0 | .33 | .28 | | .04 | .28 | | .04 | .04 | | 2.731 | 5 | Jenkins *et al* (1970) |
| 0 | 0 | 0 | 0 | 0 | .429 | .333 | | .027 | .163 | | .048 | | | .94 | 2 | Steinberg *et al* (1975) |
| 0 | 4 | 8 | 0 | 0 | .391 | .307 | .111 | .006 | .017 | .158 | .007 | .003 | | 14.92 | 4 | Steinberg *et al* (1975) |
| 50 | 0 | | 0 | 57 | .006 | .070 | .378 | | | | .173 | .357 | .015 | 1.473 | 3 | Jenkins *et al* (1970) |
| | 0 | 5 | 9 | 0 | .211 | .284 | .160 | | | .107 | .196 | .042 | | 15.08 | 8 | Steinberg *et al* (1975) |
| 6 | 0 | | 0 | 2 | .231 | .150 | .139 | .058 | .057 | | .061 | .304 | | 7.96 | 8 | Jenkins *et al* (1970) |

# Table 11. Inv System

| Population | n | Phenotype Inv(1) | Gene frequencies $Inv^1$ & $Inv^3$ $Inv^{1,2}$ | | Reference |
|---|---|---|---|---|---|
| **Bantu-speaking Negroes:** | | | | | |
| <u>Nguni</u>: | | | | | |
| Swati | 126 | 78 | .378 | .622 | Jenkins (1972) |
| Zulu | 131 | 77 | .358 | .642 | Jenkins (1972) |
| Ndebele | 129 | 75 | .353 | .647 | Jenkins (1972) |
| Xhosa | 214 | 129 | .370 | .630 | Jenkins (1972) |
| <u>Sotho/Tswana</u>: | | | | | |
| Tswana | 155 | 85 | .328 | .672 | Jenkins (1972) |
| S. Sotho | 144 | 78 | .309 | .691 | Jenkins (1972) |
| Pedi (Urban) | 147 | 43 | .159 | .841 | Jenkins (1972) |
| Kgalagadi | 48 | 32 | .425 | .575 | Jenkins (1972) |
| <u>Venda</u>: | 80 | 53 | .419 | .581 | Jenkins (1972) |
| <u>Kavango</u>: | | | | | |
| Mbukushu | 115 | 66 | .347 | .653 | Jenkins (1972) |
| Gciriku | 59 | 34 | .349 | .651 | Jenkins (1972) |
| Sambyu | 98 | 49 | .293 | .707 | Jenkins (1972) |
| Kwangali | 47 | 22 | .271 | .729 | Jenkins (1972) |
| <u>Ambo</u>: | | | | | |
| Kwambi | 119 | 62 | .308 | .692 | Jenkins (1972) |
| Kwanyama | 118 | 62 | .311 | .689 | Jenkins (1972) |
| <u>Herero</u>: | | | | | |
| Botswana | 62 | 38 | .349 | .651 | Jenkins (1972) |
| Himba | 54 | 27 | .293 | .707 | Jenkins (1972) |
| **Khoisan-speaking Negroes:** | | | | | |
| Dama | 52 | 42 | .562 | .438 | Jenkins (1972) |
| **Khoisan:** | | | | | |
| <u>Khoi</u>: | | | | | |
| Keetmanshoop Nama | 150 | 90 | .367 | .633 | Steinberg *et al* (1975) |
| Topnaars (Kuiseb) | 57 | 17 | .162 | .838 | Jenkins *et al.* (1970) |
| Topnaars (Sesfontein) | 42 | 25 | .364 | .636 | Steinberg *et al.* (1975) |
| <u>San</u>: | | | | | |
| <u>Northern</u>: | | | | | |
| Tsumkwe !Kung | 198 | 136 | .440 | .560 | Nurse & Jenkins (1977a) |
| Dobe " | 394 | 253 | .402 | .598 | " |
| /Ai/ai " | 62 | 35 | .340 | .660 | " |
| /Du/da " | 100 | 60 | .367 | .633 | " |
| //Au//en | 263 | 150 | .344 | .656 | " |
| <u>Central</u>: | | | | | |
| Nharo | 140 | 82 | .357 | .643 | Nurse & Jenkins (1977a) |
| **Hybrid Peoples:** | | | | | |
| <u>Baster-like</u>: | | | | | |
| !Kuboes | 132 | 49 | .207 | .793 | Jenkins (1972) |
| 'Cape Coloureds': | | | | | |
| Johannesburg | 112 | 60 | .198 | .802 | Jenkins *et al* (1970) |

# Table 12. Gc System

| Population | n | Phenotypes | | | | | Gene Frequencies | | | $\chi^2$[1] | References |
|---|---|---|---|---|---|---|---|---|---|---|---|
| | | 1 | 2-1 | 2 | 1-Ab | 2-Ab | $Gc^1$ | $Gc^2$ | $Gc^{Ab}$ | | |
| **Bantu-speaking Negroes:** | | | | | | | | | | | |
| *Nguni:* | | | | | | | | | | | |
| Swati | 43 | 33 | 9 | 1 | | | .872 | .128 | | .152 | Hitzeroth & Hummel (1978) |
| Zulu | 119 | 97 | 22 | 0 | | | .908 | .092 | | | Hitzeroth & Hummel (1978) |
| Zulu | 169 | 121 | 34 | 2 | 10 | 2 | .846 | .118 | .036 | .504* | McDermid & Vos (1971) |
| Ndebele | 169 | 144 | 24 | 1 | | | .923 | .077 | | .008 | Hitzeroth & Hummel (1978) |
| Xhosa | 136 | 107 | 27 | 2 | | | .886 | .114 | | .039 | Weissman *et al* (1982) |
| *Sotho/Tswana:* | | | | | | | | | | | |
| Pedi | 269 | 224 | 43 | 2 | | | .913 | .087 | | .003 | Hitzeroth & Hummel (1978) |
| Tswana | 153[v] | 131 | 20 | 2 | | | .922 | .078 | | 1.417 | Hitzeroth & Hummel (1978) |
| Ngwato | 91 | 77 | 7 | 0 | 4 | 0 | .923 | .038 | .022 | | Present study |
| *Venda:* | | | | | | | | | | | |
| Urban | 43 | 33 | 9 | 1 | | | .872 | .128 | | .164 | Hitzeroth & Hummel (1978) |
| Rural | 104 | 82 | 17 | 0 | 5 | 0 | .894 | .082 | .024 | | Present study |
| *Tsonga/Ronga:* | | | | | | | | | | | |
| Ronga | 41 | 38 | 3 | 0 | | | .963 | .037 | | | Matznetter & Spielmann (1969) |
| Shangaan | 115 | 97 | 16 | 2 | | | .913 | .087 | | 1.763 | Matznetter & Spielmann (1969) |
| Tsonga | 228 | 216 | 12 | 0 | | | .974 | .026 | | | Matznetter & Spielmann (1969) |
| Shangaan/Tsonga | 155 | 131 | 22 | 2 | | | .916 | .084 | | .902 | Hitzeroth & Hummel (1978) |
| **Khoisan-speaking Negroes:** | | | | | | | | | | | |
| Dama | 445 | 345 | 93 | 7 | | | .880 | .120 | | .065 | Knussman & Knussman (1976) |
| **Khoisan:** | | | | | | | | | | | |
| *San:* | | | | | | | | | | | |
| Sekele | 117[♂] | 72 | 13 | 0 | 24 | 5 | .773 | .077 | .150 | 3.178* | Present study |
| !Xo | 72 | 54 | 18 | 0 | | | .875 | .125 | | | Nurse & Jenkins (1977a) |

[v] includes three individuals with phenotypes 1-Va (V representing an unidentified anodal variant), giving a frequency of Va = .017

[♂] including three individuals homozygous for the Ab phenotype

* for 2 degrees of freedom

# Table 13. Haemoglobin Systems

| Population | n | β Phenotypes A | AS | AD | $Hb\beta^A$ | $Hb\beta^S$ | $Hb\beta^D$ | $A_2$ | $A_2B_2$ | $B_2$ | $Hb\delta^{A_2}$ | $Hb\delta^{B_2}$ | References |
|---|---|---|---|---|---|---|---|---|---|---|---|---|---|
| **Bantu-speaking Negroes:** | | | | | | | | | | | | | |
| **Nguni:** | | | | | | | | | | | | | |
| Swati | 92 | 92 | 0 | 0 | 1.000 | .000 | .000 | 92 | 0 | 0 | 1.000 | .000 | Nurse *et al* (1974) |
| Zulu | 123 | 123 | 0 | 0 | 1.000 | .000 | .000 | 123 | 0 | 0 | 1.000 | .000 | Jenkins (1972) |
| Zulu | 102 | 102 | 0 | 0 | 1.000 | .000 | .000 | 99 | 3 | 0 | .985 | .015 | Present study |
| **Sotho/Tswana:** | | | | | | | | | | | | | |
| S.Sotho(Highlands) | 90 | 90 | 0 | 0 | 1.000 | .000 | .000 | 85 | 5 | 0 | .972 | .028 | Jenkins (1972) |
| S.Sotho(Lowlands) | 91 | 91 | 0 | 0 | 1.000 | .000 | .000 | 88 | 3 | 0 | .983 | .017 | " |
| Pedi(rural) | 98 | 98 | 0 | 0 | 1.000 | .000 | .000 | 95 | 3 | 0 | .985 | .015 | Present study |
| Pedi(urban) | 205 | 205 | 0 | 0 | 1.000 | .000 | .000 | 204 | 1 | 0 | .998 | .002 | Jenkins (1972) |
| Fokeng | 118 | 118 | 0 | 0 | 1.000 | .000 | .000 | 118 | 0 | 0 | 1.000 | .000 | " |
| Ngwato | 59 | 59 | 0 | 0 | 1.000 | .000 | .000 | 58 | 1 | 0 | .992 | .008 | Present study |
| Kgalagadi | 124 | 124 | 0 | 0 | 1.000 | .000 | .000 | 115 | 8 | 1 | .960 | .040 | " |
| Kgalagadi | 52 | 52 | 0 | 0 | 1.000 | .000 | .000 | 52 | 0 | 0 | 1.000 | .000 | Jenkins (1972) |
| **Venda:** | | | | | | | | | | | | | |
| Urban | 254 | 473 | 2 | | .996 | .004 | | 252 | 2 | 0 | .996 | .004 | " |
| Rural | 221 | | | | | | | 218 | 3 | 0 | .996 | .004 | " |
| Rural | 104 | 103 | 1 | | .995 | .005 | | 103 | 1 | 0 | .995 | .005 | Present study |
| **Kavango:** | | | | | | | | | | | | | |
| Mbukushu | 54 | 54 | 0 | 0 | 1.000 | .000 | .000 | 52 | 2 | 0 | .981 | .019 | Nurse & Jenkins (1977b) |
| Gciriku | 110 | 99 | 11 | 0 | .901 | .099 | .000 | 109 | 0 | 0 | .995 | .005 | " |
| Sambyu | 92 | 78 | 14 | 0 | .848 | .152 | .000 | 92 | 0 | 0 | 1.000 | .000 | " |
| Kwangali | 107 | 100 | 7 | 0 | .967 | .033 | .000 | 107 | 0 | 0 | 1.000 | .000 | " |
| **Ambo:** | | | | | | | | | | | | | |
| Kwambi | 107 | 107 | 0 | 0 | 1.000 | .000 | .000 | 104 | 3 | 0 | .986 | .014 | Nurse *et al* (1983) |
| Kwaluudhi | 69 | 67 | 2 | 0 | .985 | .015 | .000 | 69 | 0 | 0 | 1.000 | .000 | " |
| Ndonga | 112 | 110 | 2 | 0 | .991 | .009 | .000 | 110 | 2 | 0 | .991 | .009 | " |
| Mbalantu | 70 | 70 | 0 | 0 | 1.000 | .000 | .000 | 70 | 0 | 0 | 1.000 | .000 | " |
| Kwanyama | 125 | 121 | 4 | 0 | .984 | .016 | .000 | 124 | 1 | 0 | .996 | .004 | " |
| Ngandjera | 68 | 68 | 0 | 0 | 1.000 | .000 | .000 | 67 | 1 | 0 | .993 | .007 | " |
| Nkolonkadhi | 32 | 31 | 1 | 0 | .985 | .015 | .000 | 32 | 0 | 0 | 1.000 | .000 | " |

Column groups: Haemoglobin β-chain — *Phenotypes* (A, AS, AD) and *Gene Frequencies* ($Hb\beta^A$, $Hb\beta^S$, $Hb\beta^D$); Haemoglobin δ-chain — *Phenotypes* ($A_2$, $A_2B_2$, $B_2$) and *Gene Frequencies* ($Hb\delta^{A_2}$, $Hb\delta^{B_2}$).

| Population | $n$ | $n$ | var | $p$ | $q$ | $n_{AA}$ | $n_{AB}$ | $n_{BB}$ | $p$ | $q$ | Reference |
|---|---|---|---|---|---|---|---|---|---|---|---|
| **Herero:** | | | | | | | | | | | |
| Botswana | 62 | 62 | 0 | 1.000 | .000 | 52 | 10 | 0 | .919 | .081 | Present study |
| SWA/Namibia | 87 | 87 | 0 | 1.000 | .000 | 78 | 9 | 0 | .948 | .052 | " |
| Mbanderu | 98 | 98 | 0 | 1.000 | .000 | 80 | 18 | 0 | .908 | .092 | " |
| Himba | 67 | 67 | 0 | 1.000 | .000 | 51 | 15 | 1 | .873 | .127 | " |
| Chimba | 34 | 34 | 0 | 1.000 | .000 | 22 | 11 | 1 | .804 | .191 | " |
| **Khoisan-speaking Negroes:** | | | | | | | | | | | |
| Sarwa | 77 | 77 | 0 | 1.000 | .000 | 77 | 0 | 0 | 1.000 | .000 | Chasko et al (1979) |
| Kwengo | 35 | 35 | 0 | 1.000 | .000 | 35 | 0 | 0 | 1.000 | .000 | Nurse et al (1977a) |
| Dama | 92 | 92 | 0 | 1.000 | .000 | 92 | 0 | 0 | 1.000 | .000 | Nurse et al (1976) |
| **Khoisan:** | | | | | | | | | | | |
| **Khoi:** | | | | | | | | | | | |
| Keetmanshoop Nama | 146 | 146 | 0 | 1.000 | .000 | 146 | 0 | 0 | 1.000 | .000 | Jenkins (1972) |
| Kuiseb Topnaars | 58 | 58 | 0 | 1.000 | .000 | | | | | | " |
| Sesfontein Nama | 42 | 42 | 0 | 1.000 | .000 | 42 | 0 | 0 | 1.000 | .000 | " |
| **San:** | | | | | | | | | | | |
| **Northern:** | | | | | | | | | | | |
| Tsumkwe !Kung | 215 | 215 | 0 | 1.000 | .000 | 215 | 0 | 0 | 1.000 | .000 | Jenkins (1972) |
| Dobe " | 424* | 422 | 1 | .999 | .001 | 422 | 2 | 0 | .998 | .002 | " |
| /Ai/ai " | 64 | 62 | 2 | .992 | .008 | 64 | 0 | 0 | 1.000 | .000 | " |
| /Du/da " | 100 | 99 | 1 | .995 | .005 | 100 | 0 | 0 | 1.000 | .000 | " |
| //Au//en | 267 | 251 | 16 | .971 | .029 | 267 | 0 | 0 | 1.000 | .000 | " |
| **Central:** | | | | | | | | | | | |
| Nharo | 142 | 141 | 1 | .996 | .004 | 142 | 0 | 0 | 1.000 | .000 | " |
| **Southern:** | | | | | | | | | | | |
| !Xõ | 72 | 72 | 0 | 1.000 | .000* | | | | | | " |
| **Caucasoids:** | | | | | | | | | | | |
| **Afrikaans-speaking:** | | | | | | | | | | | |
| Johannesburg | 160 | 160 | 0 | 1.000 | .000 | 160 | 0 | 0 | 1.000 | .000 | Present study |
| **Hybrid Peoples:** | | | | | | | | | | | |
| **Baster-like:** | | | | | | | | | | | |
| !Kuboes | 137 | 137 | 0 | 1.000 | .000 | 137 | 0 | 0 | 1.000 | .000 | Jenkins (1972) |
| **"Cape-Coloureds":** | | | | | | | | | | | |
| Johannesburg | 205+/117 | 204 | 0 | 1.000 | .000 | 114 | 3 | 0 | .987 | .013 | Jenkins (1972) |
| Orange River Valley | 205+ | 204 | 0 | 1.000 | .000 | 205 | 0 | 0 | 1.000 | .000 | " |

\* includes one individual with an unidentified phenotype

+ includes one individual with the variant "HbJα Cape Town"

# Table 14. Acid Phosphatase System

| Population | n | A | AB | B | AR | BR | R | AC | BC | C | $p^a$ | $p^b$ | $p^r$ | $p^c$ | $\chi^2$ | df | Reference |
|---|---|---|---|---|---|---|---|---|---|---|---|---|---|---|---|---|---|
| **Bantu-speaking Negroes:** | | | | | | | | | | | | | | | | | |
| **Nguni:** | | | | | | | | | | | | | | | | | |
| Swati | 72 | 3 | 19 | 45 | 0 | 5 | 0 | 0 | 0 | 0 | .173 | .792 | .035 | .000 | .825 | 3 | Jenkins (1972) |
| Swati | 52 | 1 | 12 | 33 | 0 | 6 | 1 | 0 | 0 | 0 | .135 | .808 | .058 | .000 | 1.284 | 3 | Hitzeroth *et al* (1981) |
| Zulu/Tonga | 121 | 6 | 25 | 78 | 1 | 10 | 0 | 0 | 0 | 0 | .157 | .789 | .054 | .000 | 6.07 | 3 | Nurse *et al* (1974) |
| Zulu | 89 | 0 | 21 | 62 | 0 | 6 | 0 | 0 | 0 | 0 | .118 | .848 | .034 | .000 | 3.036 | 3 | Jenkins (1974) |
| Zulu | 102 | 5 | 24 | 66 | 1 | 6 | 0 | 0 | 0 | 0 | .172 | .794 | .034 | .000 | 2.067 | 3 | Present study |
| Durban | 304 | 4 | 67 | 212 | 4 | 16 | 1 | 0 | 0 | 0 | .129 | .834 | .036 | .000 | 1.961 | 3 | Kirk *et al* (1971) |
| Zulu | 108 | 1 | 22 | 68 | 2 | 11 | 1 | 0 | 0 | 0 | .148 | .782 | .070 | .000 | 2.048 | 3 | Hitzeroth *et al* (1981) |
| Xhosa | 100 | 1 | 18 | 66 | 2 | 11 | 2 | 0 | 0 | 0 | .109 | .807 | .084 | .000 | 3.115 | 3 | Jenkins (1972) |
| Cape Town | 198 | | | | | | | | | | .140 | .840 | .020 | .000 | | – | Harris *et al* (1967) |
| Baca | 54 | 1 | 11 | 34 | 1 | 7 | 0 | 0 | 0 | 0 | .130 | .796 | .074 | .000 | .012 | – | Jenkins (1972) |
| Ndebele | 141 | 3 | 35 | 80 | 3 | 18 | 2 | 0 | 0 | 0 | .156 | .755 | .089 | .000 | 1.111 | 3 | Hitzeroth *et al* (1981) |
| **Sotho/Tswana:** | | | | | | | | | | | | | | | | | |
| Tswana | 150 | 3 | 39 | 75 | 6 | 25 | 2 | 0 | 0 | 0 | .131 | .762 | .107 | .000 | .611 | 3 | Jenkins (1972) |
| S.Sotho | 110 | 4 | 19 | 81 | 0 | 6 | 0 | 0 | 0 | 0 | .123 | .850 | .027 | .000 | 4.993 | 3 | " |
| Pedi(Urban) | 76 | 0 | 14 | 54 | 3 | 5 | 0 | 0 | 0 | 0 | .112 | .836 | .052 | .000 | 6.502 | 3 | " |
| Pedi(Rural) | 187 | 6 | 40 | 126 | 0 | 13 | 2 | 0 | 0 | 0 | .139 | .816 | .045 | .000 | 10.80 | 3 | " |
| Pedi | 227 | 6 | 55 | 122 | 6 | 35 | 3 | 0 | 0 | 0 | .161 | .736 | .103 | .000 | .500 | 3 | Hitzeroth *et al* (1981) |
| Fokeng | 191 | 6 | 36 | 114 | 2 | 27 | 6 | 0 | 0 | 0 | .173 | .792 | .035 | .000 | 11.814 | 3 | Jenkins (1972) |
| Ngwato | 76 | 2 | 14 | 50 | 3 | 5 | 0 | 0 | 0 | 0 | .164 | .783 | .053 | .000 | 6.299 | 3 | Present study |
| Tswana | 143 | 2 | 34 | 82 | 2 | 23 | 0 | 0 | 0 | 0 | .140 | .773 | .087 | .000 | 3.121 | 3 | Hitzeroth *et al* (1981) |
| Kgalagadi | 124 | 6 | 40 | 48 | 7 | 19 | 0 | 0 | 4 | 0 | .238 | .641 | .105 | .016 | 4.455 | 3 | Present study |
| **Venda:** | | | | | | | | | | | | | | | | | |
| Urban | 35 | 1 | 8 | 23 | 0 | 3 | 0 | 0 | 0 | 0 | .143 | .814 | .043 | .000 | .736 | 3 | Hitzeroth *et al* (1981) |
| Rural | 104 | 3 | 16 | 74 | 0 | 11 | 0 | 0 | 0 | 0 | .106 | .841 | .053 | .000 | 5.001 | 3 | Present study |
| **Tsonga/Ronga:** | | | | | | | | | | | | | | | | | |
| Shangaan | 141 | 3 | 39 | 82 | 4 | 13 | 0 | 0 | 0 | 0 | .174 | .766 | .060 | .000 | 1.323 | 3 | Hitzeroth *et al* (1981) |
| **Kavango:** | | | | | | | | | | | | | | | | | |
| Mbukushu | 56 | 4 | 14 | 35 | 0 | 3 | 0 | 0 | 0 | 0 | .196 | .777 | .027 | .000 | 2.89 | 2 | Nurse *et al* (1977) |
| Gciriku | 110 | 7 | 23 | 73 | 0 | 6 | 1 | 0 | 0 | 0 | .168 | .800 | .032 | .000 | 6.640 | 3 | " |
| Sambyu | 92 | 8 | 26 | 40 | 3 | 13 | 2 | 0 | 0 | 0 | .244 | .647 | .109 | .000 | 3.00 | 3 | " |
| Kwangali | 101 | 4 | 18 | 57 | 2 | 19 | 1 | 0 | 0 | 0 | .139 | .747 | .114 | .000 | 3.44 | 3 | " |
| **Ambo:** | | | | | | | | | | | | | | | | | |
| Kwambi | 105 | 5 | 27 | 66 | 2 | 3 | 1 | 0 | 1 | 0 | .186 | .776 | .033 | .005 | 2.47 | 2 | Nurse *et al* (1983) |
| Kwaluudhi | 69 | 3 | 21 | 41 | 3 | 1 | 0 | 0 | 0 | 0 | .217 | .754 | .029 | .000 | 6.56 | 3 | " |
| Ndonga | 111 | 5 | 33 | 66 | 1 | 6 | 0 | 0 | 0 | 0 | .198 | .770 | .032 | .000 | 1.31 | 2 | " |
| Mbalantu | 71 | 6 | 23 | 41 | 0 | 1 | 0 | 0 | 0 | 0 | .247 | .746 | .007 | .000 | 1.10 | 1 | " |
| Kwanyama | 127 | 5 | 33 | 76 | 2 | 8 | 1 | 0 | 2 | 0 | .177 | .768 | .047 | .008 | 2.76 | 4 | " |
| Ngandjera | 68 | 3 | 20 | 42 | 0 | 2 | 0 | 0 | 1 | 0 | .191 | .787 | .015 | .007 | .28 | 1 | " |
| Nkolonkadhi | 32 | 0 | 8 | 23 | 0 | 0 | 0 | 0 | 1 | 0 | .125 | .859 | .000 | .016 | – | 1 | " |
| **Herero:** | | | | | | | | | | | | | | | | | |
| Botswana | 59 | 3 | 11 | 42 | 1 | 2 | 0 | 0 | 0 | 0 | .153 | .822 | .025 | .000 | 3.778 | 3 | Present study |
| SWA/Namibia | 88 | 1 | 24 | 61 | 1 | 1 | 0 | 0 | 0 | 0 | .153 | .835 | .011 | .000 | 2.567 | 3 | " |
| Mbanderu | 98 | 8 | 32 | 56 | 0 | 1 | 1 | 0 | 0 | 0 | .250 | .740 | .010 | | 2.545 | 3 | " |
| Himba | 86 | 5 | 28 | 47 | 2 | 4 | 0 | 0 | 0 | 0 | .233 | .732 | .035 | .000 | .498 | 3 | " |
| Chimba | 28 | 1 | 16 | 11 | 0 | 0 | 0 | 0 | 0 | 0 | .321 | .679 | .000 | .000 | 2.693 | 1 | " |

| Population | N | 1 | 2 | 3 | 4 | 5 | 6 | 7 | 8 | 9 | | | | | | | Reference |
|---|---|---|---|---|---|---|---|---|---|---|---|---|---|---|---|---|---|
| **Khoisan-speaking Negroes:** | | | | | | | | | | | | | | | | | |
| Sarwa | 77 | 3 | 16 | 51 | 0 | 6 | 1 | 0 | 0 | 0 | .143 | .805 | .052 | .000 | 5.74 | 3 | Chasko *et al* (1979) |
| Kwengo | 35 | 2 | 17 | 11 | 1 | 3 | 1 | 0 | 0 | 0 | .314 | .600 | .086 | .000 | 4.54 | 3 | Nurse & Jenkins (1977b) |
| Dama | 92 | 2 | 30 | 58 | 0 | 1 | 0 | 1 | 0 | 0 | .190 | .799 | .005 | .006 | .66 | 1 | Nurse *et al* (1976) |
| **Khoisan:** | | | | | | | | | | | | | | | | | |
| **Khoi:** | | | | | | | | | | | | | | | | | |
| Keetmanshoop Nama | 150 | 1 | 37 | 60 | 27 | 10 | 14 | 0 | 1 | 0 | .164 | .617 | .218 | .003 | 15.07 | | Jenkins (1972) |
| **San:** | | | | | | | | | | | | | | | | | |
| **Northern:** | | | | | | | | | | | | | | | | | |
| Glang!ai | 36 | 1 | 7 | 15 | 6 | 5 | 2 | 0 | 0 | 0 | .208 | .583 | .208 | .000 | 5.48 | 3 | Nurse & Jenkins (1977a) |
| Tsumkwe !Kung | 271 | 17 | 45 | 101 | 31 | 64 | 13 | 0 | 0 | 0 | .203 | .574 | .223 | .000 | 10.42 | 3 | " " |
| Samanlaika | 33 | 3 | 5 | 9 | 5 | 5 | 6 | 0 | 0 | 0 | .242 | .424 | .333 | .000 | 6.18 | 3 | Nurse *et al* (1977) |
| **Central:** | | | | | | | | | | | | | | | | | |
| Hei//om | 69 | 6 | 14 | 29 | 8 | 10 | 2 | 0 | 0 | 0 | .246 | .594 | .159 | .000 | 5.55 | 3 | Nurse *et al* (1977) |
| Nharo | 36 | 3 | 8 | 12 | 1 | 8 | 4 | 0 | 0 | 0 | .208 | .556 | .236 | .000 | 5.43 | 3 | " " |
| G/wi | 94 | 11 | 18 | 13 | 15 | 25 | 10 | 0 | 2 | 0 | .292 | .367 | .333 | .010 | 2.10 | 3 | Jenkins *et al* (1975) |
| G//ana | 49 | 13 | 10 | 6 | 10 | 7 | 3 | 0 | 0 | 0 | .469 | .296 | .235 | .000 | 2.17 | 2 | " " |
| **Southern:** | | | | | | | | | | | | | | | | | |
| !Kö | 51 | 12 | 14 | 8 | 5 | 8 | 4 | 0 | 0 | 0 | .422 | .372 | .206 | .000 | .448 | 3 | Nurse & Jenkins (1977a) |
| Glaokx'ate | 33 | 1 | 2 | 7 | 3 | 11 | 9 | 0 | 0 | 0 | .106 | .379 | .515 | .000 | .260 | 3 | Nurse *et al* (1977) |
| **Uncertain:** | | | | | | | | | | | | | | | | | |
| Eastern ǂHuã | 35 | 2 | 10 | 5 | 5 | 11 | 2 | 0 | 0 | 0 | .271 | .443 | .286 | .000 | 1.77 | 3 | Nurse & Jenkins (1977a) |
| **Caucasoid:** | | | | | | | | | | | | | | | | | |
| **Afrikaans-speaking:** | | | | | | | | | | | | | | | | | |
| Johannesburg | 77 | 11 | 19 | 33 | 0 | 0 | 0 | 2 | 12 | 0 | .279 | .630 | .000 | .091 | 9.54 | 3 | Present study |
| Johannesburg | 159 | 15 | 56 | 68 | 0 | 0 | 0 | 7 | 12 | 1 | .293 | .642 | .000 | .066 | .754 | 3 | Palmhert-Keller *et al* (1983) |
| Namibia | 57 | 7 | 18 | 25 | 0 | 0 | 0 | 3 | 4 | 0 | .307 | .632 | .000 | .061 | 2.27 | 3 | Palmhert-Keller *et al* (1983) |
| **German-speaking:** | | | | | | | | | | | | | | | | | |
| Namibia | 118 | 15 | 45 | 48 | 0 | 0 | 0 | 4 | 4 | 2 | .330 | .601 | .000 | .069 | 8.16 | 3 | Present study |
| **Indian:** | | | | | | | | | | | | | | | | | |
| **Transvaal:** | | | | | | | | | | | | | | | | | |
| Hindu | 53 | 3 | 22 | 28 | 0 | 0 | 0 | 0 | 0 | 0 | .264 | .736 | .018 | | 0.24 | 1 | Present study |
| Moslem | 56 | 5 | 25 | 23 | 0 | 1 | 0 | 1 | 1 | 0 | .321 | .652 | | | 0.40 | 2 | " " |
| **Hybrid Peoples:** | | | | | | | | | | | | | | | | | |
| **Baster-like:** | | | | | | | | | | | | | | | | | |
| Rehoboth | 119 | 15 | 49 | 37 | 3 | 11 | 2 | 0 | 2 | 0 | .334 | .572 | .008 | .076 | 2.85 | 3 | Nurse *et al* (1982) |
| !Kuboes | 133 | 1 | 19 | 66 | 2 | 40 | 5 | 0 | 0 | 0 | .087 | .718 | .196 | .004 | 2.048 | 3 | Jenkins (1972) |
| Riemvaasmaak | 89 | 15 | 20 | 53 | 2 | 9 | 2 | 0 | 1 | 0 | .146 | .764 | .084 | .006 | 3.860 | 4 | Nurse & Jenkins (1978) |
| **Griqua:** | 266 | 15 | 64 | 114 | 12 | 56 | 4 | 0 | 1 | 0 | .199 | .656 | .143 | .002 | 4.60 | 4 | Nurse & Jenkins (1975) |
| **"Cape-Coloureds":** | | | | | | | | | | | | | | | | | |
| Johannesburg | 96 | 6 | 27 | 43 | 5 | 13 | 2 | 0 | 1 | 0 | .229 | .667 | .094 | .010 | 1.544 | 3 | Jenkins (1972) |
| Cape Town | 174 | | | | | | | | | | .28 | .70 | .00 | .02 | - | - | Harris *et al* (1967) |

↓ presumably Zulu   ↓ presumably Xhosa

# Table 15. Glucose-6-Phosphate Dehydrogenase System

| Population | n (m) | n (f) | Males A | Males A⁻ | Males B | Females A | Females A⁻ | Females AB+A⁻B | Females B | $Gd^A$ | $Gd^{A^-}$ | $Gd^B$ | $\chi^2_{[1]}$ | References |
|---|---|---|---|---|---|---|---|---|---|---|---|---|---|---|
| **Bantu-speaking Negroes:** | | | | | | | | | | | | | | |
| **Nguni:** | | | | | | | | | | | | | | |
| Swati | 23 | 69 | 5 | 0 | 18 | 5 | 0 | 18 | 46 | .162 | .041 | .797 | 3.216 | Hitzeroth & Bender (1980) |
| Zulu/Ronga | 46 | 100 | 7 | 2 | 37 | 7 | 1 | 18 | 74 | .130 | .087 | .783 | 12.510 | Nurse et al (1974) |
| Zulu | 101 | 53 | 18 | 2 | 81 | 1 | 0 | 9 | 43 | .115 | .015 | .870 | 5.203 | Present study |
| Zulu | 100 | | 15 | 1 | 84 | | | | | .180 | .020 | .800 | | Balinsky & Jenkins (1972) |
| Xhosa | | 119 | | | | 13 | 0 | 27 | 79 | .163 | .059 | .777 | 16.484 | Hitzeroth & Bender (1980) / Jenkins (1972) |
| Ndebele | | 172 | | | | 11 | 0 | 41 | 120 | .140 | .043 | .817 | 8.023 | Hitzeroth & Bender (1980) |
| **Sotho/Tswana:** | | | | | | | | | | | | | | |
| S.Sotho(Highlands) | 18 | | 1 | 1 | 16 | | | | | .060 | .060 | .880 | | Beaumont et al (1979) |
| S.Sotho(Lowlands) | 33 | | 2 | 2 | 29 | | | | | .060 | .060 | .880 | | " |
| S.Sotho | 30 | | 5 | 1 | 24 | | | | | .160 | .030 | .810 | | Balinsky & Jenkins (1967) |
| Pedi(Rural) | 116 | | 19 | 5 | 92 | | | | | .160 | .040 | .800 | | Jenkins (1972) |
| Pedi(Urban) | 84 | | 14 | 3 | 67 | | | | | .170 | .040 | .790 | | " |
| Pedi(Urban) | 46 | 269 | | | | 17 | 0 | 50 | 201 | .118 | .039 | .843 | 25.999 | Hitzeroth & Bender (1980) |
| Fokeng | 46 | | 5 | 1 | 40 | | | | | .110 | .020 | .870 | | Jenkins (1972) |
| Ngwato | 60 | | 5 | 4 | 51 | | | | | .067 | .083 | .850 | | Present study |
| Tswana | 28 | 152 | 7 | 1 | 20 | 10 | 0 | 34 | 108 | .131 | .047 | .822 | 9.879 | Hitzeroth & Bender (1980) |
| Kgalagadi | | 96 | | | | 14 | 0 | 15 | 67 | .216 | .016 | .768 | 12.020 | Present study |
| **Venda:** | | | | | | | | | | | | | | |
| Rural | 47 | 56 | 15 | 4 | 28 | 10 | 0 | 10 | 36 | .319 | .085 | .596 | 24.308 | Present study |
| Urban | | 47 | | | | 6 | 0 | 9 | 32 | .150 | .074 | .776 | 11.274 | Hitzeroth & Bender (1980) |
| **Tsonga/Ronga:** | | | | | | | | | | | | | | |
| Shangaan/Tsonga | | 153 | | | | 9 | 0 | 46 | 97 | .158 | .053 | .789 | 2.022 | Hitzeroth & Bender (1980) |
| **Kavango:** | | | | | | | | | | | | | | |
| Mbukushu | 30 | 23 | 7 | 1 | 22 | 3 | 1 | 6 | 13 | .212 | .077 | .711 | 9.910 | Nurse & Jenkins (1977b) |
| Gciriku | 48 | 62 | 17 | 6 | 25 | 12 | 4 | 11 | 35 | .234 | .142 | .624 | 28.400 | " |
| Sambyu | 39 | 52 | 10 | 7 | 21 | 12 | 0 | 17 | 21 | .305 | .103 | .572 | 7.480 | " |
| Kwangali | 63 | 38 | 17 | 3 | 43 | 4 | 0 | 11 | 23 | .241 | .040 | .719 | 3.580 | " |
| **Ambo:** | | | | | | | | | | | | | | |
| Kwambi | 104 | | 21 | 7 | 76 | | | | | .202 | .067 | .731 | | Nurse et al (1983) |
| Kwaluudhi | 69 | | 18 | 4 | 47 | | | | | .261 | .058 | .681 | | " |
| Ndonga | 110 | | 35 | 10 | 65 | | | | | .318 | .091 | .591 | | " |
| Mbalantu | 72 | | 17 | 7 | 48 | | | | | .236 | .097 | .667 | | " |
| Kwanyama | 124 | | 19 | 16 | 88 | | | | | .153 | .129 | .711 | | " |
| Ngandjera | 67 | | 17 | 5 | 45 | | | | | .254 | .074 | .672 | | " |
| Nkolonkadhi | 32 | | 4 | 1 | 27 | | | | | .125 | .031 | .844 | | " |
| **Herero:** | | | | | | | | | | | | | | |
| Botswana | 22 | | 4 | 2 | 16 | | | | | .180 | .090 | .740 | | Jenkins (1972) |
| Mbanderu | 49 | 43 | 8 | 2 | 39 | 2 | 0 | 10 | 31 | .158 | .020 | .822 | .997 | Present study |

| Population | n | | | | | | | f1 | f2 | f3 | χ² | Reference |
|---|---|---|---|---|---|---|---|---|---|---|---|---|
| **Khoisan-speaking Negroes:** | | | | | | | | | | | | |
| Sarwa | 36 | 41 | 2 | 34 | 4 | 6 | 31 | .136 | .000 | .864 | 16.140 | Chasko et al (1979) |
| Kwengo | 20 | 15 | 7 | 12 | 3 | 3 | 9 | .301 | .039 | .660 | 4.160 | Nurse & Jenkins (1977b) |
| Dama | 63 | 56 | 11 | 51 | 0 | 12 | 44 | .126 | .011 | .863 | 2.900 | Nurse et al (1976) |
| **Khoi:** | | | | | | | | | | | | |
| Keetmanshoop Nama | 59 | | | 59 | | 3 | 20 | .000 | .000 | 1.000 | | Jenkins (1972) |
| Kuiseb Topnaars | 25 | 23 | 0 | 25 | 0 | 0 | 22 | .042 | .000 | .958 | 1.840 | Nurse & Jenkins (1977a) |
| Sesfontein Nama | 20 | 22 | 0 | 20 | 0 | 0 | 22 | .000 | .000 | 1.000 | | " |
| **San:** | | | | | | | | | | | | |
| **Northern:** | | | | | | | | | | | | |
| G/anigai | 14 | 22 | 1 | 13 | 0 | 1 | 21 | .034 | .000 | .966 | .700 | Nurse & Jenkins (1977a) |
| Tsumkwe !Kung | 69 | 77 | 5 | 64 | 2 | 2 | 73 | .049 | .000 | .951 | 20.910 | Nurse et al (1977) |
| Saman!aiika | 4 | 5 | 0 | 4 | 0 | 0 | 5 | .000 | .000 | 1.000 | | Nurse & Jenkins (1977a) |
| Dobe | 94 | 110 | 4 | 89 | 0 | 10 | 100 | .038 | .010 | .952 | .200 | Nurse et al (1977) |
| /Du/da | 27 | 20 | 1 | 26 | 0 | 0 | 20 | .000 | .037 | .963 | | " |
| //Au//en | 43 | 20 | 1 | 42 | 0 | 0 | 20 | .023 | .000 | .977 | .150 | Nurse et al (1977) |
| **Central:** | | | | | | | | | | | | |
| Hei//om | 42 | 16 | 2 | 39 | 0 | 0 | 16 | .027 | .013 | .960 | 1.240 | " |
| Nharo | 28 | 47 | 1 | 25 | 1 | 0 | 46 | .071 | .036 | .893 | | Jenkins et al (1975) |
| G/wi | 47 | 47 | 1 | 46 | 1 | 0 | 24 | .000 | .021 | .979 | | " |
| G//ana | 25 | 25 | 0 | 24 | 1 | 0 | 24 | .040 | .000 | .960 | | " |
| **Southern:** | | | | | | | | | | | | |
| !Xo | 28 | 23 | 0 | 28 | 0 | 0 | 23 | .000 | .000 | 1.000 | | Nurse & Jenkins (1977a) |
| G!aokx'ate | 18 | 15 | 0 | 18 | 0 | 0 | 15 | .000 | .000 | 1.000 | | Nurse et al (1977) |
| **Uncertain:** | | | | | | | | | | | | |
| Eastern ≠Huã | 12 | 23 | 1 | 10 | 0 | 1 | 22 | .026 | .026 | .948 | 8.920 | Nurse & Jenkins (1977a) |
| **Caucasoids** | | | | | | | | | | | | |
| **Afrikaans-speaking:** | | | | | | | | | | | | |
| Johannesburg | 77 | | | | | | | .000 | .000 | 1.000 | | Present study |
| Johannesburg | 160 | | | | | | | .000 | .000 | 1.000 | | Present study |
| **German-speaking:** | | | | | | | | | | | | |
| SWA/Namibia | 64 | | 0 | | 0 | 2 | 62 | .011 | .000 | .989 | | Palmhert-Keller et al (1983) |
| **Indian:** | | | | | | | | | | | | |
| **Transvaal:** | | | | | | | | | | | | |
| Hindu | 27 | 26 | 0 | 27 | 0 | 0 | 26 | .000 | .000 | 1.000 | | Present study |
| Moslem | 27 | 29 | 1 | 25 | 0 | 0 | 29 | .012 | .012 | .976 | | Present study |
| **Hybrid Peoples:** | | | | | | | | | | | | |
| **Baster-like:** | | | | | | | | | | | | |
| Rehoboth | 52 | 67 | 2 | 50 | 0 | 1 | 66 | .016 | .000 | .984 | 2.420 | Nurse et al (1982) |
| !Kuboes | 22 | | 0 | 22 | 0 | 0 | | .000 | .000 | 1.000 | | Jenkins (1972) |
| Riemvaasmaak | 38 | 51 | 2 | 36 | 5 | 6 | 40 | .129 | .010 | .871 | 26.600 | Nurse & Jenkins (1978) |
| Griqua | 102 | 164 | 11 | 90 | 13 | 16 | 132 | .108 | .010 | .882 | 505.07 | Nurse & Jenkins (1975) |
| **Cape-Coloured:** | | | | | | | | | | | | |
| Johannesburg | 33 | 31 | 2 | 31 | | | | .000 | .060 | .940 | | Balinsky & Jenkins (1967) |
| Johannesburg | 80 | 74 | 5 | 74 | | | | .060 | .010 | .930 | | Jenkins (1972) |
| Cape Town | 100 | 90 | 8 | 90 | | | | .080 | .020 | .900 | | Gordon et al (1966) |

§ one individual with a rare variant
‡ three individuals with a rare variant giving a frequency of .220
§ one individual with a rare variant giving a frequency of .008

## Table 16. 6-Phosphogluconate Dehydrogenase System

| Population | n | Phenotypes | | | | Gene Frequencies 6-phosphogluconate dehydrogenase | | | $\chi^2_{[1]}$ | References |
|---|---|---|---|---|---|---|---|---|---|---|
| | | A | AC | C | AR | $PGD^A$ | $PGD^C$ | $PGD^R$ | | |
| **Bantu-speaking Negroes:** | | | | | | | | | | |
| **Nguni:** | | | | | | | | | | |
| Swati | 90 | 79 | 11 | 0 | 0 | .939 | .061 | .000 | | Jenkins (1972) |
| Swati | 69 | 61 | 8 | 0 | 0 | .942 | .058 | .000 | | Hitzeroth & Bender (1980) |
| Zulu/Tonga | 123 | 115 | 8 | 0 | 0 | .967 | .033 | .000 | | Nurse et al (1974) |
| Zulu ↓ | 101 | 83 | 18 | 0 | 0 | .906 | .094 | .000 | .017 | Present study |
| Durban | 304 | 242 | 57 | 1 | 1 | .893 | .104 | .003 | 4.164* | Kirk et al (1971) |
| Zulu | 120 | 96 | 21 | 1 | 2 | .896 | .096 | .008 | .255* | Hitzeroth & Bender (1980) |
| Ndebele | 174 | 138 | 35 | 1 | 0 | .894 | .106 | .000 | .599 | " " |
| **Sotho/Tswana:** | | | | | | | | | | |
| S.Sotho(Highland) | 90 | 78 | 12 | 0 | 0 | .933 | .067 | .000 | | Jenkins (1972) |
| S.Sotho(Lowland) | 92 | 73 | 19 | 0 | 0 | .897 | .103 | .000 | | " " |
| Pedi(urban) | 80 | 71 | 7 | 2 | 0 | .931 | .069 | .000 | | " " |
| Pedi(rural) | 116 | 99 | 17 | 0 | 0 | .927 | .073 | .000 | | " " |
| Pedi | 275 | 235 | 40 | 0 | 0 | .927 | .073 | .000 | 7.966 | Hitzeroth & Bender (1980) |
| Fokeng | 117 | 103 | 14 | 0 | 0 | .940 | .060 | .000 | | Jenkins (1972) |
| Ngwato | 76 | 72 | 4 | 0 | 0 | .974 | .026 | .000 | | Present study |
| Tswana | 155 | 133 | 18 | 2 | 2 | .923 | .071 | .006 | 1.589 | Hitzeroth & Bender (1980) |
| Kgalagadi | 124 | 74 | 46 | 3 | 1 | .786 | .210 | .004 | 2.660 | Present study |
| **Venda:** | | | | | | | | | | |
| Urban | 47 | 37 | 9 | 1 | 0 | .883 | .117 | .000 | .253 | Hitzeroth & Bender (1980) |
| Rural | 104 | 83 | 19 | 2 | 0 | .889 | .111 | .000 | .525 | Present study |
| **Tsonga/Ronga:** | | | | | | | | | | |
| "Shangaan"/Tsonga | 158 | 128 | 28 | 1 | 1 | .899 | .098 | .003 | .181 | Hitzeroth & Bender (1980) |
| **Kavango:** | | | | | | | | | | |
| Mbukushu | 57 | 54 | 3 | 0 | 0 | .974 | .026 | .000 | | Nurse & Jenkins (1977b) |
| Gciriku | 110 | 100 | 9 | 1 | 0 | .950 | .050 | .000 | | " " |
| Sambyu | 94 | 91 | 3 | 0 | 0 | .984 | .016 | .000 | | " " |
| Kwangali | 103 | 93 | 9 | 0 | 1 | .951 | .044 | .005 | 2.118 | " " |
| **Ambo:** | | | | | | | | | | |
| Kwambi | 104 | 90 | 14 | 0 | 0 | .933 | .067 | .000 | | Nurse et al (1983) |
| Kwaluudhi | 69 | 58 | 10 | 1 | 0 | .920 | .080 | .000 | | " " |
| Ndonga | 110 | 100 | 10 | 0 | 0 | .954 | .046 | .000 | | " " |
| Mbalantu | 72 | 68 | 4 | 0 | 0 | .972 | .028 | .000 | | " " |
| Kwanyama | 124 ↓ | 115 | 8 | 0 | 1 | .964 | .032 | .008 | | " " |
| Ngandjera | 66 | 59 | 6 | 0 | 1 | .946 | .046 | .008 | | " " |
| Nkolonkadhi | 32 | 28 | 2 | 0 | 2 | .937 | .032 | .031 | .711 | " " |
| **Herero:** | | | | | | | | | | |
| Botswana | 39 | 39 | 0 | 0 | 0 | 1.000 | .000 | .000 | | Jenkins (1972) |
| SWA/Namibia | 88 | 85 | 3 | 0 | 0 | .983 | .017 | .000 | | Present study |
| Mbanderu | 99 | 95 | 2 | 2 | 0 | .980 | .010 | .010 | | " " |
| Himba | 86 | 86 | 0 | 0 | 0 | 1.000 | .000 | .000 | | " " |
| Chimba | 29 | 29 | 0 | 0 | 0 | 1.000 | .000 | .000 | | " " |
| **Khoisan-speaking Negroes:** | | | | | | | | | | |
| Sarwa | 77 | 73 | 4 | 0 | 0 | .974 | .026 | .000 | | Chasko et al (1979) |
| Kwengo | 35 | 35 | 0 | 0 | 0 | 1.000 | .000 | .000 | | Nurse & Jenkins (1977b) |
| Dama | 119 | 112 | 4 | 0 | 3 | .970 | .017 | .013 | | Nurse et al (1976) |

334

Khoisan:

| | N | | | | | $P^A$ | $P^B$ | $P^C$ | $\chi^2$ | Reference |
|---|---|---|---|---|---|---|---|---|---|---|
| **Khoi:** | | | | | | | | | | |
| Keetmanshoop Nama | 118 | 117 | 1 | 0 | 0 | .996 | .004 | .000 | | Jenkins (1972) |
| Sesfontein Topnaars | 42 | 41 | 0 | 1 | 1 | .988 | .000 | .012 | | Jenkins (1972) |
| **San:** | | | | | | | | | | |
| **Northern:** | | | | | | | | | | |
| Glanglai | 35 | 35 | 0 | 0 | 0 | 1.000 | .000 | .000 | | Nurse & Jenkins (1977a) |
| Tsumkwe !Kung | 147 | 147 | 0 | 0 | 0 | 1.000 | .000 | .000 | | " |
| Dobe " | 328 | 328 | 0 | 0 | 0 | 1.000 | .000 | .000 | | " |
| /Du/da " | 74 | 74 | 0 | 0 | 0 | 1.000 | .000 | .000 | | " |
| //Au//en | 82 | 81 | 1 | 0 | 0 | .994 | .006 | .000 | | " |
| **Central:** | | | | | | | | | | |
| Hei//om | 62 | 61 | 1 | 0 | 0 | .992 | .008 | .000 | | " |
| Nharo | 58 | 57 | 1 | 0 | 0 | .991 | .009 | .000 | | " |
| G/wi | 94 | 93 | 1 | 0 | 0 | .995 | .005 | .000 | | " |
| G//ana | 50 | 44 | 6 | 0 | 0 | .940 | .060 | .000 | | " |
| **Southern:** | | | | | | | | | | |
| !Xõ | 51 | 49 | 2 | 0 | 0 | .980 | .020 | .000 | | " |
| Glaokx'ate | 33 | 33 | 0 | 0 | 0 | 1.000 | .000 | .000 | | " |
| **Uncertain:** | | | | | | | | | | |
| Eastern ǂHuã | 36 | 36 | 0 | 0 | 0 | 1.000 | .000 | .000 | | " |
| **Caucasoids:** | | | | | | | | | | |
| **Afrikaans-speaking:** | | | | | | | | | | |
| Johannesburg | 77 | 74 | 3 | 0 | 0 | .980 | .020 | .000 | | Present study |
| Johannesburg | 160 | 157 | 3 | 0 | 0 | .991 | .009 | .000 | | " |
| SWA/Namibia | 57 | 56 | 1 | 0 | 0 | .991 | .009 | .000 | | Palmhert-Keller et al (1983) |
| **German-speaking:** | | | | | | | | | | |
| SWA/Namibia | 118 | 114 | 3 | 0 | 1 | .983 | .013 | .004 | | " |
| **Indians:** | | | | | | | | | | |
| **Transvaal:** | | | | | | | | | | |
| Hindu | 53 | 52 | 1 | 0 | 0 | .991 | .009 | .000 | | Present study |
| Moslem | 56 | 55 | 1 | 0 | 0 | .991 | .009 | .000 | | " |
| **Hybrid Peoples:** | | | | | | | | | | |
| **Baster-like:** | | | | | | | | | | |
| Rehoboth | 119 | 111 | 8 | 0 | 0 | .966 | .034 | .000 | | Nurse et al (1982) |
| !Kuboes | 135 | 135 | 0 | 0 | 0 | 1.000 | .000 | .000 | | Jenkins (1972) |
| Riemvaasmaak | 89 | 87 | 2 | 0 | 0 | .989 | .011 | .000 | | Nurse & Jenkins (1978) |
| **Griqua:** | | | | | | | | | | |
| Griqua | 249 | 239 | 8 | 1 | 1 | .987 | .010 | .003 | 8.45 | Nurse & Jenkins (1975) |
| **"Cape-Coloureds":** | | | | | | | | | | |
| Johannesburg | 109 | 91 | 18 | 0 | 0 | .955 | .045 | .000 | | Jenkins (1972) |
| Cape Town | 200 | 182 | 18 | 0 | 0 | .917 | .083 | .000 | | Gordon et al (1968) |

§ presumably Zulu
† including one individual with 6PGD (CR)
‡ including one individual with 6PGD(AS) giving a frequency of .004
* 2 d.f.

335

# Table 17. Phosphoglucomutase System: First and Second Loci

| Population | n | First locus Phenotypes 1 | 2-1 | 2 | 6-1 | 7-1 | Gene freq. $PGM_1^1$ | $PGM_1^2$ | $PGM_1^6$ | $PGM_1^7$ | $\chi^2_{[1]}$ | Second locus Phenotypes 1 | 2-1 | 2 | Gene freq. $PGM_2^1$ | $PGM_2^2$ | $\chi^2_{[1]}$ | References |
|---|---|---|---|---|---|---|---|---|---|---|---|---|---|---|---|---|---|---|
| **Bantu-speaking Negroes:** | | | | | | | | | | | | | | | | | | |
| **Nguni:** | | | | | | | | | | | | | | | | | | |
| Swati | 91 | 66 | 22 | 2 | | 1 | .852 | .143 | | .005 | 1.326* | 84 | 7 | 0 | .962 | .038 | - | Jenkins (1972) |
| Swati | 63 | 40 | 22 | 1 | | | .810 | .190 | | | 1.109* | | | | | | - | Hitzeroth et al (1981) |
| Zulu/Ronga | 123 | 83 | 34 | 5 | 1 | | .817 | .179 | .004 | | .628 | 115 | 8 | 0 | .967 | .033 | - | Nurse et al (1974) |
| Zulu | 304 | 219 | 82 | 1 | | | .855 | .145 | | | 2.377 | 277 | 27 | 0 | .955 | .045 | - | Kirk et al (1971) |
| Zulu | 102 | 77 | 24 | 1 | | | .873 | .127 | | | 2.738 | | | | | | - | Present Study |
| Zulu | 116/114 | 83 | 31 | 2 | | | .849 | .151 | | | .215 | 107 | 7 | 0 | .969 | .031 | - | Hitzeroth et al (1981) |
| Ndebele | 166/167 | 104 | 54 | 8 | | | .789 | .211 | | | .084 | 150 | 17 | 0 | .949 | .051 | - | Hitzeroth et al (1981) |
| **Sotho/Tswana:** | | | | | | | | | | | | | | | | | | |
| S.Sotho(Highland) | 90 | 67 | 21 | 2 | | | .861 | .139 | | | .066* | 83 | 7 | 0 | .961 | .039 | - | Beaumont et al (1979) |
| S.Sotho(Lowland) | 92 | 60 | 26 | 5 | 1 | | .799 | .195 | .006 | | .944* | 87 | 5 | 0 | .973 | .027 | - | " |
| Pedi(urban) | 110 | 73 | 33 | 3 | | 1 | .818 | .177 | | .005 | 0.114* | 105 | 5 | 0 | .977 | .023 | - | Jenkins (1972) |
| Pedi(rural) | 116 | 64 | 43 | 9 | | | .737 | .263 | | | .230 | 111 | 5 | 0 | .978 | .022 | - | Jenkins (1972) |
| Pedi | 260/259 | 165 | 83 | 12 | | | .794 | .206 | | | .141 | 242 | 17 | 0 | .967 | .033 | - | Hitzeroth et al (1981) |
| Fokeng | 121 | 84 | 32 | 5 | | | .825 | .175 | | | .786 | 114 | 7 | 0 | .972 | .028 | - | Jenkins (1972) |
| Ngwato | 60 | 29 | 28 | 3 | | | .717 | .283 | | | 1.335 | 55 | 5 | 0 | .958 | .042 | - | Present study |
| Tswana | 150 | 91 | 53 | 6 | | | .783 | .217 | | | .251* | 136 | 13 | 1 | .950 | .050 | - | Hitzeroth et al (1981) |
| Kgalagadi | 124 | 83 | 31 | 9 | | 1 | .798 | .198 | | .004 | 5.709* | 107 | 17 | 0 | .931 | .069 | 1.154 | Present study |
| **Venda:** | | | | | | | | | | | | | | | | | | |
| Urban | 45 | 23 | 20 | 2 | | | .733 | .267 | | | .835 | 43 | 2 | 0 | .978 | .022 | - | Hitzeroth et al (1981) |
| Rural | 104 | 74 | 29 | 1 | | | .851 | .149 | | | 1.026 | 96 | 8 | 0 | .962 | .038 | - | Present study |
| **Tsonga/Ronga:** | | | | | | | | | | | | | | | | | | |
| "Shangaan" | 150 | 101 | 42 | 7 | | | .813 | .187 | | | .907 | 136 | 14 | 0 | .953 | .047 | - | Hitzeroth et al (1981) |
| **Kavango:** | | | | | | | | | | | | | | | | | | |
| Mbukushu | 57 | 39 | 16 | 2 | | | .825 | .175 | | | .070 | 55 | 2 | 0 | .982 | .018 | - | Nurse & Jenkins (1977b) |
| Gciriku | 111 | 63 | 43 | 5 | | | .761 | .239 | | | .290 | 89 | 6 | 0 | .973 | .027 | - | " |
| Sambyu | 94 | 57 | 31 | 6 | | | .771 | .229 | | | .410 | 89 | 5 | 0 | .973 | .027 | - | " |
| Kwangali | 103 | 57 | 42 | 4 | | | .757 | .243 | | | 1.260 | 101 | 2 | 0 | .990 | .010 | - | " |
| **Ambo:** | | | | | | | | | | | | | | | | | | |
| Kwambi | 41 | 33 | 8 | 0 | | | .902 | .098 | | | - | 41 | 0 | 0 | 1.000 | .000 | - | Nurse et al (1983) |
| Kwaluudhi | 40 | 22 | 18 | 0 | | | .775 | .225 | | | 4.07 | 39 | 1 | 0 | .987 | .013 | - | " |
| Ndonga | 59 | 43 | 12 | 4 | | | .831 | .169 | | | .28 | 58 | 1 | 0 | .991 | .009 | - | " |
| Mbalantu | 67 | 42 | 23 | 2 | | | .798 | .202 | | | 1.52 | 65 | 2 | 0 | .985 | .015 | - | " |
| Kwanyama | 119 | 75 | 36 | 8 | | | .781 | .219 | | | .04 | 118 | 1 | 0 | .996 | .004 | - | " |
| Ngandjera | 42 | 29 | 12 | 1 | | | .833 | .167 | | | .04 | 42 | 0 | 0 | 1.000 | .000 | - | " |
| Nkolonkadhi | 21 | 15 | 5 | 1 | | | .833 | .167 | | | - | 21 | 0 | 0 | 1.000 | .000 | - | " |
| **Herero:** | | | | | | | | | | | | | | | | | | |
| Botswana | 39 | 23 | 16 | 0 | | | .795 | .205 | | | 4.560 | 38 | 1 | 0 | .987 | .013 | - | Jenkins (1972) |
| S.W.A./Namibia | 88 | 38 | 46 | 4 | | | .673 | .327 | | | 2.619 | 85 | 3 | 0 | .983 | .017 | - | Present study |
| Mbanderu | 98 | 61 | 29 | 8 | | | .770 | .230 | | | 2.546 | 96 | 2 | 0 | .990 | .010 | - | " |
| Himba | 86 | 39 | 40 | 7 | | | .686 | .314 | | | 4.975 | 85 | 1 | 0 | .994 | .006 | - | " |
| Chimba | 29 | 14 | 8 | 7 | | | .621 | .379 | | | | 29 | 0 | 0 | 1.000 | .000 | - | " |
| **Khoisan-speaking Negroes:** | | | | | | | | | | | | | | | | | | |
| Sarwa | 77 | 40 | 30 | 7 | | | .714 | .286 | | | .70 | 63 | 13 | 1 | .903 | .097 | .03 | Chasko et al (1979) |
| Kwengo | 35 | 18 | 16 | 1 | | | .743 | .257 | | | 1.35 | 35 | 0 | 0 | 1.000 | .000 | - | Nurse & Jenkins (1977a) |
| Dama | 119 | 67 | 45 | 7 | | | .752 | .248 | | | .024 | 119 | 0 | 0 | 1.000 | .000 | - | Nurse et al (1976) |

Khoisan:

| Population | N | (1) | (2-1) | (2) | (other) | f | f | f | f | $\chi^2$ | N | (2-1) | (2) | f | f | Reference |
|---|---|---|---|---|---|---|---|---|---|---|---|---|---|---|---|---|
| **Khoi:** | | | | | | | | | | | | | | | | |
| Keetmanshoop Nama | 150/149ᵛ | 106 | 29 | 10 | 5 | .803 | .163 | .034 | | 12.619* | 148 | 1 | 0 | .996 | .004 | Jenkins (1972) |
| Topnaars(Sesfontein) | 42ᵛ | 31 | 7 | 0 | 3 | .856 | .096 | .048 | | | 42 | 0 | 0 | 1.000 | .000 | " |
| **San:** | | | | | | | | | | | | | | | | |
| **Northern:** | | | | | | | | | | | | | | | | |
| Glang!ai | 36 | 35 | 0 | 0 | 1 | .986 | .000 | .014 | | | 36 | 0 | 0 | 1.000 | .000 | Nurse et al (1977) |
| Tsumkwe !Kung | 169 | 149 | 15 | 3 | 1 | .906 | .086 | .004 | .004 | 8.29* | 167 | 2 | 0 | .994 | .006 | Jenkins (1972) |
| Samangaigai | 29 | 26 | 2 | 1 | 0 | .948 | .052 | | | 8.33 | 28 | 1 | 0 | .983 | .017 | Nurse et al (1977) |
| Dobe | 328 | 315 | 13 | 0 | 0 | .980 | .020 | | | | 311 | 17 | 0 | .974 | .026 | Jenkins (1972) |
| /Du/da | 74 | 73 | 1 | 0 | 0 | .993 | .007 | | | | 69 | 5 | 0 | .966 | .034 | " |
| //Au//en | 82 | 76 | 6 | 0 | 0 | .963 | .037 | | | | 78 | 4 | 0 | .976 | .024 | Nurse et al (1977) |
| **Central:** | | | | | | | | | | | | | | | | |
| Hei//om | 91 | 74 | 15 | 1 | 1 | .901 | .093 | .006 | | .17* | 81 | 10 | 0 | .941 | .059 | Nurse et al (1977) |
| Nharo | 58 | 55 | 3 | 0 | 0 | .974 | .026 | | | | 54 | 4 | 0 | .965 | .035 | " |
| G/wi | 93 | 67 | 23 | 3 | 0 | .844 | .156 | | | .31 | 85 | 7 | 1 | .947 | .053 | Jenkins et al (1975) |
| G//ana | 50 | 32 | 18 | 0 | 0 | .820 | .180 | | | | 47 | 3 | 0 | .970 | .030 | " |
| **Southern:** | | | | | | | | | | | | | | | | |
| !Xõ | 51 | 42 | 7 | 2 | 0 | .856 | .144 | | | 3.72 | 46 | 5 | 0 | .952 | .048 | Jenkins & Nurse (1976) |
| G!aokx'ate | 33 | 32 | 1 | 0 | 0 | .985 | .015 | | | | 31 | 2 | 0 | .970 | .030 | Nurse et al (1977) |
| **Uncertain:** | | | | | | | | | | | | | | | | |
| Eastern ≠Huã | 36 | 36 | 0 | 0 | 0 | 1.000 | .000 | | | | 36 | 0 | 0 | 1.000 | .000 | Nurse & Jenkins (1977a) |
| **Caucasoid:** | | | | | | | | | | | | | | | | |
| **Afrikaans-speaking:** | | | | | | | | | | | | | | | | |
| Johannesburg | 77 | 56 | 18 | 3 | 0 | .844 | .156 | | | .95 | 77 | 0 | 0 | 1.000 | .000 | Present study |
| Johannesburg | 158 | 89 | 57 | 12 | 0 | .744 | .256 | | | .457 | 158 | 0 | 0 | 1.000 | .000 | " |
| Namibia | 57 | 49 | 4 | 4 | 0 | .895 | .105 | | | 23.72 | | | | | | Palmhert-Keller et al (1983) |
| **German-speaking:** | | | | | | | | | | | | | | | | |
| Namibia | 118 | 74 | 35 | 9 | 0 | .775 | .225 | | | 2.54 | 118 | 0 | 0 | 1.000 | .000 | Palmhert-Keller et al (1983) |
| **Indian:** | | | | | | | | | | | | | | | | |
| **Transvaal:** | | | | | | | | | | | | | | | | |
| Hindu | 53 | 26 | 22 | 5 | 0 | .698 | .302 | .000 | | .737 | 53 | 0 | 0 | 1.000 | .000 | Present study |
| Moslem | 56 | 25 | 28 | 3 | 0 | .696 | .304 | .000 | | 1.302 | 56 | 0 | 0 | 1.000 | .000 | " |
| **Hybrid Peoples:** | | | | | | | | | | | | | | | | |
| **Baster-like:** | | | | | | | | | | | | | | | | |
| Rehoboth | 119 | 73 | 41 | 5 | 0 | .786 | .214 | | | .07* | 118 | 1 | 0 | .996 | .004 | Nurse et al (1982) |
| !Kuboes | 135 | 96 | 33 | 2 | 4 | .848 | .137 | .015 | | .203* | 135 | 0 | 0 | 1.000 | .000 | Jenkins (1972) |
| Riemvaasmaak | 86/87 | 54 | 28 | 4 | | .791 | .209 | | | .020 | 86 | 1 | 0 | .994 | .006 | Nurse & Jenkins (1978) |
| **Griqua:** | | | | | | | | | | | | | | | | |
| Griqua | 262 | 188 | 67 | 7 | | .845 | .155 | | | .12 | 252 | 10 | 0 | .981 | .019 | Nurse & Jenkins (1975) |
| **"Cape-Coloureds":** | | | | | | | | | | | | | | | | |
| Cape Town | 200 | 128 | 62 | 10 | | .795 | .205 | | | .482 | 194 | 6 | 0 | .985 | .015 | Gordon et al (1968) |
| Johannesburg | 109 | 66 | 33 | 10 | | .757 | .243 | | | 3.478 | 107 | 2 | 0 | .991 | .009 | Jenkins (1972) |

* individuals with $PGM_1$(6-1) and $PGM_1$(7-1) phenotypes have been excluded from $\chi^2$ calculations; ᵛ includes one individual $PGM_1$(6-2)

# Table 18. Phosphoglucomutase System: First Locus Subtyping

| Population | n | Phenotypes | | | | | | | | | | Gene Frequencies | | | | | $\chi^2$ | df | Reference |
|---|---|---|---|---|---|---|---|---|---|---|---|---|---|---|---|---|---|---|---|
| | | $1^+1^+$ | $1^+1^-$ | $1^-1^-$ | $1^+2^+$ | $1^+2^-$ | $1^-2^+$ | $1^-2^-$ | $2^+2^+$ | $2^+2^-$ | $2^-2^-$ | $PGM_1^{1+}$ | $PGM_1^{1-}$ | $PGM_1^{2+}$ | $PGM_1^{2-}$ | $PGM_1^6$ | | | |
| **Bantu-speaking Negroes:** | | | | | | | | | | | | | | | | | | | |
| **Nguni:** | | | | | | | | | | | | | | | | | | | |
| Zulu | 102 | 55 | 20 | 2 | 14 | 5 | 5 | 0 | 0 | 1 | 0 | .730 | .142 | .098 | .029 | | 2.738 | 3 | Tippler et al (1982) |
| **Sotho/Tswana:** | | | | | | | | | | | | | | | | | | | |
| Pedi | 98 | 43 | 18 | 6 | 18 | 2 | 5 | 1 | 4 | 0 | 1 | .633 | .184 | .158 | .026 | | 5.207↓ | 4 | Tippler et al (1982) |
| Ngwato | 77 | 32 | 10 | 1 | 28 | 2 | 1 | 0 | 3 | 0 | 0 | .675 | .085 | .227 | .013 | | 3.132↓ | 3 | Present study |
| Kgalagadi | 81 | 31 | 15 | 5 | 16 | 1 | 7 | 1 | 3 | 0 | 1 | .586 | .204 | .179 | .025 | .006 | 3.323 | 4 | Tippler et al (1982) |
| Venda | 104 | 49 | 22 | 3 | 23 | 3 | 3 | 0 | 1 | 0 | 0 | .702 | .149 | .135 | .014 | | 2.876 | 4 | Present study |
| Kavango | 64 | 32 | 10 | 1 | 13 | 3 | 2 | 1 | 2 | 0 | 0 | .703 | .117 | .148 | .032 | | 0.325 | 3 | Tippler et al (1982) |
| **Herero:** | | | | | | | | | | | | | | | | | | | |
| Mbanderu | 72 | 38 | 7 | 0 | 13 | 7 | 1 | 0 | 2 | 2 | 2 | .715 | .056 | .139 | .090 | | 1.457 | 4 | " |
| **Khoisan:** | | | | | | | | | | | | | | | | | | | |
| **Khoi:** | | | | | | | | | | | | | | | | | | | |
| Nama (Rehoboth) | 78 | 38 | 12 | 2 | 11 | 5 | 3 | 0 | 1 | 1 | 1 | .680 | .134 | .109 | .051 | .026 | 2.543 | 4 | " |
| **San:** | | | | | | | | | | | | | | | | | | | |
| /Ai/ai !Kung | 84 | 42 | 36 | 3 | 2 | 0 | 0 | 0 | 0 | 0 | 0 | .732 | .250 | .012 | .000 | .006 | 2.739 | 1 | " |
| **Caucasoid:** | | | | | | | | | | | | | | | | | | | |
| **Indian:** | | | | | | | | | | | | | | | | | | | |
| Johannesburg | 104 | 37 | 9 | 2 | 37 | 5 | 6 | 2 | 2 | 4 | 0 | .595 | .110 | .243 | .052 | | 10.147 | 6 | " |
| **Afrikaans-speaking:** | | | | | | | | | | | | | | | | | | | |
| Johannesburg | 157 | 55 | 31 | 3 | 33 | 18 | 4 | 1 | 3 | 7 | 2 | .612 | .134 | .159 | .096 | | 6.385 | 6 | " |

→ $\chi^2$ obtained by pooling expected values less than one

↓ includes one individual $PGM_1(1^+6)$

→ $\chi^2$ obtained by pooling expected values less than one

↓ includes two individuals $PGM_1(1^+6)$ and two individuals $PGM_1(1^-6)$

↓ includes one individual $PGM_1(1^+6)$

# Table 19. Adenylate Kinase System

| Population | n | 1 | 2-1 | 2 | $AK^1$ | $AK^2$ | $\chi^2_{[1]}$ | References |
|---|---|---|---|---|---|---|---|---|
| **Bantu-speaking Negroes:** | | | | | | | | |
| **Nguni:** | | | | | | | | |
| Swati | 91 | 89 | 2 | 0 | .989 | .011 | | Jenkins (1972) |
| Swati | 62 | 59 | 3 | 0 | .976 | .024 | | Hitzeroth *et al* (1981) |
| Zulu | 102 | 101 | 1 | 0 | .995 | .005 | | Present study |
| Zulu | 304 | 298 | 6 | 0 | .990 | .010 | | Kirk *et al* (1971) |
| Zulu | 116 | 109 | 7 | 0 | .970 | .030 | | Hitzeroth *et al* (1981) |
| Xhosa | 100 | 98 | 2 | 0 | .990 | .010 | | Gordon *et al* (1966) |
| Ndebele | 164 | 159 | 5 | 0 | .985 | .015 | | Hitzeroth *et al* (1981) |
| **Sotho/Tswana:** | | | | | | | | |
| S.Sotho(Highlands) | 90 | 87 | 3 | 0 | .983 | .017 | | Beaumont *et al* (1979) |
| S.Sotho(Lowlands) | 92 | 88 | 4 | 0 | .978 | .022 | | "       " |
| Pedi(urban) | 88 | 85 | 3 | 0 | .983 | .017 | | Jenkins (1972) |
| Pedi(rural) | 116 | 113 | 3 | 0 | .987 | .013 | | "       " |
| Pedi | 261 | 256 | 5 | 0 | .990 | .010 | | Hitzeroth *et al* (1981) |
| Fokeng | 121 | 118 | 3 | 0 | .988 | .012 | | Jenkins (1972) |
| Ngwato | 77 | 74 | 3 | 0 | .981 | .019 | | Present study |
| Tswana | 150 | 146 | 4 | 0 | .987 | .013 | | Hitzeroth *et al* (1981) |
| Kgalagadi | 124 | 122 | 2 | 0 | .992 | .008 | | Present study |
| **Venda:** | | | | | | | | |
| Urban | 45 | 44 | 1 | 0 | .989 | .011 | | Hitzeroth *et al* (1981) |
| Rural | 104 | 104 | 0 | 0 | 1.000 | .000 | | Present study |
| **Tsonga/Ronga:** | | | | | | | | |
| "Shangaan"/Tonga | 150 | 150 | 0 | 0 | 1.000 | .000 | | Hitzeroth *et al* (1981) |
| **Ambo:** | | | | | | | | |
| Kwambi | 104 | 104 | 0 | 0 | 1.000 | .000 | | Nurse *et al* (1983) |
| Kwaluudhi | 69 | 69 | 0 | 0 | 1.000 | .000 | | "       " |
| Ndonga | 110 | 110 | 0 | 0 | 1.000 | .000 | | "       " |
| Mbalantu | 72 | 72 | 0 | 0 | 1.000 | .000 | | "       " |
| Kwanyama | 124 | 124 | 0 | 0 | 1.000 | .000 | | "       " |
| Ngandjera | 66 | 66 | 0 | 0 | 1.000 | .000 | | "       " |
| Nkolonkadhi | 32 | 32 | 0 | 0 | 1.000 | .000 | | "       " |
| **Herero:** | | | | | | | | |
| Botswana | 39 | 37 | 2 | 0 | .974 | .026 | | Jenkins (1972) |
| Mbanderu | 99 | 89 | 10 | 0 | .949 | .051 | | Present study |
| **Khoisan-speaking Negroes:** | | | | | | | | |
| Sarwa | 77 | 77 | 0 | 0 | 1.000 | .000 | | Chasko *et al* (1979) |
| Dama | 119 | 98 | 21 | 0 | .912 | .088 | | Nurse *et al* (1976) |
| **Khoisan:** | | | | | | | | |
| **Khoi:** | | | | | | | | |
| Keetmanshoop Nama | 118 | 109 | 9 | 0 | .962 | .038 | | Jenkins (1972) |
| Sesfontein Topnaars | 42 | 34 | 8 | 0 | .905 | .095 | | Nurse & Jenkins (1977a) |
| **San:** | | | | | | | | |
| **Northern:** | | | | | | | | |
| Tsumkwe !Kung | 147 | 131 | 16 | 0 | .946 | .054 | | Nurse & Jenkins (1977a) |
| Saman!aika " | 13 | 11 | 2 | 0 | .923 | .077 | | Nurse *et al* (1977) |
| Dobe " | 328 | 312 | 16 | 0 | .976 | .024 | | Nurse & Jenkins (1977a) |
| /Du/da " | 74 | 63 | 10 | 1 | .919 | .081 | .590 | "     "     " |
| //Au//en " | 82 | 68 | 14 | 0 | .915 | .085 | | "     "     " |
| **San:** | | | | | | | | |
| **Central:** | | | | | | | | |
| Hei//um | 51 | 47 | 4 | 0 | .960 | .040 | | Nurse & Jenkins (1977a) |
| Nharo | 58 | 50 | 8 | 0 | .931 | .069 | | "     "     " |
| **Southern:** | | | | | | | | |
| G!aokx'ate | 33 | 33 | 0 | 0 | 1.000 | .000 | | "     "     " |
| **Caucasoids:** | | | | | | | | |
| **Afrikaans-speaking:** | | | | | | | | |
| Johannesburg | 77 | 73 | 4 | 0 | .974 | .026 | | Present study |
| Johannesburg | 160 | 148 | 12 | 0 | .963 | .037 | | "       " |
| SWA/Namibia | 57 | 51 | 6 | 0 | .947 | .053 | | Palmhert-Keller *et al* (1983) |
| **German-speaking:** | | | | | | | | |
| SWA/Namibia | 118 | 111 | 7 | 0 | .970 | .030 | | "       "       " |
| **Indian:** | | | | | | | | |
| **Transvaal:** | | | | | | | | |
| Hindu | 51 | 41 | 10 | 0 | .902 | .098 | | Present study |
| Moslem | 55 | 47 | 8 | 0 | .927 | .073 | | "       " |
| **Hybrid Peoples:** | | | | | | | | |
| **Baster-like:** | | | | | | | | |
| Rehoboth | 120 | 110 | 9 | 1 | .954 | .046 | 1.850 | Nurse *et al* (1982) |
| !Kuboes | 116 | 109 | 7 | 0 | .970 | .030 | | Jenkins (1972) |
| **Griqua:** | | | | | | | | |
| Griqua | 87 | 84 | 3 | 0 | .983 | .017 | | Nurse & Jenkins (1975) |
| **"Cape-Coloureds":** | | | | | | | | |
| Johannesburg | 109 | 104 | 5 | 0 | .977 | .023 | | Jenkins (1972) |
| Cape Town | 100 | 90 | 10 | 0 | .950 | .050 | | "       " |

# Table 20. Peptidase A System

| Population | n | Phenotypes | | | | Gene Frequencies | | | $\chi^2_{[1]}$ | References |
|---|---|---|---|---|---|---|---|---|---|---|
| | | 1 | 2-1 | 2 | 8-2 | $PepA^1$ | $PepA^2$ | $PepA^8$ | | |
| **Bantu-speaking Negroes:** | | | | | | | | | | |
| Nguni: | | | | | | | | | | |
| Zulu/Tonga | 123 | 97 | 26 | 0 | 0 | .894 | .106 | .000 | | Jenkins (1972) |
| Zulu | 303 | 249 | 50 | 4 | 0 | .904 | .096 | .000 | .662 | Kirk *et al* (1971) |
| Zulu | 102 | 72 | 27 | 0 | 3 | .833 | .147 | .020 | | Present study |
| Sotho/Tswana: | | | | | | | | | | |
| Pedi(rural) | 98 | 86 | 12 | 0 | 0 | .939 | .061 | .000 | | Present study |
| Ngwato | 75 | 63 | 9 | 0 | 3 | .900 | .080 | .020 | | " " |
| Kgalagadi | 124 | 113 | 10 | 0 | 1 | .952 | .044 | .004 | | " " |
| Venda: | | | | | | | | | | |
| Rural | 104 | 76 | 25 | 2 | 1 | .851 | .139 | .010 | .013 | " " |
| Kavango: | | | | | | | | | | |
| Mbukushu | 54 | 40 | 14 | 0 | 0 | .870 | .130 | .000 | | Nurse & Jenkins (1977b) |
| Gciriku | 105 | 84 | 18 | 3 | 0 | .886 | .114 | .000 | 2.370 | " " " |
| Sambyu | 91 | 62 | 28 | 1 | 0 | .835 | .165 | .000 | 1.270 | " " " |
| Kwangali | 101 | 80 | 20 | 1 | 0 | .891 | .109 | .000 | .040 | " " " |
| Ambo: | | | | | | | | | | |
| Kwambi | 41 | 37 | 4 | 0 | 0 | .951 | .049 | .000 | | Nurse *et al* (1983) |
| Kwaluudhi | 40 | 36 | 3 | 1 | 0 | .937 | .063 | .000 | 5.101 | " " |
| Ndonga | 58 | 50 | 8 | 0 | 0 | .931 | .069 | .000 | | " " |
| Mbalantu | 66 | 59 | 5 | 2 | 0 | .932 | .068 | .000 | 9.418 | " " |
| Kwanyama | 119 | 102 | 17 | 0 | 0 | .929 | .071 | .000 | | " " |
| Ngandjera | 42 | 40 | 2 | 0 | 0 | .976 | .024 | .000 | | " " |
| Nkolonkadhi | 21 | 19 | 1 | 1 | 0 | .929 | .071 | .000 | 50.840 | " " |
| Herero: | | | | | | | | | | |
| SWA/Namibia | 88 | 72 | 16 | 0 | 0 | .909 | .091 | .000 | | Present study |
| Mbanderu | 99 | 92 | 6 | 0 | 1 | .960 | .035 | .005 | | " " |
| Himba | 85 | 70 | 15 | 0 | 0 | .912 | .088 | .000 | | " " |
| Chimba | 16 | 16 | 0 | 0 | 0 | 1.000 | .000 | .000 | | " " |
| **Khoisan-speaking Negroes:** | | | | | | | | | | |
| Sarwa | 77 | 72 | 3 | 2 | 0 | .955 | .045 | .000 | 18.200 | Chasko *et al* (1979) |
| Kwengo | 35 | 29 | 4 | 2 | 0 | .886 | .114 | .000 | 6.590 | Jenkins & Nurse (1977b) |
| Dama | 92 | 88 | 4 | 0 | 0 | .978 | .022 | .000 | | Nurse *et al* (1976) |
| **Khoisan:** | | | | | | | | | | |
| Khoi: | | | | | | | | | | |
| Keetmanshoop Nama | 175 | 168 | 7 | 0 | 0 | .980 | .020 | .000 | | Jenkins (1972) |
| San: | | | | | | | | | | |
| Northern: | | | | | | | | | | |
| G!ang!ai | 35 | 35 | 0 | 0 | 0 | 1.000 | .000 | .000 | | Nurse & Jenkins (1977a) |
| Tsumkwe !Kung | 278 | 277 | 1 | 0 | 0 | .998 | .002 | .000 | | " " " |
| Central: | | | | | | | | | | |
| Hei//om | 68 | 65 | 2 | 1 | 0 | .967 | .033 | .000 | .936 | " " " |
| G/wi | 79 | 78 | 1 | 0 | 0 | .994 | .006 | .000 | | " " " |
| G//ana | 50 | 49 | 1 | 0 | 0 | .990 | .010 | .000 | | " " " |
| San: | | | | | | | | | | |
| Southern: | | | | | | | | | | |
| !Kõ | 51 | 51 | 0 | 0 | 0 | 1.000 | .000 | .000 | | Nurse & Jenkins (1977a) |
| G!aokx'ate | 33 | 33 | 0 | 0 | 0 | 1.000 | .000 | .000 | | " " " |
| Uncertain: | | | | | | | | | | |
| Eastern ǂHuǎ | 36 | 31 | 4 | 1 | 0 | .917 | .083 | .000 | 3.630 | " " " |
| **Caucasoids:** | | | | | | | | | | |
| Afrikaans-speaking: | | | | | | | | | | |
| Johannesburg | 160 | 160 | 0 | 0 | 0 | 1.000 | .000 | .000 | | Present study |
| Johannesburg | 77 | 76 | 1 | 0 | 0 | .994 | .006 | .000 | | " " |
| **Hybrid Peoples:** | | | | | | | | | | |
| Baster-like: | | | | | | | | | | |
| Riemvaasmaak | 88 | 74 | 14 | 0 | 0 | .921 | .079 | .000 | | Nurse & Jenkins (1978) |
| Griqua: | | | | | | | | | | |
| Griqua | 266 | 250 | 15 | 1 | 0 | .968 | .032 | .000 | 2.11 | Nurse & Jenkins (1975) |
| "Cape-Coloureds": | | | | | | | | | | |
| Johannesburg | 95 | 89 | 4 | 2 | 0 | .958 | .042 | .000 | 17.036 | Jenkins (1972) |

# Table 21. Adenosine Deaminase System

| Population | n | Phenotypes | | | Gene Frequencies | | $\chi^2_{[1]}$ | References |
|---|---|---|---|---|---|---|---|---|
| | | 1 | 2-1 | 2 | $ADA^1$ | $ADA^2$ | | |
| **Bantu-speaking Negroes:** | | | | | | | | |
| _Nguni:_ | | | | | | | | |
| Swati | * | | | | 1.000 | .000 | – | Hitzeroth _et al_ (1981) |
| Zulu | * | | | | 1.000 | .000 | – | " " |
| Zulu | 102 | 102 | 0 | 0 | 1.000 | .000 | – | Present study |
| Zulu/Tonga | 257 | 254 | 3 | 0 | .994 | .006 | – | Jenkins (1972) |
| Xhosa | 138 | 137 | 1 | 0 | .996 | .004 | – | Weissmann _et al_ (1981) |
| Ndebele | 168 | 166 | 2 | 0 | .994 | .006 | – | Hitzeroth _et al_ (1981) |
| _Sotho/Tswana:_ | | | | | | | | |
| Tswana | * | | | | 1.000 | .000 | – | " " |
| Pedi(urban) | 261 | 260 | 1 | 0 | .998 | .002 | – | " " |
| Pedi(rural) | 98 | 98 | 0 | 0 | 1.000 | .000 | – | Present study |
| Ngwato | 77 | 77 | 0 | 0 | 1.000 | .000 | – | " " |
| Kgalagadi | 124 | 124 | 0 | 0 | 1.000 | .000 | – | " " |
| _Venda:_ | | | | | | | | |
| Urban | * | | | | 1.000 | .000 | – | Hitzeroth _et al_ (1981) |
| Rural | 104 | 104 | 0 | 0 | 1.000 | .000 | – | Present study |
| _Tsonga/Ronga:_ | | | | | | | | |
| "Shangaan"/Tsonga | * | | | | 1.000 | .000 | – | Hitzeroth _et al_ (1981) |
| _Herero:_ | | | | | | | | |
| Mbanderu | 99 | 99 | 0 | 0 | 1.000 | .000 | – | Present study |
| **Khoisan-speaking Negroes:** | | | | | | | | |
| Sarwa | 77 | 77 | 0 | 0 | 1.000 | .000 | – | Chasko _et al_ (1979) |
| Kwengo | 35 | 35 | 0 | 0 | 1.000 | .000 | – | Nurse & Jenkins (1977a) |
| Dama | 92 | 91 | 1 | 0 | .995 | .005 | – | Nurse _et al_ (1976) |
| **Khoisan:** | | | | | | | | |
| _Khoi:_ | | | | | | | | |
| Keetmanshoop Nama | 134 | 133 | 0 | 1 | .993 | .007 | – | Jenkins (1972) |
| _San:_ | | | | | | | | |
| _Northern:_ | | | | | | | | |
| G!ang!ai | 33 | 33 | 0 | 0 | 1.000 | .000 | – | Jenkins _et al_ (1979) |
| Tsumkwe !Kung | 278 | 277 | 1 | 0 | .998 | .002 | – | Jenkins (1972) |
| _Central:_ | | | | | | | | |
| Hei//om | 75 | 75 | 0 | 0 | 1.000 | .000 | – | Jenkins _et al_ (1979) |
| G/wi | 71 | 71 | 0 | 0 | 1.000 | .000 | – | " " " |
| G//ana | 21 | 21 | 0 | 0 | 1.000 | .000 | – | " " " |
| _Southern:_ | | | | | | | | |
| !Xõ | 51 | 51 | 0 | 0 | 1.000 | .000 | – | " " " |
| G!aokx'ate | 33 | 33 | 0 | 0 | 1.000 | .000 | – | " " " |
| _Uncertain:_ | | | | | | | | |
| Eastern ≠Huã | 36 | 36 | 0 | 0 | 1.000 | .000 | – | " " " |
| **Caucasoids:** | | | | | | | | |
| _Afrikaans-speaking:_ | | | | | | | | |
| Johannesburg | 76 | 69 | 7 | 0 | .954 | .046 | – | Present study |
| Johannesburg | 160 | 151 | 9 | 0 | .972 | .028 | – | " " |
| SWA/Namibia | 57 | 54 | 3 | 0 | .974 | .026 | – | Palmhert Keller _et al_ (1983) |
| _German-speaking:_ | | | | | | | | |
| SWA/Namibia | 118 | 103 | 15 | 0 | .936 | .064 | – | " " " |
| _Indian:_ | | | | | | | | |
| _Transvaal:_ | | | | | | | | |
| Hindu | 53 | 46 | 7 | 0 | .934 | .066 | – | Present study |
| Moslem | 56 | 45 | 10 | 1 | .893 | .107 | .031 | " " |
| **Hybrid Peoples:** | | | | | | | | |
| _Baster-Like:_ | | | | | | | | |
| Rehoboth | 120 | 114 | 6 | 0 | .975 | .025 | – | Nurse _et al_ (1982) |
| _Griqua:_ | | | | | | | | |
| Griqua | 263 | 257 | 5 | 1 | .987 | .013 | 6.210 | Nurse & Jenkins (1975) |
| _"Cape-Coloureds":_ | | | | | | | | |
| Johannesburg | 144 | 141 | 3 | 0 | .990 | .010 | – | Jenkins (1972) |

\* numbers tested not specified

# Table 22. Glutamate Pyruvate Transaminase System

| Population | n | Phenotypes 1 | Phenotypes 2-1 | Phenotypes 2 | Gene Frequencies $GPT^1$ | Gene Frequencies $GPT^2$ | $x^2_{[1]}$ | References |
|---|---|---|---|---|---|---|---|---|
| **Bantu-speaking Negroes:** | | | | | | | | |
| _Nguni_: | | | | | | | | |
| Swati | 65 | 52 | 13 | 0 | .900 | .100 | | Hitzeroth _et al_ (1981) |
| Zulu | 102 | 70 | 31 | 1 | .838 | .162 | 1.483 | Present study |
| Zulu | 120 | 102 | 15 | 3 | .913 | .087 | 5.713 | Hitzeroth _et al_ (1981) |
| Ndebele | 171 | 133 | 34 | 4 | .877 | .123 | 1.014 | " " |
| _Sotho/Tswana_: | | | | | | | | |
| Tswana | 155 | 128 | 26 | 1 | .910 | .090 | .068 | Hitzeroth _et al_ (1981) |
| Pedi(urban) | 261 | 194 | 62 | 5 | .862 | .138 | .001 | " " |
| Pedi(rural) | 98 | 75 | 21 | 2 | .872 | .128 | .135 | Present study |
| Ngwato | 75 | 47 | 27 | 1 | .807 | .193 | 1.783 | " " |
| Kgalagadi | 124 | 105 | 15 | 4 | .907 | .093 | 9.810 | " " |
| _Venda_: | | | | | | | | |
| Urban | 47 | 39 | 7 | 1 | .904 | .096 | .915 | Hitzeroth _et al_ (1981) |
| Rural | 104 | 79 | 22 | 3 | .865 | .135 | .878 | Present study |
| _Tsonga/Ronga_: | | | | | | | | |
| Shangaan-Tsonga | 154 | 124 | 28 | 2 | .896 | .104 | .085 | Hitzeroth _et al_ (1981) |
| _Ambo_: | | | | | | | | |
| Kwambi | 76 | 50 | 24 | 2 | .816 | .184 | .210 | Marks _et al_ (1977) |
| Kwaluudhi | 60 | 35 | 21 | 4 | .758 | .242 | .120 | " " |
| Ndonga | 98 | 58 | 35 | 5 | .770 | .230 | .010 | " " |
| Mbantu | 56 | 37 | 14 | 5 | .786 | .214 | 3.610 | " " |
| Kwanyama | 115 | 75 | 36 | 4 | .809 | .191 | .020 | " " |
| Ngandjera | 56 | 30 | 21 | 5 | .723 | .277 | .220 | " " |
| Nkolonkadhi | 29 | 18 | 10 | 1 | .793 | .207 | .100 | " " |
| _Herero_: | | | | | | | | |
| Mbanderu | 97 | 51 | 42 | 4 | .742 | .258 | 1.681 | Present study |
| **Khoisan-speaking Negroes:** | | | | | | | | |
| Sarwa | 77 | 56 | 19 | 2 | .851 | .149 | .070 | Chasko _et al_ (1979) |
| **Khoisan:** | | | | | | | | |
| _San_: | | | | | | | | |
| _Southern_: | | | | | | | | |
| G!aokx'ate | 33 | 5 | 20 | 8 | .455 | .545 | 1.60 | Nurse & Jenkins(1977a) |
| **Caucasoids:** | | | | | | | | |
| _Afrikaans-speaking_: | | | | | | | | |
| Johannesburg | 77 | 9 | 41 | 27 | .383 | .617 | 1.230 | Present study |
| Johannesburg | 160* | 32 | 76 | 51 | .440 | .560 | .210 | " " |
| SWA/Namibia | 57 | 13 | 22 | 22 | .421 | .579 | 2.480 | Palmhert-Keller _et al_ (1983) |
| _German-speaking_: | | | | | | | | |
| SWA/Namibia | 118 | 32 | 55 | 31 | .504 | .496 | .540 | Palmhert-Keller _et al_ (1983) |

\* excluding 1 individual with phenotype 3-1

# Table 23.  Carbonic Anhydrase: Second Locus System

| Population | n | Phenotypes 1 | Phenotypes 2-1 | Phenotypes 2 | Gene Frequencies $CA^1_{II}$ | Gene Frequencies $CA^2_{II}$ | $\chi^2_{[1]}$ | References |
|---|---|---|---|---|---|---|---|---|
| Bantu-speaking Negroes: | | | | | | | | |
| Nguni: | | | | | | | | |
| Zulu | 102 | 87 | 13 | 2 | .917 | .083 | 2.790 | Present study |
| Sotho/Tswana: | | | | | | | | |
| Pedi(rural) | 98 | 82 | 15 | 1 | .913 | .087 | .112 | "    " |
| Ngwato | 77 | 63 | 14 | 0 | .909 | .091 | | "    " |
| Kgalagadi | 124 | 115 | 9 | 0 | .964 | .036 | | "    " |
| Venda: | | | | | | | | |
| Rural | 104 | 81 | 19 | 4 | .870 | .130 | 3.796 | "    " |
| Ambo: | | | | | | | | |
| Kwambi | 76 | 65 | 10 | 1 | .921 | .079 | .510 | Marks et al (1977) |
| Kwaluudhi | 60 | 42 | 17 | 1 | .842 | .158 | .240 | "    " |
| Ndonga | 98 | 84 | 14 | 0 | .929 | .071 | | "    " |
| Mbalantu | 56 | 38 | 16 | 2 | .821 | .179 | 1.820 | "    " |
| Kwanyama | 115 | 93 | 18 | 4 | .887 | .113 | 5.370 | "    " |
| Ngandjera | 56 | 44 | 11 | 1 | .884 | .116 | .150 | "    " |
| Nkolonkadhi | 29 | 17 | 10 | 2 | .759 | .241 | .090 | "    " |
| Herero: | | | | | | | | |
| Mbanderu | 99 | 87 | 9 | 3 | .924 | .076 | 12.159 | Present study |
| Khoisan-speaking Negroes: | | | | | | | | |
| Sarwa | 77 | 68 | 9 | 0 | .942 | .058 | | Chasko et al (1979) |
| Khoisan: | | | | | | | | |
| San: | | | | | | | | |
| Southern: | | | | | | | | |
| G!aokx'ate | 33 | 33 | 0 | 0 | 1.000 | .000 | | Nurse & Jenkins (1977a) |
| Caucasoids: | | | | | | | | |
| Afrikaans-speaking: | | | | | | | | |
| Johannesburg | 77 | 77 | 0 | 0 | 1.000 | .000 | | Present study |
| Johannesburg | 160 | 160 | 0 | 0 | 1.000 | .000 | | "    " |

# Table 24. Esterase D System

| Population | n | Phenotypes | | | Gene Frequencies | | $\chi^2_{[1]}$ | References |
|---|---|---|---|---|---|---|---|---|
| | | 1 | 2-1 | 2 | $EsD^1$ | $EsD^2$ | | |
| **Bantu-speaking Negroes:** | | | | | | | | |
| **Nguni:** | | | | | | | | |
| Swati | 69 | 69 | 0 | 0 | 1.000 | .000 | | Hitzeroth *et al* (1976) |
| Zulu | 119 | 111 | 8 | 0 | .966 | .034 | | " " |
| Zulu | 102 | 95 | 7 | 0 | .966 | .034 | | Present study |
| Ndebele | 174 | 166 | 8 | 0 | .977 | .023 | | Hitzeroth *et al* (1976) |
| **Sotho/Tswana:** | | | | | | | | |
| Tswana | 155 | 148 | 6 | 1 | .964 | .026 | 8.074 | Hitzeroth *et al* (1976) |
| Pedi(urban) | 275 | 254 | 20 | 1 | .960 | .040 | .773 | " " |
| Pedi(rural) | 98 | 90 | 7 | 1 | .964 | .046 | 3.364 | Present study |
| Ngwato | 77 | 73 | 4 | 0 | .974 | .026 | | " " |
| Kgalagadi | 124 | 107 | 17 | 0 | .931 | .069 | | " " |
| **Venda:** | | | | | | | | |
| Urban | 47 | 43 | 4 | 0 | .957 | .043 | | Hitzeroth *et al* (1976) |
| Rural | 104 | 96 | 8 | 0 | .962 | .038 | | Present study |
| **Tsonga/Ronga:** | | | | | | | | |
| Shangaan-Tsonga | 158 | 147 | 11 | 0 | .965 | .034 | | Hitzeroth *et al* (1976) |
| **Ambo:** | | | | | | | | |
| Kwambi | 76[v] | 60 | 12 | 3 | .773 | .112 | .110 | Marks *et al* (1977) |
| Kwaluudhi | 60 | 54 | 5 | 1 | .942 | .058 | 3.680 | " " |
| Ndonga | 98 | 89 | 9 | 0 | .954 | .046 | | " " |
| Mbalantu | 56 | 48 | 7 | 1 | .920 | .080 | 1.080 | " " |
| Kwanyama | 115 | 106 | 9 | 0 | .961 | .039 | | " " |
| Ngandjera | 56 | 48 | 7 | 1 | .920 | .080 | 1.080 | " " |
| Nkolonkadhi | 29 | 26 | 3 | 0 | .948 | .052 | | " " |
| **Herero:** | | | | | | | | |
| Mbanderu | 98 | 89 | 9 | 0 | .954 | .046 | | Present study |
| **Khoisan-speaking Negroes:** | | | | | | | | |
| Sarwa | 77 | 74 | 3 | 0 | .980 | .020 | . | Chasko *et al* (1979) |
| **Khoisan:** | | | | | | | | |
| **San:** | | | | | | | | |
| **Southern:** | | | | | | | | |
| !Xõ | 51 | 46 | 5 | 0 | .951 | .049 | | Nurse & Jenkins (1977a) |
| G!aokx'ate | 33 | 33 | 0 | 0 | 1.000 | .000 | | " " " |
| **Uncertain:** | | | | | | | | |
| Eastern ≠Huã | 35 | 34 | 1 | 0 | .986 | .014 | | " " " |
| **Caucasoids:** | | | | | | | | |
| **Afrikaans-speaking:** | | | | | | | | |
| Johannesburg | 77 | 54 | 21 | 2 | .838 | .162 | .001 | Present study |
| Johannesburg | 160 | 112 | 46 | 2 | .844 | .156 | 1.310 | " " |
| SWA/Namibia | 57 | 46 | 10 | 1 | .895 | .105 | .320 | Palmhert-Keller *et al* (1983) |
| **German-speaking:** | | | | | | | | |
| SWA/Namibia | 118 | 88 | 30 | 0 | .873 | .127 | | Palmhert-Keller *et al* (1983) |
| **Hybrid Peoples:** | | | | | | | | |
| **Baster-like:** | | | | | | | | |
| Riemvasmaak | 88 | 83 | 5 | 0 | .972 | .028 | | Nurse & Jenkins (1978) |

[v] excluding one individual of phenotype EsD0, giving an $EsD^0$ frequency of .115

# Table 25. Glyoxalase I System

| Population | n | Phenotypes | | | Gene Frequencies | | $\chi^2_{[1]}$ | References |
|---|---|---|---|---|---|---|---|---|
| | | 1 | 2-1 | 2 | $GLO^1$ | $GLO^2$ | | |
| **Bantu-speaking Negroes:** | | | | | | | | |
| Nguni: | | | | | | | | |
|   Swati | 57 | 2 | 16 | 39 | .175 | .825 | .051 | Bender *et al* (1977) |
|   Zulu | 103 | 5 | 45 | 53 | .267 | .733 | 1.391 | " " |
|   Zulu | 102 | 13 | 39 | 50 | .319 | .681 | 1.458 | Present study |
|   Ndebele | 142 | 6 | 53 | 83 | .229 | .771 | .467 | Bender *et al* (1977) |
| Sotho/Tswana: | | | | | | | | |
|   Tswana | 131 | 9 | 45 | 77 | .241 | .759 | .464 | Bender *et al* (1977) |
|   Pedi (urban) | 238 | 18 | 102 | 118 | .290 | .710 | .398 | " " |
|   Pedi (rural) | 98 | 14 | 38 | 46 | .337 | .663 | 1.704 | Present study |
|   Ngwato | 75 | 8 | 29 | 38 | .300 | .700 | .472 | " " |
|   Kgalagadi | 114 | 18 | 53 | 43 | .390 | .609 | .061 | " " |
| Venda: | | | | | | | | |
|   Urban | 43 | 6 | 16 | 21 | .326 | .674 | 1.001 | Bender *et al* (1977) |
|   Rural | 104 | 8 | 41 | 55 | .274 | .726 | .009 | Present study |
| Tsonga/Ronga: | | | | | | | | |
|   Shangaan/Tsonga | 129 | 6 | 55 | 68 | .260 | .740 | 1.526 | Bender *et al* (1977) |
| Herero: | | | | | | | | |
|   Mbanderu | 98 | 5 | 28 | 65 | .194 | .806 | .537 | Present study |
| **Caucasoids:** | | | | | | | | |
| Afrikaans-speaking: | | | | | | | | |
|   Johannesburg | 160 | 30 | 83 | 47 | .447 | .553 | .390 | Present study |


# Table 26. Phenylthiocarbamide Tasting System

| Population | n | Phenotypes | | Gene Frequencies | | Reference |
|---|---|---|---|---|---|---|
| | | Tasters | non-Tasters | $T$ | $t$ | |
| **Bantu-speaking Negroes:** | | | | | | |
| Nguni: | | | | | | |
|   Zulu | 86 | 84 | 2 | .849 | .151 | Jenkins (1972) |
|   Xhosa | 100 | 94 | 6 | .749 | .251 | Gordon (1965) |
| Sotho/Tswana: | | | | | | |
|   S.Sotho | 149 | 146 | 3 | .859 | .141 | Beaumont *et al* (1979) |
|   Pedi (rural) | 111 | 99 | 12 | .671 | .329 | Jenkins (1972) |
|   Kgalagadi | 38 | 36 | 2 | .772 | .228 | Jenkins (1965) |
| Venda: | 348 | 328 | 20 | .761 | .239 | de Villiers (1970) |
| **Khoisan-speaking Negroes:** | | | | | | |
|   Dama | 6 | 5 | 1 | .582 | .408 | Nurse *et al* (1976) |
| **Khoisan:** | | | | | | |
| Khoi: | | | | | | |
|   Keetmanshoop Nama | 115 | 111 | 4 | .814 | .186 | Jenkins (1972) |
|   Kuiseb Topnaars | 29 | 23 | 6 | .545 | .455 | Jenkins (1972) |
| San: | 85 | 79 | 6 | .735 | .265 | Jenkins (1965) |
| **Hybrid Peoples:** | | | | | | |
| Baster-like: | | | | | | |
|   !Kuboes | 114 | 102 | 12 | .676 | .324 | Jenkins (1972) |
| "Cape-Coloureds": | | | | | | |
|   Johannesburg | 103 | 92 | 11 | .663 | .337 | Jenkins (1972) |
|   Cape Town | 200 | 160 | 40 | .549 | .451 | Gordon (1965) |

## Table 27. Acetylator System

| Population | n | Slow Acetylators n | Slow Acetylators % | Slow Acetylator gene frequency $Ac^S$ | Reference |
|---|---|---|---|---|---|
| Bantu-speaking Negroes: | | | | | |
| Nguni: | | | | | |
| Swati | 16 | 8 | 50 | .707 | Eidus *et al* (1979) |
| Zulu | 40 | 17 | 43 | .652 | Eidus *et al* (1979) |
| Zulu | 83 | 35 | 42 | .649 | Bach *et al* (1976) |
| Zulu | 69 | 12 | 17 | .417 | Present study |
| Xhosa | 32 | 16 | 50 | .707 | Eidus *et al* (1979) |
| Xhosa | 10 | 5 | 50 | .707 | Bach *et al* (1976) |
| Pondo | 8 | 2 | 25 | .500 | Bach *et al* (1976) |
| Ndebele | 12 | 3 | 25 | .500 | Eidus *et al* (1979) |
| Sotho/Tswana: | | | | | |
| Sotho | 39 | 18 | 46 | .679 | Eidus *et al* (1979) |
| Pedi | 18 | 8 | 44 | .667 | Eidus *et al* (1979) |
| Pedi | 98 | 31 | 32 | .562 | Present study |
| Tswana | 43 | 15 | 35 | .591 | Eidus *et al* (1979) |
| Kgalagadi | 12 | 4 | 33 | .577 | Present study |
| Venda: | | | | | |
| Rural | 60 | 40 | 67 | .816 | Present study |
| Tsonga/Ronga: | | | | | |
| Shangaan | 11 | 2 | 18 | .426 | Eidus *et al* (1979) |
| Mixed: | 100 | 27 | 27 | .520 | Buchanan *et al* (1976) |
| Khoisan: | | | | | |
| San: | | | | | |
| !Kung | 30 | 1 | 3 | .183 | Jenkins *et al* (1974) |

## Table 28. Primary Adult Lactase Deficiency

| Population | n | PHILA negative n | PHILA negative % | $PHILA^-$ gene frequency | Reference |
|---|---|---|---|---|---|
| Bantu-speaking Negroes: | | | | | |
| Nguni: | | | | | |
| Zulu | 45 | 42 | 93 | .97 | Jenkins (1982) |
| Sotho/Tswana: | | | | | |
| Pedi | 29 | 29 | 100 | 1.00 | Jenkins (1982) |
| Kgalagadi | 16 | 16 | 100 | 1.00 | Jenkins (1982) |
| Venda: | | | | | |
| Rural | 27 | 27 | 100 | 1.00 | Jenkins (1982) |
| Herero: | | | | | |
| Namibia | 91 | 86 | 95 | .97 | Currie *et al* (1978) |
| Mixed: | 30 | 30 | 100 | 1.00 | Jersky & Kinsley (1967) |
| Khoisan-speaking Negroes: | | | | | |
| Sarwa: | 50 | 50 | 100 | 1.00 | Chasko *et al* (1979) |
| Khoisan: | | | | | |
| Khoi: | | | | | |
| Keetmanshoop Nama (adults) | 18 | 9 | 50 | .71 | Nurse & Jenkins (1974) |
| Keetmanshoop Nama (children) | 21 | 17 | 81 | .90 | Nurse & Jenkins (1974) |
| San: | | | | | |
| Northern: | | | | | |
| Tsumkwe !Kung | 40 | 39 | 98 | .98 | Jenkins *et al* (1974) |
| Uncertain: | | | | | |
| ≠Huã | 25 | 23 | 92 | .96 | Nurse & Jenkins (1974) |
| Caucasoids: | | | | | |
| English-speaking: | | | | | |
| Johannesburg | 42 | 24 | 57 | .76 | Jenkins (1982) |
| Hybrid Peoples: | | | | | |
| Baster-like: | | | | | |
| Rehoboth: | 20 | 13 | 65 | .81 | Nurse (1977a) |
| 'Cape-Coloureds': | | | | | |
| Johannesburg | 57 | 39 | 68 | .83 | Jenkins (1982) |

# BIBLIOGRAPHY

Abraham, D. P. (1964). Ethno-history of the Empire of Mutapa: problems and methods. In *The historian in Central Africa* (eds. J. Vansina, R. Mauny and L. V. Thomas), pp. 104-26. Oxford University Press for International African Institute, London.

Acocks, J. P. H. (1953). *The veld types of South Africa*. Government Printer, Pretoria.

Ady, P. H. (1965). *Oxford regional economic atlas: Africa*. Clarendon Press, Oxford.

Alexander, J. E. (1838). *An expedition of discovery into the interior of Africa*. 2 Vols. Colburn, London.

Alexander, R. A. (1951). Rift Valley Fever in the Union. *Journal of the South African Veterinary Medical Association* 22, 105-11.

Allan, W. (1967). *The African husbandman*. Oliver and Boyd, Edinburgh.

Allen, J. W. (1877). The influence of physical conditions on the genesis of species. *Radical Review* 1, 108-40.

de Almeida, A. (1964). *Bushmen and other non-Bantu peoples of Angola*. Witwatersrand University Press, Johannesburg.

Altmann, A. (1945). Sickle cell anaemia in a South African-born European. *Clinical Proceedings* 4, 1-10.

Andersson, C. J. (1856). *Lake Ngami*. Hurst and Blackett, London.

Andrews, P. and Cronin, J. E. (1982). The relationships of *Sivapithecus* and *Ramapithecus* and the evolution of the orang-utan. *Nature* 297, 541-6.

Arkell, A. J. (1949). *Early Khartoum*. Oxford University Press, Oxford.

— (1966). *Early Khartoum* (2nd edn). Oxford University Press, Oxford.

Ashton, H. (1952). *The Basuto*. Oxford University Press, London.

Bach, P. H., Higgins–Opitz, S. B., Bima, B. and Leary, W. P. (1976). Isoniazid acetylator status of Black South African tuberculosis patients. *South African Medical Journal* 50, 1132-4.

Baker, P. and Weiner, J. S. (eds.) (1966). *The biology of human adaptability*. Oxford University Press, Oxford.

Balinsky, D. and Jenkins, T. (1967). Electrophoretic variants of glucose-6-phosphate dehydrogenase and phosphoglucomutase in Bantu and Coloured subjects. *South African Journal of Medical Science* 32, 96.

Barnard, A. (1975). Australian models in the South West African Highlands. *African Studies* 34, 9–18.

Barnes, J. A. (1954a). *Politics in a changing society*. Oxford University Press, London.

— (1954b). Class and committees in a Norwegian island parish. *Human Relations* 7, 39–58.

Barnicot, N. A., Garlick, J. P., Singer, R. and Weiner, J. S. (1959). Haptoglobin and transferrin variants in Bushmen and some other South African peoples. *Nature* 184, 2042.

Bartholomew, G. A. Jnr. and Birdsell, J. B. (1953). Ecology and the proto-hominids. *American Anthropologist* 55, 481–98.

Batchelor, J. R. and Morris, P. J. (1978). HLA and disease. In *Histocompatibility testing 1977* (ed. W. F. Bodmer), pp. 205–6. Munksgaard, Copenhagen.

Bayless, T. M. (1971). Junior, why don't you drink your milk? *Gastroenterology* 60, 479–80.

Beaumont, B., Nurse, G. T. and Jenkins, T. (1979). Highland and lowland populations in Lesotho. *Human Heredity* 29, 42–9.

Beaumont, P. B., de Villiers, H., and Vogel, J. C. (1978). Modern man in sub-Saharan Africa prior to 49 000 years B.P.: a review and evaluation with particular reference to Border Cave. *South African Journal of Science* 74, 409–19.

Beet, E. A. (1946). Sickle cell disease in the Balovale District of Northern Rhodesia. *East African Medical Journal* 23, 75–86.

— (1949). The genetics of the sickle-cell trait in a Bantu tribe. *Annals of Eugenics* 14, 279–84.

Beighton, P., Durr, L. and Hamersma, H. (1976). The clinical features of sclerosteosis. *Annals of Internal Medicine* 84, 393–7.

Bender, K., Frank, R. and Hitzeroth, H. W. (1977). Glyoxalase I polymorphism in South African Bantu-speaking Negroids. *Human Genetics* 38, 223–6.

Benedict, B. (1978). Social regulation of fertility. In *The Structure of Human Populations* (eds. G. A. Harrison and A. J. Boyce), pp. 73–89. Clarendon Press, Oxford.

Benyon, J. A. (1974). The process of political incorporation. In *The Bantu-speaking peoples of Southern Africa* (ed. W. D. Hammond-Tooke), pp. 367–96. Routledge and Kegan Paul, London and Boston.

Bergmann, C. (1847). Über die Verhaltnisse der Wärmeökonomie der Thiere zu ihrer Grösse. *Göttinger Studien* 3, 595–708.

Berman, C. (1935). Malignant disease in the Bantu of Johannesburg and the Witwatersrand gold mines. *South African Journal of Medical Science* 1, 12–30.

Beyers, C. (1967). *Die Kaapse Patriotte*. Tweede, hersiende, vermeerde en geïllustreerde uitgawe. Van Schaik, Pretoria.

Bienzle, U., Agent, O., Lucas, A. O. and Luzzatto, L. (1972). Glucose-6-phosphate dehydrogenase and malaria. Greater resistance of females heterozygous for enzyme deficiency and of males with non-deficient variant. *Lancet* i, 107–10.

Bishop, W. W. and Clark, J. D. (eds.) (1967). *Background to evolution in Africa*. Chicago University Press, Chicago.

Black, F. L. (1975). Infectious disease in primitive societies. *Science* 187, 515–18.

Bleek, D. (1927). The distribution of Bushman languages in South Africa. In *Festschrift Meinhof*, pp. 55–64, cited in Schapera, I: *The Khoisan peoples of South Africa*, pp. 419–20. Routledge and Kegan Paul, London.

Bley, H. (1971). *South West Africa under German rule*, 1894–1914. Heinemann, London, Ibadan and Nairobi.

Bodmer, W. F. (ed.) (1978). *Histocompatibility testing 1977*. Munksgaard, Copenhagen.

Botha, C. G. (1919). *The French refugees at the Cape*. Cape Times, Cape Town.

Botha, M. C. (1972). Blood group gene frequencies. An indication of the genetic constitution of population samples in Cape Town. *South African Medical Journal* 46, supplement 1, 1–27.

— du Toit, E. D., Jenkins, T., van Leeuwen, A., D'Amaro, J., Meera Khan, P., van der Steen, G., van Rood, J. J. and van der Does, J. A. (1973). The HL-A system in Bushmen (San) and Hottentot (Khoikhoi) populations of South West Africa. In *Histocompatibility testing 1972* (eds. J. J. Dausset and J. Colombani), pp. 422–32. Munksgaard, Copenhagen.

— and van Zyl, L. J. (1966). Abnormal haemoglobins in Cape Town. *South African Medical Journal* 40, 753–6.

Bott, E. (1957). *Family and social network*. Tavistock Press, London.

Boucher, M. (1981). *French speakers at the Cape in the first hundred years of Dutch East India Company rule*. University of South Africa, Pretoria.

Boule, M. and Vallois, H. (1932). L'homme fossil d'Asselar, Sahara. *Archives de l'Institut de Paléontologie Humaine* 9, 1–90.

Boulle, G. J., Peisach, M. and Jacobson, L. (1979). Archaeological significance of trace element analysis of South West African potsherds. *South African Journal of Science* 75, 215–17.

Bovill, E. W. (1968). *The golden trade of the Moors*. Oxford University Press, London.

Boyden, S. (1972). Ecology in relation to urban population structure. In *The structure of human populations* (eds. G. A. Harrison and A. J. Boyce), pp. 411–41. Clarendon Press, Oxford.

Bradshaw, E. and Harrington, J. S. (1976). Temporal changes in primary liver cancer in Black goldminers from Moçambique. *South African Medical Journal* 50, 2022.

Brain, P. (1966). Subgroups of A in the South African Bantu. *Vox Sanguinis* 11, 686–98.

— and Hammond, M. G. (1973). Frequency of HL-A antigens in South African Bantu, Indians and Caucasians. In *Histocompatibility testing 1972* (eds. J. J. Dausset and J. Colombani), pp. 433–9. Munksgaard, Copenhagen.

Brandel-Syrier, M. (1971). *Reeftown elite*. Routledge and Kegan Paul, London.

Bräuer, G. (1976). Morphological and multivariate analysis of human skeletons from Iron Age raves northwreast of Lake Eyasi (Tanzania). *Homo* 27, 185–96.

Brincker, P. H. (1886). *Wörterbuch und kurzgefasste Grammatik des Otji-Herero*. Weigel, Leipzig.

de Brito, G. (1735). *História tragico-maritima*. Lisboa.

Brontë-Stewart, B., Botha, M. C. and Krut, L. H. (1962). ABO blood groups in relation to ischaemic heart disease. *British Medical Journal* 1, 1646–50.

— Budtz-Olsen, O. E., Hickley, J. M. and Brock, J. F. (1960). The health and nutritional status of the Kung Bushmen of South West Africa. *South African Journal of Laboratory and Clinical Medicine* 6, 187–216.

Brothwell, D. R. (1963). Evidence of early population change in Central and Southern Africa: doubts and problems. *Man o.s.* 63, 101–4.

— and Shaw, T. (1971). A late Upper Pleistocene proto-West African Negro from Nigeria. *Man* 6, 221–7.

Bryant, A. T. (1929). *Olden times in Zululand and Natal*. Longman, London.

Buchanan, N., Strickwold, B. and Shuenyane, E. (1976). Isoniazid inactivation in Black patients with tuberculosis. *South African Medical Journal* 50, 463-5.

Buckwalter, J. A., Kark, A. E. and Knowler, L. A. (1961). A study in human genetics. The ABO blood groups and disease in South Africa. *Archives of Internal Medicine* 107, 558-67.

Burrell, R. J. W. (1957). Oesophageal cancer in the Bantu. *South African Medical Journal* 31, 401-9.

Bush, G. L., Case, S. M., Wilson, A. C. and Patton, S. L. (1977). Rapid speciation and chromosomal evolution in mammals. *Proceedings of the National Academy of Sciences of the U.S.A.* 74, 3942-6.

Campbell, G. D. (1963). Diabetes in Asians and Africans in and around Durban. *South African Medical Journal* 37, 1195-208.

Campbell, J. (1815). *Travels in South Africa, undertaken at the request of the Missionary Society*. Black and Parry, London.

Carstens, W. P. (1966). *The social structure of a Cape Coloured Reserve*. Oxford University Press, Cape Town.

Cary, M. and Warmington, E. H. (1929). *The ancient explorers*. Methuen, London.

Cavalli-Sforza, L. L. (1972). Pygmies, an example of hunter–gatherers, and genetic consequences for man of domestication of plants and animals. In *Human genetics* (eds. J. de Grouchy, F. J. G. Ebling and I. W. Henderson), pp. 79-95. Excerpta Medica, Amsterdam.

— (1974). The genetics of human populations. In *The human population*, a *Scientific American* book. pp. 80-9. W. H. Freeman, San Francisco.

— and Feldman, M. W. (1981). *Cultural Transmission and Evolution: a Quantitative Approach*. Princeton University Press, Princeton.

— Zonta, L. A., Nuzzo, F., Bernini, L., de Jong, W. W. W., Meera Khan, P., Ray, A. K., Went, L. N., Siniscalco, M., Nijenhuis, L. E., van Loghem, E. and Modiano, G. (1969). Studies on African Pygmies in the Central African Republic (with an analysis of genetic distance). *American Journal of Human Genetics* 21, 252-74.

Chafulumira, E. W. (1948). *Mbiri ya Amang'anja*. Zomba: Nyasaland Education Department.

Chalmers, J. N. M., Ikin, E. W. and Mourant, A. E. (1953). A study of two unusual blood-group antigens in West Africans. *British Medical Journal* 2, 175-7.

Chamla, M. C. (1968). Les populations anciennes du Sahara et des régions limitrophes. Études des restes osseux humains néolithiques et protohistoriques. *Mémoires due Centre des Récherches Anthropologiques, Préhistoriques et Ethnographiques en Algérie* 9, 15-99.

Chapman, J. (1971). *Travels in the interior of South Africa, 1849-1863*. Edited by E. C. Tabler. 2 vols. A. A. Balkema, Cape Town.

Chasko, W. J., Nurse, G. T., Harpending, H. C. and Jenkins, T. (1979). Serogenetic studies on a 'Masarwa' population of north-eastern Botswana. *Botswana Notes and Records* 11, 15-23.

Clark, J. D. (1950). The newly discovered Nachikufu culture of Northern Rhodesia and the possible origin of certain elements of the South African Smithfield culture. *South African Archaeological Bulletin* 5, 86-98.

— (1955). A note on the early river-craft and fishing practices in South-East Africa. *South African Archaeological Bulletin* 15, 77-9.

Clark, J. D. (1959). *The prehistory of Southern Africa.* Penguin Books, Harmondsworth.

— (1963). *Prehistoric cultures of Northeast Angola and their significance in tropical Africa.* Publicacões do Museu de Dundo, Lisboa.

— (1970). *The prehistory of Africa.* Thames and Hudson, London.

Cohen, Y. A. (ed.) (1974). *Man in adaptation: the biosocial background.* Aldine Publishing Co., Chicago.

Cole, S. (1954). *The prehistory of East Africa.* Penguin Books, Harmondsworth.

— (1970). *The Neolithic revolution* (5th edn). British Museum, London.

Colenbrander, H. T. (1902). *De Afkomst der Boeren.* Het Algemeen Nederlandsch Verbond, Kaapstad.

Collins, K. J. and Weiner, J. S. (eds.) (1977). *Human adaptability.* Taylor and Francis, London.

Conroy, G. C., Jolly, C. J., Cramer, D. and Kalb, J. E. (1978). Newly discovered fossil hominid skull from the Afar depression, Ethiopia. *Nature* 276, 67-70.

Cook, G. C. and Kajubi, S. K. (1966). Tribal incidence of lactase deficiency in Uganda. *Lancet* i, 725-30.

— and Nurse, G. T. (1980). The intestinal lactase polymorphism in Papua New Guinea, *Papua New Guinea Medical Journal* 23, 141-5.

Cooke, C. K. (1965). Evidence of human migrations from the rock art of Southern Rhodesia. *Africa* 35, 263-85.

Coon, C. S. (1963). *The origin of races.* Jonathan Cape, London.

Cronin, J. E., Boaz, N. T., Stringer, C. B. and Rak, Y. (1981). Tempo and mode in hominid evolution. *Nature* 292, 113-22.

Cummins, H. (1955). Dermatoglyphs of Bushmen (South Africa). *American Journal of Physical Anthropology* 13, 699-709.

Cunha, A. Xavier da. (1968). Étude séro-anthropologique d'une population métissée au Mozambique, 8th *International Congress of Anthropological and Ethnological Sciences*, Tokyo and Kyoto, 191-5.

Currie, B., Jenkins, T. and Nurse, G. T. (1978). Low frequency of persistent high intestinal lactose activity (lactose tolerance) in the Herero. *South African Journal of Science* 74, 227-8.

Dart, R. A. (1937). The physical characters of the /?auni-≠Khomani Bushmen. In *Bushmen of the Southern Kalahari* (eds. J. D. Rheinallt Jones and C. M. Doke), pp. 117-88. Witwatersrand University Press, Johannesburg.

Dausset, J. J. and Colombani, J. (eds.) (1973). *Histocompatibility testing 1972.* Munksgaard, Copenhagen.

Davies, J. N. P. (1979). *Pestilence and disease in the history of Africa.* Witwatersrand University Press, Johannesburg.

Davis D. N. S. (ed.) (1964). *Ecological studies in Southern Africa.* Junk, The Hague.

Davis, K. (1974). The migrations of human populations. In *The human population,* a *Scientific American* book. pp. 92-105. W. H. Freeman, San Francisco.

Dean, G. (1963). *The Porphyrias.* Pitman, London.

Deevy, D. E. (1949). Report on the human remains. In *Early Khartoum* (ed. A. J. Arkell), pp. 31-33. Oxford University Press, Oxford.

DeVore, I. and Lee, R. (eds.) (1976). *Kalahari hunter-gatherers.* Harvard University Press, Cambridge, Mass.

Doke, C. M. (1945). *Bantu.* Oxford University Press for International African Institute, London.

—— (1954). *The southern Bantu languages.* Oxford University Press for International African Institute, London.

Dornan, S. S. (1925). *Pygmies and Bushmen of the Kalahari.* Seeley, Service, London.

Douglas, M. (1964). Matriliny and pawnship in Central Africa. *Africa* 34, 301–13.

Downing, B. H. (1978). Environmental consequences of agricultural expansion in South Africa since 1850. *South African Journal of Science* 74, 420–2.

Drascher, W. and Rust, H. J. (eds.) (1961). *Ein Leben für Südwestafrika: Festschrift Dr. h.c. H. Vedder.* SWA Wissenchaftliche Gesellschaft, Windhoek.

Drennan, M. R. (1929). The dentition of a Bushman tribe. *Annals of the South African Museum* 24, 61–87.

Drury, J. and Drennan, M. R. (1926). The pudendal parts of the South African Bush race. *Medical Journal of South Africa* 22, 113–17.

Dubb, A. A. (1974). The impact of the city. In *The Bantu-speaking peoples of Southern Africa* (ed. W. D. Hammond-Tooke), pp. 441–72. Routledge and Kegan Paul, London and Boston.

Dutton, T. P. (1970). Iron smelting furnace in the Ndumu Game Reserve. *Lammergeyer* 12, 37–40.

Dyer, K. F. (1976). Patterns of gene flow between Negroes and Whites in the U.S. *Journal of Biosocial Science* 8, 309–33.

Edginton, M. E., Hodkinson, J. and Seftel, H. C. (1972). Disease patterns in a South African rural Bantu population. *South African Medical Journal* 46, 968–76.

Edwards, A. W. F. and Cavalli-Sforza, L. L. (1972). Affinity as revealed by differences in gene frequencies. In *The assessment of population affinities in man* (eds. J. S. Weiner and J. Huizinga), pp. 37–47. Oxford University Press, Oxford.

Ehret, C. (1967). Cattle keeping and milking in Eastern and Southern African history: the linguistic evidence. *Journal of African History* 8, 1–17.

Eibl-Eibesfeldt, I. (1972). *Die !Ko-Buschmann-Gesellschaft.* Piper Verlag, München.

Eidus, L., Glathaar, E., Hodgkin, M. M., Nel, E. E. and Kleeberg, H. H. (1979). Comparison of isoniazid phenotyping of Black and White patients with emphasis on South African Blacks. *International Journal of Clinical Pharmacology and Biopharmacy* 17, 311–16.

Ellenberger, D. F. (1912). *History of the Basuto, ancient and modern.* Translated and edited by J. C. MacGregor. Caxton, London.

Elphick, R. (1977). *Kraal and Castle: Khoikhoi and the founding of White South Africa.* Yale University Press, New Haven and London.

Elsdon-Dew, R. (1939). Blood groups in Africa. *Publications of The South African Institute for Medical Research* 9, 29–94.

Engelbrecht, J. A. (1936). *The Korana.* Maskew Miller, Cape Town.

Erickson, C. (1976). *Emigration from Europe 1815–1914.* A. and C. Black, London.

Estermann, C. (1956). *Ethnográfia do sudoeste de Angola,* vol. 1. *Os povos não-bantos e o grupo étnico dos Ambos.* Ministero do Ultramar, Porto.

Facer, C. A. and Brown, J. (1979). ABO blood groups and *falciparum* malaria. *Transactions of the Royal Society of Tropical Medicine and Hygiene* 73, 599–600.

Fagan B. M. (1965). *Southern Africa during the Iron Age.* Thames and Hudson, London.

Fagan, B. M. (1967a). The Iron Age peoples of Zambia and Malawi. In *Background to evolution in Africa* (eds. W. W. Bishop and J. D. Clark), pp. 659–86. Chicago University Press, Chicago.

— (1967b). Radiocarbon dates for sub-Saharan Africa. *Journal of African History* 8, 525.

Fagg, B. E. B. (1969). Recent work in West Africa: new light on the Nok Culture. *World Archaeology* 1, 41–50.

Feely, J. M. (1980). Did Iron Age man have a role in the history of Zululand's wilderness landscapes? *South African Journal of Science* 76, 150–2.

Festenstein, H., Adams, E., Brown, J., Burke, J., Lincoln, P., Oliver, R. T. D., Rondiak, G., Sachs, J. A., Welch, S. G. and Wolf, E. (1973). The distribution of HL-A antigens and other polymorphisms in Bantu-speaking Negroids living in Zambia. In *Histocompatibility testing 1972* (eds. J. J. Dausset and J. Colombani), pp. 397–407. Munksgaard, Copenhagen.

Findlay, G. (1936). *Miscegenation*. Pretoria News, Pretoria.

Findlay, G. H., Nurse, G. T., Heyl, T., Hull, P. R., Jenkins, T., Klevansky, H., Morrison, G. L., Sher, J., Schultz, E. J., Swart, E., Venter, I. J. and Whiting, D. A. (1977). Keratolytic winter erythema or "Oudtshoorn Skin": a newly recognized inherited dermatosis prevalent in South Africa. *South African Medical Journal* 52, 871–4.

Fischer, E. (1913). *Die Rehobother Bastards und das Bastardierungsproblem beim Menschen*. Gustav Fischer, Jena.

Flatz, G. and Rotthauwe, H. W. (1977). The human lactase polymorphism: physiology and genetics of lactose absorption and malabsorption. In *Progress in medical genetics, n.s.* (eds. A. G. Steinberg, A. G. Bearn, A. G. Motulsky and B. Childs) vol. 2, pp. 205–49. W. B. Saunders, Philadelphia.

Flight, C. (1970). Excavations at Kintampo. *West African Archaeological Newsletter* 12, 71–3.

Ford, E. G. (1964). *Ecological genetics*. Chapman and Hall, London.

Fouché, L. (1937). *Mapungubwe*. Cambridge University Press, Cambridge.

Fourie, L. (1925). The Bushmen of South West Africa. In *The native tribes of South West Africa*. pp. 81–105. Cape Times, Cape Town.

Fox, R. H. (1958). Heat stress and nutrition. *Proceedings of the Nutrition Society* 17, 173–9.

Freedman, J. and Goldberg, L. (1976). Incidence of retinoblastoma in the Bantu of South Africa. *British Journal of Ophthalmology* 60, 655–6.

Freeman-Grenville, G. S. P. (1962). *The East African coast: select documents from the first to the early nineteenth century*. Oxford University Press, Oxford.

Frisch, R. E. (1978). Population, food intake and fertility. *Science* 199, 22–30.

— and McArthur, J. W. (1974). Menstrual cycles: fatness as a determinant of minimum weight for height necessary for their maintenance or onset. *Science* 185, 949–51.

Fritsch, G. (1872). *Die Eingeborenen Süd-Afrikas*. Hirt, Breslau.

Galloway, A. (1959). *The skeletal remains of Bambandyanalo*. Witwatersrand University Press, Johannesburg.

Galton, F. (1853). *The narrative of an explorer in tropical South Africa*. John Murray, London.

— (1869). *Hereditary genius: an enquiry into its laws and consequences*. Macmillan, London.

Gardner, G. A. (1963). *Mapungubwe*. Vol. 2. Van Schaik, Pretoria.

Gershater, G. (1955). From Lithuania to South Africa. In *The Jews in South Africa: a history* (eds. G. Saron and L. Hotz), pp. 59-84. Oxford University Press, London.

Giblett, E. R., Motulsky, A. G. and Fraser, G. R. (1966). Population genetic studies in the Congo. IV. Haptoglobin and transferrin serum groups in the Congo and in other African populations. *American Journal of Human Genetics*, 18, 553-8.

Glass, B. and Li, C. C. (1953). The dynamics of racial intermixture: an analysis based on the American Negro. *American Journal of Human Genetics* 5, 1-20.

Gloger, C. L. (1833). *Das Abändern der Vögel durch Einfluss des Klimas.* Schulz, Breslau.

Godber, M. Kopeć, A. C., Mourant, A. E., Teesdale, P., Tills, D., Weiner, J. S., El-Niel, H., Wood, C. H. and Barley, S. (1976). The blood groups, serum groups, red cell isoenzymes and haemoglobins of the Sandawe and Nyaturu in Tanzania. *Annals of Human Biology* 3, 463-73.

Goldberg, L. (1977). The rising incidence of retinoblastoma in Blacks. *South African Medical Journal* 51, 368.

Goldblatt, J. and Beighton, P. (1979). Gaucher's disease in South Africa. *Journal of Medical Genetics* 16, 302-5.

Gomperts, E., Geefhuysen, J., Katz, J. and Metz, J. (1969). Von Willebrand's disease in the Bantu. *South African Medical Journal* 43, 1107-9.

Goodwin, A. J. H. and van Riet Lowe, C. (1929). *The Stone Age cultures of South Africa.* South African Museum, Cape Town.

Gordon, H. (1965a). Genetics and race. Part I. *South African Medical Journal* 39, 553.

—— (1965b). Genetics and race. Part II. *South African Medical Journal* 39, 543.

—— Gordon, W. and Botha, V. (1969). Lipoid proteinosis in an inbred Namaqualand community. *Lancet* i, 1032-5.

—— —— Botha, V. and Edelstein, I. (1971). Lipoid proteinosis. Birth defects: *Original Article Series*, VII, No. 8, 164-77.

—— Robertson, M., Blair, J. M. and Vooijs, M. (1964). Genetic markers in liver disease. *South African Medical Journal* 78, 734-5.

—— Vooijs, M. and Keraan, M. M. (1966). Genetical variation in some human red cell enzymes: An interracial study. *South African Medical Journal* 40, 1031-2.

Gould, S. J. and Eldredge, N. (1977). Punctuated equilibria: the tempo and mode of evolution reconsidered. *Palaeobiology* 3, 115-51.

Govaerts, A., Massart, T. H., Rivat, L. and Brocteur, J. (1973). Serological characters of a Bantu population. In *Histocompatibility testing 1972* (eds. J. J. Dausset and J. Colombani), pp. 409-13. Munksgaard, Copenhagen.

Gramly, R. M. and Rightmire, G. P. (1973). A fragmentary cranium and dated Later Stone Age assemblage from Lukenya Hill, Kenya. *Man* 8, 571-9.

Greenacre, M. J. and Degos, L. (1977). Correspondence analysis of HLA frequency data from 124 population samples. *American Journal of Human Genetics* 29, 60-75.

Greenberg, J. H. (1955). *Studies in African linguistic classification.* Compass Publishing Company, New Haven, Connecticut.

—— (1963). *The languages of Africa.* Indiana University Press, Bloomington.

Grobbelaar, C. S. (1956). The physical characters of the Korana. *South African Journal of Science* 53, 97-159.

te Groen, L. H. T. and Rose, E. F. (1974). A preliminary investigation into the incidence of cancer of the cervix. *South African Medical Journal* 48, 2341-5.

de Grouchy, J., Ebling, F. J. G. and Henderson, I. W. (eds.) (1972). *Human genetics*. Excerpta Medica, Amsterdam.

— Turleau, C., Roubin, M. and Chavin, Colin F. (1973). Chromosomal evolution of man and the primates (*Pan troglodytes, Gorilla gorilla, Pongo pygmaeus*). *Nobel* 23, 124-30.

Grové, S. S. and Ledger, J. A. (1975). Leishmania from a hyrax in South West Africa. *Transactions of the Royal Society of Tropical Medicine and Hygiene* 69, 523.

Guerreiro, M. V. (1968). *Bochimanes !Khũ de Angola*. Junta de Investigações do Ultramar, Lisboa.

Guthrie, M. (1962). Some developments in the prehistory of the Bantu languages. *Journal of African History* 3, 273-82.

— (1967). *The classification of the Bantu languages*. Dawsons for the International African Institute, London.

Hahn, C. H. L. (1928). The Ovambo. In *The native tribes of South West Africa*. pp. 1-36. Cape Times, Cape Town.

Hahn, T. (1881). *Tsuni-//Goam, the Supreme Being of the Khoi-Khoi*. Trübner and Co., London.

Halford, S. J. (undated). *The Griquas of Griqualand*. Juta, Cape Town and Johannesburg.

Hall, M. and Vogel, J. C. (1978). Enkwazini: fourth-century Iron Age site on the Zululand coast. *South African Journal of Science* 74, 70-1.

Hammond-Tooke, W. D. (1965). Segmentation and fission in Cape Nguni political units. *Africa* 35, 143-66.

— (ed.) (1974a). The *Bantu-speaking peoples of Southern Africa*. Routledge and Kegan Paul, London and Boston.

— (1974b). World-view: i. A system of beliefs. ii. A system of action. In *The Bantu-speaking peoples of Southern Africa* (ed. W. D. Hammond-Tooke), pp. 318-63. Routledge and Kegan Paul, London and Boston.

Hance, W. A. (1964). *The geography of modern Africa*. Columbia University Press, New York and London.

Harinck, G. (1969). Interaction between Xhosa and Khoi: emphasis on the period 1620-1750. In *African societies in Southern Africa* (ed. L. Thompson), pp. 145-70. Heinemann, London, Ibadan and Nairobi.

Harington, J. S., McGlashan, N. D., Bradshaw, E., Geddes, E. W. and Purves, L. R. (1975). A spatial and temporal analysis of four cancers in African gold miners from Southern Africa. *British Journal of Cancer* 31, 665-78.

Harris, D. R. (ed.) (1980). *Human ecology in savannah environments*. Academic Press, London.

Harris, H., Hopkinson, D. A., Luffman, J. E., and Rapley, S. (1967). Personal observation, quoted by Hopkinson, D. A. (1968).

Harrison, G. A. and Boyce, A. J. (eds.) (1972a). *The structure of human populations*. Clarendon Press, Oxford.

— — (1972b). Migration, exchange and the genetic structure of populations. In *The structure of human populations* (eds. G. A. Harrison and A. J. Boyce), pp. 128-45. Clarendon Press, Oxford.

— Kuchemann, C. F., Moore, M. A. S., Boyce, A. J., Mourant, A. E., Godber, M. J., Glasgow, B. J., Kopeć, A. C., Tills, D. and Clegg, E. J. (1969). The effects of altitudinal variation in Ethiopian populations. *Philosophical Transactions of the Royal Society, London* 256, 147–82.

Harrison, G. A. (ed.) (1977). *Population structure and human variation.* Cambridge University Press, Cambridge.

— Weiner, J. S., Tanner, J. M. and Barnicot, N. A. (1977). *Human biology* (2nd edn). Oxford University Press, Oxford.

Hartwig, G. W. and Patterson, K. D. (eds.) (1978). *Disease in African history.* Duke University Press, Durham, N.C.

el Hassan, A. M., Godber, M. G., Kopeć, A. C., Mourant, A. E., Tills, D. and Lehmann, H. (1968). The heredity of blood factors of the Beja of Sudan. *Man* 3, 272–83.

Hayden, M. R. and Beighton, P. (1982). Genetic aspects of Huntington's Chorea. *American Journal of Medical Genetics* 11, 135–41.

— MacGregor, J. M. and Beighton, P. H. (1980). The prevalence of Huntingdon's Chorea in South Africa. *South African Medical Journal* 58, 193–6.

Heese, J. J. (1971). *Die Herkoms van die Afrikaner.* A. A. Balkema, Cape Town.

Hellman, E. (1948). *Rooiyard: a sociological survey of an urban native slum.* Oxford University Press, Cape Town.

Helman, J. (1957). Some diseases amongst the Hottentots of South West Africa. *Central African Journal of Medicine* 3, 143–4.

Herodotus (1920). *The histories.* With a translation by A. D. Godley. 4 vols. Loeb's Classical Library. Heinemann, London.

Herzog, P., Bohatova, J. and Drdova, A. (1970). Genetic polymorphisms in Kenya. *American Journal of Human Genetics* 22, 287–91.

Heyl, T. (1970). Genealogical study of lipoid proteinosis in South Africa. *British Journal of Dermatology* 83, 338–40.

Hiernaux, J. (1968). *La diversité humanine en Afrique sub-saharienne: Récherches biologiques.* Institut de Sociologie, Université Libre, Bruxelles.

— (1976). Blood polymorphism frequencies in the Sara Majingay of Chad. *Annals of Human Biology* 3, 129–40.

Higginson, J. and Oettlé, A. G. (1960). Cancer incidence in the Bantu and 'Cape Colored' races of South Africa: report of a cancer survey in the Transvaal (1953–55). *Journal of the National Cancer Institute* 24, 589–671.

Hirsch, H. (1958). Blood group distribution in Natal: preliminary note. *Transactions of the Royal Society of Tropical Medicine and Hygiene* 52, 408–10.

Hitzeroth, H. W. (1972). *Fisiese Antropologie van die Inheemse Mense in Suidelike Afrika.* Afrika-Instituut, Pretoria.

— and Bender, K. (1980). Erythrocyte G-6-PD and 6-PGD genetic polymorphisms in South African Negroes, with a note on G-6-PD and the malaria hypothesis. *Human Genetics* 54, 233–42.

— and Hummel, K. (1978). Serum protein polymorphisms Hp, Tf, Gc, Gm, Inv and Pt in Bantu speaking South African Negroids. *Anthropologische Anzeiger* 36, 127–41.

— Bender, K., and Frank, R. (1981). South African Negroes: Isoenzyme polymorphisms (GPT, $PGM_1$, $PGM_2$, ACP, AK and ADA) and tentative genetic distances. *Anthropologische Anzeiger* 39, 20–35.

— — and Wolfswinkel, J. M. (1976). Esterase D polymorphisms in South African Negroids. *South African Journal of Science* 72, 301–3.

— Walter, H. and Hilling, M. (1978). Genetic markers and leprosy in South

African Negroes. I. Serum protein polymorphisms. *South African Medical Journal* 54, 653–8.

—— —— and Munderlow, W. (1979). Genetic markers and leprosy in South African Negroes, II. Erythrocyte enzyme polymorphisms. *South African Medical Journal* 56, 507–10.

Hopkinson, D. A. (1968). Genetically determined polymorphisms of erythrocyte enzymes in man. In *Advances in clinical chemistry.* (eds. Bodansky, O. and Stewart, C. P.) 2, 21–79.

Horner, R. and Lanzkowsky, P. (1966). Incidence of congenital abnormalities in Cape Town. *South African Medical Journal* 40, 171.

Houghton, D. H. (1974). The process of economic incorporation. In *The Bantu-speaking peoples of Southern Africa* (ed. W. D. Hammond-Tooke), pp. 397–414. Routledge and Kegan Paul, London and Boston.

—— and Walton, E. M. (1952). *The economy of a native reserve: Keiskammahoek rural survey.* Vol. 2. Shuter and Shooter, Pietermaritzburg.

Howell, N. (1979). *Demography of the Dobe !Kung.* Academic Press, New York.

Howells, W. W. (ed.) (1962). *Ideas on human evolution.* Harvard University Press, Cambridge, Mass.

Inskeep, R. R. (1967). The Late Stone Age in Southern Africa. In *Background to evolution in Africa* (eds. W. W. Bishop and J. D. Clark), pp. 557–82. Chicago University Press, Chicago.

Isaacson, C., Selzer, G., Kaye, V., Greenberg, M., Woodruff, J. D., Davies, J. N. P., Ninin, D., Vetten, D. and Andrew, M. (1978). Cancer in the urban Blacks of South Africa. *South African Cancer Bulletin* 22, 49–84.

Jackson, W. P. U. (1978). Epidemiology of diabetes in South Africa. *Advances in Metabolic Disorders* 9, 111–46.

Jacobson, L. and Vogel, J. C. (1979). Radiocarbon dates for two Khoi ceramic vessels from Conception Bay, South West Africa/Namibia. *South African Journal of Science* 75, 230–1.

Jaffey, A. J. E. (1966). A reappraisal of the history of the Rhodesian Iron Age up to the fifteenth century. *Journal of African History* 7, 187–95.

Jeffreys, K. (1928). *Kaapse Argiefstukken.* Cape Times, Cape Town.

Jeffreys, M. D. W. (1966). Pre-Colombian maize in Southern Africa. *Nature* 215, 695–7.

—— (1968). *Some Semitic influences in Hottentot culture.* (4th Raymond Dart Lecture). Witwatersrand University Press, Johannesburg.

Jenkins, T. (1972). *Genetic polymorphisms of Man in Southern Africa.* M.D. Thesis, University of London.

—— (1974). Blood group A$^{bantu}$ population and family studies. *Vox Sanguinis* 26, 537–50.

—— (1977). Defining sero-genetic parameters in various South African Negro populations stratified according to their degree of urbanization. In *Human adaptability* (eds. K. O. Collins and J. S. Weiner), pp. 253–4. Taylor and Francis, London.

—— (1982). Human evolution in Southern Africa. In *Human genetics. Part A: The unfolding genome* (ed. B. Bonné-Tamir). Alan R. Liss Inc., New York.

—— and Brain, C. P. (1967). The peoples of the lower Kuiseb Valley South West Africa. *Scientific papers of the Namib Desert Research Station* 35.

—— and Corfield, V. (1972). Red cell acid phosphatase polymorphism in southern Africa: population data and studies on the R, RA and RB phenotypes. *Annals of Human Genetics* 35, 379–91.

— and Dunn, D. S. (1981). Haematological genetics in the tropics: I. Africa. In *Haematology in tropical areas* (ed. L. Luzzatto), *Clinics in haematology*. Vol. 10 no. 3. pp. 1029–50. W. B. Saunders, London and Philadelphia.

— Grové, S. S., Dunn, D. S., and Nurse, G. T. (1983). Serogenetic studies on the Herero and Herero-related peoples of Namibia and Botswana. In press.

— Harpending, H. C., Gordon, H., Keraan, M. M. and Johnston, S. (1971). Red-cell-enzyme polymorphisms in the Khoisan peoples of Southern Africa. *American Journal of Human Genetics* 23, 513–532.

— — and Nurse, G. T. (1978). Genetic distance among certain Southern African populations. In *Evolutionary models and studies in human diversity* (eds. R. J. Meier, C. M. Otten and F. Abdel-Hameed), pp. 227–43. Mouton, The Hague.

— Lane, A. B., Hopkinson, D. A. and Nurse, G. T. (1979). The red cell adenosine deaminase polymorphism in Southern Africa, with special reference to deficiency among the !Kung. *Annals of Human Genetics* 42, 425–33.

— — Nurse, G. T. and Tanaka, J. (1975). Serogenetic studies on the G/wi and G//ana San of Botswana. *Human Heredity* 25, 318–28.

— Lehmann, H. and Nurse, G. T. (1974). Public health and the genetic constitution of the San ('Bushmen'): Carbohydrate metabolism and acetylator status of the !Kung of Tsumkwe in the north-western Kalahari. *British Medical Journal* 2, 23–26.

— and Nurse, G. T. (1972). Blood group gene frequencies. *South African Medical Journal* 46, 560.

— — (1974a). The red cell 6-phosphogluconate dehydrogenase polymorphism in certain Southern African populations; with the first report of a new phenotype. *Annals of Human Genetics* 58, 19–29.

— — (1974b). Genetic studies on Botswana peoples. *Botswana Notes and Records* 6, 221–2.

— — (1976). Biochemical studies on the desert-dwelling hunter–gatherers of Southern Africa. In *Progress in medical genetics, n.s.* (eds. A. G. Steinberg and A. G. Bearn) Vol. 1, pp. 211–81. W. B. Saunders, Philadelphia.

— Rabson, A., Lane, A. B., Hopkinson, D. A. and Nurse, G. T. (1976). Deficiency of adenosine deaminase not associated with combined immune deficiency. *Journal of Paediatrics* 89, 732–6.

— and Steinberg, A. G. (1966). Some serum polymorphisms in Kalahari Bushmen and Bantu: Gamma globulins, haptoglobins and transferrins. *American Journal of Human Genetics* 18, 399–407.

— Zoutendyk, A. and Steinberg, A. G. (1970). Gammaglobulin groups (Gm and Inv) of various Southern African populations. *American Journal of Physical Anthropology* 32, 197–218.

Jenny, H. (1967). *Südwestafrika: Land zwischen den Extremen*. W. Kohlhammer Verlag, Stuttgart.

Jersky, J. and Kinsley, R. H. (1967). Lactase deficiency in the South African Bantu. *South African Medical Journal* 41, 1194–6.

Jessop, J. P. (1974). The ecological setting. In *The Bantu-speaking peoples of Southern Africa* (ed. W. D. Hammond-Tooke), pp. 46–55. Routledge and Kegan Paul, London and Boston.

Joffe, B. I., Jackson, W. P. U., Thomas, M. E., Toyer, M. G., Keller, P., Pimstone, B. L. and Zamit, R. (1971). Metabolic responses to oral glucose in the Kalahari Bushmen. *British Medical Journal* 4, 206–8.

Johanson, D. C. and Edey, M. A. (1981). *Lucy: the beginnings of humankind.* Granada, London, Toronto, Sydney and New York.

Johnson, J. D., Kretchmer, N. and Simoons, F. J. (1974). Lactose malabsorption: its biology and history. *Advances in Pediatrics* 21, 197-237.

Junod, H. A. (1927). *The life of a South African tribe* (2nd edn, 2 vols). Macmillan, London.

Junod, H. P. (1936). Notes on the ethnographical situation in Portuguese East Africa. *Bantu Studies* 10, 293-311.

Kaminer, B. and Lutz, W. P. W. (1960). Blood pressure in Bushmen of the Kalahari Desert. *Circulation* 22, 289-95.

Kew, M. C., Geddes, E. W., MacNab, G. M. and Bersohn, I. (1974). Hepatitis-B antigen and cirrhosis in Bantu patients with primary liver cancer. *Cancer* 34, 539-41.

Kimambo, I. (1968). The rise of the Congolese state systems. In *Aspects of Central African history* (ed. T. D. Ranger), pp. 29-48. Heinemann, London, Ibadan and Nairobi.

Kirk, R. L., Blake, N. M. and Vos, G. H. (1971). The distribution of enzyme group systems in a sample of South African Bantu. *South African Medical Journal* 45, 69-72.

Klapwijk, M. (1973). An early Iron Age site near Tzaneen, N. E. Transvaal. *South African Journal of Science* 69, 324.

— (1974). A preliminary report on pottery from the north-eastern Transvaal, South Africa. *South African Archaeological Bulletin* 29, 19-23.

Klein, R. G. (1977). The ecology of early man in Southern Africa. *Science* 197, 115-26.

Knodel, J. (1977). Breast-feeding and population growth. *Science* 198, 111-15.

Knussmann, R. (1969). Bericht über eine anthropologische Forschungsreise zu den Dama in Südwestafrika. *Homo* 20, 34-66.

— and Knussmann, R. (1969/1970). Die Dama—eine Altschicht in Südwestafrika? *Journal der Südestafrikanische Wissenschaftlige Gessellschaft* 24, 9-32.

de Kock, V. (1963). *Those in bondage.* Juta, Cape Town.

Köhler, O. (1961). Die Sprachforschung in Südwestafrika. In *Ein Leben fur Südwestafrika: Festschrift Dr. H. Vedder* (eds. W. Drascher and H. J. Rust), pp. 61-77. SWA Wissenschaftliche Gesellschaft, Windhoek.

Kokernot, R. H., Szlamp, E. L., Levitt, J. and MacIntosh, B. M. (1965). Survey for antibodies against arthropod-borne viruses in the sera of indigenous residents of the Caprivi Strip and Bechuanaland Protectorate. *Transactions of the Royal Society of Tropical Medicine and Hygiene* 59, 553-62.

Kolata, G. B. (1974). !Kung hunter-gatherers: feminism, diet and birth control. *Science* 185, 932-4.

Konner, M. and Worthman, C. (1980). Nursing frequency, gonadal function and birth spacing among !Kung hunter-gatherers. *Science* 207, 788-91.

Krige, E. J. and Krige, J. D. (1943). *The realm of a rain queen.* International African Institute, London.

Kroenlein, J. G. (1889). *Wortschatz der Khoi-Khoin.* Deutsche Kolonialgesellschaft, Berlin.

Kromberg, J. and Jenkins T. (1982*a*). Prevalence of albinism in the South African Negro. *South African Medical Journal* 61, 383-6.

— — (1982*b*). Common birth defects in South African Blacks. *South African Medical Journal* 62, 599-602.

Krut, L. H. and Singer, R. (1963). Steatopygia: the fatty acid composition of subcutaneous adipose tissue in the Hottentot. *American Journal of Physical Anthropology* 21, 181-7.

Kurth, G. (ed.) (1962). *Evolution and hominisation.* Gustav Fischer, Stuttgart.

Law, R. C. C. (1967). The Garamantes and Trans-Saharan enterprise in classical times. *Journal of African History* 8, 181-200.

Leakey, L. S. B. (1935). *The Stone Age races of Kenya.* Oxford University Press, Oxford.

Leakey, R. E. F. (1976). New hominid fossils from the Koobi Fora formation in northern Kenya. *Nature* 261, 574-6.

Lebzelter, V. (1931). Zur Anthropologie der Kung-Buschleute. *Anzeichnungen der Akademie der Wissenschaft, Wien* 68, 24-6.

Lee, R. (1965). *Subsistence ecology of !Kung Bushmen.* Doctoral dissertation presented to the University of California, Berkeley.

— (1976). !Kung spatial organization. In *Kalahari hunter-gathers* (eds. I. DeVore and R. Lee), pp. 73-97. Harvard University Press, Cambridge, Mass.

— (1979) *!Kung San: men, women and work in a foraging society.* Cambridge University Press, New York.

— and DeVore, I. (eds.) (1968a). *Man the hunter.* Aldine Press, Chicago.

— — (1968b). Problems in the study of hunters and gatherers. In *Man the hunter* (eds. R. Lee and I. DeVore), pp. 3-12. Aldine Press, Chicago.

Legassick, M. (1969). The Sotho-Tswana peoples before 1800. In *African societies in Southern Africa* (ed. L. Thompson), pp. 86-125. Heinemann, London, Ibadan and Nairobi.

Le Gros Clark, W. E. (1967). *Man-apes or ape-men?* Holt, Rinehart and Wilson, New York.

— (1971). *The antecedents of man* (3rd edn). University Press, Edinburgh.

Lévi-Strauss, C. (1952). La notion d'archaisme en ethnologie. *Cahiers internationaux de Sociologie* 12, 32-5.

Lewis, S. M., Anderson, C. G. and Baskind, E. (1957). Homozygous haemoglobin-C disease in a White family, with special reference to blood autolysis studies. *British Journal of Haematology* 3, 68-76.

Linden, I. (1972). 'Mwali' and the Luba origin of the Chewa. *Society of Malaŵi Journal* 25, 1, 11-19.

Livingstone, D. (1856). *Missionary travels and researches in South Africa.* John Murray, London.

Livy (Titus Livius) (1965). *The war with Hannibal.* Translation by Aubrey de Sélincourt of Books XXI-XXX of the *History of Rome.* Penguin Books, Harmondsworth.

Lobato, A. (1954). *A expansão portuguesa em Moçambique de 1498 a 1530.* Ministerio das Colonias, Porto.

Louw, J. T. (1960). *Prehistory of the Matjes River rock shelter.* National Museum, Bloemfontein.

Lowe, R. F. (1969). Rhodesian tribal blood groups. *Central African Journal of Medicine* 15, 151-64.

Lurie, A. and Jenkins, T. (1975). The incidence of haemophilia in South Africa. In *Handbook of hemophilia, part I* (eds. K. M. Brinkhous and H. C. Hemker), pp. 49-58. Excerpta Medica, Amsterdam.

von Luschan, F. (1907). *The racial affinities of the Hottentots.* British South Africa Association, London.

Luzzatto, L. (1973). Studies of polymorphic traits for the characterization of populations. *Israel Journal of Medical Sciences* 9, 1181-94.

Lye, W. (1969). The distribution of the Sotho peoples after the Difaqane. In *African Societies in Southern Africa* (ed. L. Thompson), pp. 191-206. Heinemann, London, Ibadan and Nairobi.

McCalman, H. R. and Grobbelaar, B. J. (1965). Preliminary report of two stoneworking OvaTjimba groups in the Northern Kaokoveld of South West Africa. *Cimbebasia* 13, 1-39.

McDermid, E. M. and Vos, G. H. (1971a). Serum protein groups of South African Bantu I. Albumin ceruloplasmin, transferrin and haptoglobin. *South African Journal of Medical Science* 36, 7-14.

— — (1971b). Serum protein groups of South African Bantu II. $a^1$-antitrypsin, group specific component and further observations on haptoglobin and ceruloplasmin. *South African Journal of Medical Science* 36, 63-8.

— — (1971c). Serum protein groups of South African Indians. *South African Journal of Medical Science* 36, 57-62.

Maggs, T. M. O'C. and Michael, M. A. (1976). Ntshekane, an Early Iron Age site in the Tugela Basin, Natal. *Annals of the Natal Museum* 22, 705-59.

Mainga, M. (1966). The origin of the Lozi: some oral traditions. In *The Zambesian past* (eds. E. T. Stokes and R. Brown), pp. 238-47. Manchester University Press, Manchester.

Maingard, J. F. (1932). Physical characters of the Korana. *Bantu Studies* 6, 163-82.

— (1937). Some notes on health and disease among the Bushmen of the Southern Kalahari. In *Bushmen of the Southern Kalahari* (eds. J. D. Rheinallt Jones and C. M. Doke), pp. 227-36. Witwatersrand University Press, Johannesburg.

Maingard, L. F. (1931). The lost tribes of the Cape. *South African Journal of Science* 28, 487-504.

— (1934). The linguistic approach to South Africa prehistory and ethnology. *South African Journal of Science* 31, 117-43.

— (1964). The Korana dialects. *African Studies* 23, 57-66.

Manning, E. B., Mann, J. I. and Sophangisa, E. (1974). Dietary patterns in urbanized blacks. A study in Guguletu, Cape Town, 1971. *South African Medical Journal* 48, 485-98.

Maquet, J. J. (1961). *The premise of inequality in Rwanda*. Oxford University Press for International African Institute, London.

Marais, J. S. (1939). *The Cape Coloured Peoples 1652-1937*. Witwatersrand University Press, Johannesburg.

Marks, M., Jenkins, T. and Nurse, G. T. (1977). GPT, Ca$_{II}$ and EsD polymorphisms in South West Africa. *Human Genetics* 37, 49-54.

Marrett, J. R. (1936). *Race, sex and environment*. Hutchinson, London.

Marshall, L. J. (1959). Marriage among !Kung Bushmen. *Africa* 29, 335-65.

— (1960). !Kung Bushman bonds. *Africa* 30, 325-55.

— (1961). Sharing, talking and giving: relief of social tension among the !Kung Bushmen. *Africa* 31, 231-49.

— (1962). !Kung Bushman religious beliefs. *Africa* 32, 221-51.

Mason, R. J. (1962). *The prehistory of the Transvaal*. Witwatersrand University Press, Johannesburg.

— (1973). First Early Iron Age settlement in South Africa: Broederstroom 24/73, Brits District, Transvaal. *South African Journal of Science* 69, 324-5.

Matznetter, T. and Spielmann, W. (1969) Blutgruppen moçambiquanischer Bantustämmer, *Zeitschrift für Morphologie und Anthropologie* 61, 57–71.

Mauny, R. (1961). *Tableau géographique de l'Ouest Africain au Moyen Age.* Institut Africain de l'Afrique Noire, Dakar.

May, R. M. (1978). Human reproduction reconsidered. *Nature* 272, 491–5.

Mayer, P. (1961). *Townsmen or tribesmen: conservatism and the process of urbanization in a South African city.* Oxford University Press, Cape Town.

—— (1962). Migrancy and the study of Africans in town. *American Anthropologist* 64, 576–92.

Medawar, P. B. (1975). Technology and evolution. In *Technology and the frontiers of knowledge.* The Frank Nelson Doubleday Lectures 1972–3 (ed. S. Bellow), pp. 99–109. Doubleday and Co., New York.

Meigs, P. (1952). *Distribution of arid and homoclimates.* UNESCO, New York.

Mentzel, O. F. (1921). *A geographical and topographical description of the Cape of Good Hope.* Translated by H. J. Mandelbrote. Van Riebeeck Society, Cape Town.

van der Merwe, A. le R. (1969). Isolated homogeneous groups in nutritional studies: report on a pilot survey on the Himba and Tjimba of the Baynes Mountains. *South African Medical Journal* 43, 331–6.

Miers, S. and Kopytoff, I. (eds.) (1977). *Slavery in Africa.* University of Wisconsin Press, Madison.

Miller, C. H., Mason, S. J., Clyde, D. F. and McGinniss, M. H. (1976). The resistance factor to *Plasmodium vivax* in blacks. The Duffy blood-group genotype *FyFy. New England Journal of Medicine* 295, 302–4.

Mistry, S. D. (1965). Ethnic groups of Indians in South Africa. *South African Medical Journal* 39, 691–4.

Mitchell, J. C. (1956a). Urbanization, detribalization and stabilization in Southern Africa. In *Social implications of industrialization and urbanization in Africa south of the Sahara*, pp. 693–711. UNESCO, Paris.

—— (1956b). *The Yao village.* Manchester University Press, Manchester.

Mönnig, H. O. (1967). *The Pedi.* Van Schaik, Pretoria.

Moores, P. P. (1980). Blood groups of the Natal Indian peoples. Ph.D. Thesis, University of Natal.

Morton, N. E., Chung, C. S., and Mi, M. P. (1967). *Genetics of interracial crosses in Hawaii.* S. Karger, Basel.

Moullec, J., Mendrez, C. and Van Cong, N. (1966). Les groupes sanguins au Lessouto. *Bulletin et mémoires de la Société d'Anthropologie de Paris* 9, 363–6.

Mourant, A. E., Kopeć, A. and Domaniewska-Sobczak, K. (1976). *The distribution of the human blood groups and other polymorphisms.* Oxford University Press, Oxford.

Mudau, E. (1940). Ngoma-Lungundu and the early invaders of Venda. In *The copper miners of Musina and the early history of the Zoutpansberg* (ed. N. van Warmelo), pp. 10–32. Government Printer, Pretoria.

Murray, J. F., Freedman, M. L., Lurie, H. I. and Merriweather, A. M. (1957). Witkop: a synonym for favus. *South African Medical Journal* 31, 657–60.

—— Merriweather, A. M. and Freedman, M. L. (1956). Endemic syphilis in the Bakwena Reserve of the Bechuanaland Protectorate. A report on mass examination and treatment. *Bulletin of the World Health Organization* 15, 1975.

Muuka, L. S. (1966). The colonization of Barotseland in the seventeenth century.

In *The Zambesian past* (eds. E. T. Stokes and R. Brown), pp. 248–60. Manchester University Press, Manchester.

Mwale, E. B. (1962). *Za Acewa* (3rd edn). Macmillan, London.

Nerup, J., Platz, P., Anderson, O. O., Christy, M., Lyngsøe, J., Poulsen, J. E., Ryder, L. P., Nielsen, L. S., Thomsen, M. and Svejgaard, A. (1974). HL-A antigens and diabetes mellitus. *Lancet* 2, 864.

Nienaber, G. S. (1963). *Hottentots*. Van Schaik, Pretoria.

van Noten, F. (1965). Nouvelles fouilles à Lochinvar (Zambia), 1964. *Afrika-Tervuren* 11, 16–22.

Ntara, S. J. (1965). *Mbiri ya Achewa*. Malaŵi Literature and Publications Bureau, Blantyre.

Nurse, G. T. (1967). The name 'Akafula'. *Society of Malaŵi Journal* 20, 2, 17–22.

— (1968). Seasonal fluctuations in the body weight of African villagers. *Central African Journal of Medicine* 14, 122–7 and 147–50.

— (1970). Cognate sets determining the Maravi periphery. *Review of Ethnology* 3, 25–31 and 33–40.

— (1973). Ndwandwe and the Ngoni. *Society of Malaŵi Journal* 26, 1, 7–14.

— (1974). *The physical characters of the Maravi*. Ph.D. Thesis, University of the Witwatersrand.

— (1975a). *The origins of the northern Cape Griqua*. Institute for the Study of Man in Africa, Johannesburg.

— (1975b). Seasonal hunger among the Ngoni and Ntumba of Central Malaŵi. *Africa* 45, 1–11.

— (1976). Isonymic studies on the Griqua of the northern Cape Province. *Journal of Biosocial Science* 8, 277–86.

— (1977a). Intestinal lactose and milk tolerance in Southern Africa. *Leech* 47, 8–10.

— (1977b). The survival of the Khoisan race. *Bulletin of the International Committee on Urgent Anthropological and Ethnological Research* 19, 39–46.

— (1978). *Clanship in Central Malaŵi*. Institut für Völkerkunde, Vienna.

— (1980). Labour migration as an agent of change in African and Melanesian societies. In *The tribal world and its transformations* (eds. Bhupinder Singh and J. S. Bhandari), pp. 99–109. Concept Press, New Delhi.

— (1983). Population movement around the northern Kalahari. *African Studies* 76, 153–63.

— Bodmer, J. G., Bodmer, W. F., van Leeuwen, A., van Rood, J. J., du Toit, E. D. and Botha, M. C. (1975). A reassessment of the HL-A system in Khoisan populations of South West Africa. *Tissue Antigens* 5, 401–14.

— Botha, M. C. and Jenkins, T. (1977). Serogenetic studies on the San of South West Africa. *Human Heredity* 27, 81–98.

— Elphinstone, C. D. and Jenkins, T. (1974). Mseleni joint disease: population genetic studies. *South African Journal of Science* 70, 360–5.

— Harpending, H. and Jenkins, T. (1972). Biology and the history of Southern African populations. In *Evolutionary models and studies in human diversity* (eds. R. J. Meier, C. Otten and F. Abdel-Hameed), pp. 245-54. Mouton, The Hague.

— and Jenkins, T. (1973). Genetically determined hazards of blood transfusion within and between races. *South African Medical Journal* 47, 56–61.

— — (1974). Lactose intolerance in San populations. *British Medical Journal* iii, 809.

— — (1975). The Griqua of Campbell, Cape Province. *American Journal of Physical Anthropology* 43, 71–8.

— — (1977a). *Health and the hunter-gatherer*. Karger, Basel.

— — (1977b). Serogenetic studies on the Kavango of South West Africa. *Annals of Human Biology* 4, 465–78.

— — (1978). Riemvasmaak before resettlement. *South African Journal of Science* 74, 339–41.

— — Africa, B. J. and Stellmacher, F. F. (1982). Sero-genetic studies on the Basters of Rehoboth, South West Africa/Namibia. *Annals of Human Biology* 9, 157–66.

— Lane, A. B. and Jenkins, T. (1976). Sero-genetic studies on the Dama of South West Africa. *Annals of Human Biology* 3, 33–50.

— Rootman, A. J. and Jenkins, T. (1983). Serogenetic studies on the Ambo of Namibia. (in press).

— Santos, David J. H., Steinberg, A. G. and Jenkins, T. (1979). Serogenetic studies on the Njinga of Angola. *Annals of Human Biology* 6, 335–44.

— Tanaka, N., MacNab, G. and Jenkins, T. (1973). Non-veneral syphilis and Australia antigen among the G/wi and G//ana San ('Bushmen') of the Central Kalahari Reserve, Botswana. *Central African Journal of Medicine* 19, 207–13.

Oettlé, A. G. (1963a). Regional variations in the frequency of Bantu oesophageal cancer cases admitted to hospitals in South Africa. *South African Medical Journal* 37, 434–39.

— (1963b). Skin cancer in Africa. *National Cancer Institute Monograph* 10, 197–214.

— (1965). The aetiology of primary carcinoma of the liver in Africa: a critical appraisal of previous ideas with an outline of the mycotoxin. *South African Medical Journal* 39, 817–25.

Ogbu, J. U. (1973). Seasonal hunger in tropical Africa as a cultural phenomenon. *Africa* 43, 317–32.

Ojikutu, R. O., Nurse, G. T. and Jenkins, T. (1977). Red cell enzyme polymorphisms in the Yoruba. *Human Heredity* 27, 444–53.

Oliver, R. O. (1966). The problem of Bantu expansion. *Journal of African History* 7, 361–76.

— and Fagan, B. (1975). *Africa in the Iron Age*. Cambridge University Press, London.

Omer-Cooper, J. D. (1966). *The Zulu aftermath*. Longman, London.

— Aspects of political change in the Mfecane. In *African societies in Southern Africa* (ed. L. Thompson), pp. 207–29. Heinemann, London, Ibadan and Nairobi.

Pachai, B. (ed.) (1972). *The early history of Malaŵi*. Longman, London.

— (1971). *The international aspects of the South African Indian question, 1860–1971*. Struik, Cape Town.

Palmer, E. (1966). *The Plains of Camdeboo*. Collins, London.

Palmhert-Keller, R., Nurse, G. T. and Jenkins, T. (1983). Serogenetic studies on the Caucasoids of South West Africa/Namibia. *Human Heredity* 33, 79–87.

Pama, C./de Villiers, C. C. (1966). *Geslagsregisters van die Ou Kaapse Families*. 3 Dele. A. A. Balkema, Kaapstad en Amsterdam.

Passarge, S. (1908). *Südafrika. Eine Landes-, Volks- und Wirtschaftskunde*. Quelle and Meyer, Leipzig.

Patterson, K. D. and Hartwig, G. W. (1978). The disease factor: an introductory review. In *Disease in African history* (eds. G. W. Hartwig and K. D. Patterson), pp. 3–24. Duke University Press, Durham, N.C.

Pauw, B. A. (1963). *The second generation: a study of the family among urbanized Bantu in East London*. Oxford University Press, Cape Town.

Pearson, J. B. (1963). Carcinoma of the breast in Nigeria: a review of 100 patients. *British Journal of Cancer* 17, 559–65.

Peringuey, L. (1911). The stone ages of South Africa as represented in the collection of the South African Museum. *Annals of the South African Museum* 8, 1–218.

Phillipson, D. W. (1969). Early iron-using peoples of Southern Africa. In *African societies in Southern Africa* (ed. L. Thompson), pp. 24–49. Heinemann, London, Ibadan and Nairobi.

— (1970). Notes on the later prehistoric radiocarbon chronology of eastern and southern Africa. *Journal of African History* 11, 1–15.

— (1977). *The later prehistory of Eastern and Southern Africa*. Heinemann, London.

Pliny the Elder (C. Plinius Secundus) (1855–1857). *Natural History*. Translated by J. Bostock and H. T. Riley. London, Bohn.

Post, R. H. (1962). Population differences in red and green colour deficiency. A review and a query on selection relaxation. *Eugenics Quarterly* 9, 131–46.

Potgieter, E. F. (1955). *The disappearing Bushmen of Lake Chrissie*. Van Schaik, Pretoria.

Prinsloo, H. (1974). Early Iron Age site at Klein Afrika near Wyliespoort, Soutpansberg Mountains, South Africa. *South African Journal of Science* 70, 271–3.

Ptolemy (Claudius Ptolemaeus) (1932). *Geography*. Translated by E. L. Stevenson. New York Public Library, New York.

Race, R. R. and Sanger, R. (1968). *Blood groups in Man* (5th edn). Blackwell, Oxford.

— — (1975). *Blood Groups in Man* (6th edn). Blackwell, Oxford.

Ramsay, M. and Jenkins, T. (1984). α-Thalassaemia in Africa: the oldest malaria protective trait? *The Lancet* ii, 410.

Ranger, T. O. (ed.) (1968). *Aspects of Central African History*. Heinemann, London, Ibadan and Nairobi.

Raper, P. E. (1979). Notes on Khoekhoen place names. *South African Journal of Science* 75, 448–51.

Raven-Hart, R. (1967). *Before Van Riebeeck: callers at South Africa from 1488 to 1652*. Struik, Cape Town.

Redinha, J. (1974). *Distribução étnica de Angola* (8a edição). Cita, Luanda.

van Rensberg, S. J., Kirsipuu, A., Coutinho, L., Watt, J. J. (1975). Circumstances associated with the contamination of food by aflatoxin in a high primary liver cancer area. *South African Medical Journal* 49, 877–83.

Rensch, B. (1947). *Neuer Probleme der Abstammungslehre; die trans-spezifische Evolution*. F. Enke, Stuttgart.

Rheinallt Jones, J. D. and Doke, C. M. (eds.) (1937). *Bushmen of the Southern Kalahari*. Witwatersrand University Press, Johannesburg.

Richards, A. I. (1939). *Land, labour and diet in Northern Rhodesia*. Oxford University Press for International African Institute, London.

Rightmire, G. P. (1975). New studies of Post-Pleistocene human skeletal remains from the Rift Valley, Kenya. *American Journal of Physical Anthropology* 42, 351–70.

Robbs, J. V. and Moshal, M. G. (1979). Duodenal ulceration in Indians and Blacks in Durban. *South African Medical Journal* 55, 39–42.

Roberts, D. F. (1953). Body weight, race and climate. *American Journal of Physical Anthropology* 11, 533–58.

Robertson, M. A., Harington, J. S. and Bradshaw, E. (1971a). The cancer pattern in Africans at Baragwanath Hospital, Johannesburg. *British Journal of Cancer* 25, 377–84.

— — — (1971b). The cancer pattern in Africans of the Transvaal lowveld. *British Journal of Cancer* 25, 385–94.

— — — (1971c). The cancer pattern in African gold miners. *British Journal of Cancer* 25, 395–402.

Robinson, K. R. (1959). *Khami ruins.* Cambridge University Press, Cambridge.

Robinson, K. R. (1966). The archaeology of the Rozwi. In *The Zambesian past* (eds. E. Stokes and R. Brown), pp. 3–27. Manchester University Press Manchester.

Ross, R. (1976). *Adam Kok's Griquas.* Cambridge University Press, Cambridge.

Rudner, J. (1968). Strandloper pottery from South and South West Africa. *Annals of the South African Museum* 49, 441–663.

Saha, N., Samuel, A. P. W., Omar, A. and Ahmed, M. A. (1978). A study of some genetic characteristics of the population of the Sudan. *Annals of Human Biology* 5, 569–75.

Sahlins, M. (1968). Notes on the original affluent society. In *Man the hunter* (eds. R. Lee and I. DeVore), pp. 85–9. Aldine Press, Chicago.

Sampson, C. G. (1967). *Excavations at Zaayfontein shelter, Norvalspont, northern Cape.* National Museum, Bloemfontein.

Sansom, B. (1974). Traditional economic systems. In *The Bantu-speaking peoples of Southern Africa* (ed. W. D. Hammond-Tooke), pp. 135–76. Routledge and Kegan Paul, London and Boston.

Santachiara-Benerecetti, A. S., Beretta, M., Negri, M., Ranzani, G., Antonini, G., Barberio, C., Modiano, G. and Cavalli-Sforza, L. L. (1980). Population genetics of red cell enzymes in Pygmies: a conclusive account. *American Journal of Human Genetics* 32, 934–54.

— Ranzani, A. N. and Antonini, G. (1977). Study on African Pygmies. V. Red cell Acp in Babinga Pygmies. *American Journal of Human Genetics* 29, 635–8.

Saron, G. (1955). Jewish immigration, 1880–1913. In *The Jews in South Africa: a history* (eds. G. Saron and L. Hotz), pp. 85–104. Oxford University Press, London.

— and Hotz, L. (eds.) (1955). *The Jews in South Africa: a history.* Oxford University Press, London.

Schapera, I. (1930). *The Khoisan peoples of South Africa.* Routledge and Kegan Paul, London.

— (ed.) (1937). *The Bantu-speaking tribes of South Africa.* Routledge, London.

— (1953). *The Tswana.* International African Institute, London.

Schoffeleers, J. M. (1972). The meaning and use of the name *Malaŵi* in oral traditions and precolonial documents. In *The early history of Malaŵi* (ed. B. Pachai), pp. 91–132. Longman, London.

Schofield, J. F. (1935). Natal coastal pottery from the Durban district, Part I. *South African Journal of Science* 32, 508–27.

— (1936). Natal coastal pottery from the Durban district. Part II. *South African Journal of Science* 33, 993–1009.

— (1937). The work done in 1934: pottery. In *Mapungubwe* (ed. L. Fouché), pp. 32-55. Cambridge University Press, London.

— (1948). *Primitive pottery*. Handbook no. 3 of the South African Archaeological Society, Cape Town.

— (1958). Pottery from Natal, Zululand, Bechuanaland and South West Africa. *South African Journal of Science* 53, 382-95.

Schonland, M. and Bradshaw, E. (1968). Cancer in the Natal African and Indian 1964-66. *International Journal of Cancer* 3, 304-16.

— — (1969). Oesophageal cancer in Natal Bantu: a review of 516 cases. *South African Medical Journal* 43, 1028-31.

Schultze, L. (1907). *Aus Namaland and Kalahari*. Gustav Fischer, Jena.

— (1928). Zur Kenntnis des Körpers der Hottentoten und Buschmänner. *Jenaische Denkschriften* 17, 147-228.

Schumacher, A., Knussmann, R. and Knussmann, R. (1979). Die Transferrintypen der Dama (Südwest-Afrika). *Anthropologische Anzeiger* 37, 101-6.

Seedat, Y. K., Seedat, M. A. and Nkomo, M. N. (1978). The prevalence of hypertension in the urban Zulu. *South African Medical Journal* 53, 923-7.

Seftel, H. C. (1977). Diseases in urban and rural Black populations. *South African Medical Journal* 51, 121-3.

— (1978). The rarity of coronary heart disease in South African Blacks. *South African Medical Journal* 54, 99-105.

Segi, M., Noye, H., Hattori, H., Yamazaki, Y. and Segi, R. (1979). *Age-adjusted death rate for cancer for selected sites (A- classification) in 51 countries in 1974*. Segi Institute of Cancer Epidemiology, Nagoya, Japan.

Sensabaugh, G. P. and Golden, V. L. (1978). Phenotype dependence in the inhibition of red cell acid phosphatase (ACP) by folates. *American Journal of Human Genetics* 30, 553-60.

Shapiro, M. (1951a). The ABO, M, P and Rh blood group systems in the South African Bantu. *South African Medical Journal* 25, 165-70 and 189-92.

— (1951b). Further evidence of homogeneity of blood group distribution in the South African Bantu. *South African Medical Journal* 25, 406-11.

— (1952). Observations on the Kell-Cellano (K-k) blood group system. *South African Medical Journal* 26, 951-5.

Short, R. V. (1976). Definition of the problem: the evolution of human reproduction. *Proceedings of the Royal Society, London* series B, 195, 3-24.

Shrubsall, F. C. (1898). Crania of African Bush races. *Journal of the Royal Anthropological Institute* 27, 263-92.

— (1907). Notes on some Bushman crania and bones from the South African Museum, Cape Town. *Annals of the South African Museum* 5, 227-70.

— (1911). A note on craniology. *Annals of the South African Museum* 8, 202-8.

von Sicard, H. (1954). Rhodesian sidelights on Bechuanaland history. *NADA* 31, 68-94.

Silberbauer, G. B. (1965). *Bushman survey report*. Bechuanaland Government, Gaberones.

— and Kuper, A. J. (1966). Kgalagari masters and Bushman serfs: some observations. *African Studies* 25, 171-9.

Simoons, F. J. (1970). Primary adult lactose intolerance and the milking habit: a problem in biologic and cultural interrelations. II. A culture historical hypothesis. *American Journal of Digestive Diseases* 15, 695-710.

— (1978). The geographic hypothesis and lactose absorption: a weighing of the evidence. *American Journal of Digestive Diseases* 23, 963-80.

Sineer, H. A., Nelson, M. M. and Beighton, P. H. (1979). Spina bifida and anencephaly in the Cape. *South African Medical Journal* 53, 626-7.

Singer, R. (1958). The Boskop 'race' problem. *Man* 58, 173-8.

— and Weiner, J. S. (1963). Biological aspects of some indigenous African populations. *South-Western Journal of Anthropology* 19, 168-76.

— — and Zoutendyk, A. (1961). The blood groups of the Hottentots. In *Proceedings of the Second International Congress of Human Genetics*, pp. 884-7. Edizione Instituto 'G. Mendel', Rome.

Smith, A. K. (1973). The peoples of southern Mozambique: an historical survey. *Journal of African History* 14, 565-80.

Smithells, R. W., Sheppard, S., Schorah, C. J., Seller, M. J., Nevin, N. C., Harris, R., Read, A. P. and Fielding, D. W. (1980). Possible prevention of neural tube defects by periconceptional vitamin supplementation. *Lancet* i, 339-40.

Soga, J. H. (1930). *The South-Eastern Bantu*. Witwatersrand University Press, Johannesburg.

Soussi-Tsafrir, J. (1974). *Light-eyed Negroes and the Klein–Waardenburg Syndrome*. Macmillan, London.

Spedini, G., Capucci, E., Fuciarelli, M. and Richards, O. (1980). The AcP polymorphism frequencies in the Mbugo and Sango of Central Africa (correlations between the $P^r$ allele frequencies and some climatic factors in Africa). *Annals of Human Biology* 7, 125-8.

Splaine, M., Hayes, E. B. and Barclay, G. P. T. (1971). Calculations for changes in sickle-cell trait rates. *American Journal of Human Genetics* 23, 368-74.

Stayt, H. (1931). *The Bavenda*. Oxford University Press, London.

Steinberg, A. G., Jenkins, T., Harpending, H. C. and Nurse, G. T. (1975). Gamma globulin groups of the Khoisan peoples of Southern Africa: evidence for polymorphism for a $Gm^{1,5,13,14,21}$ haplotype among the San. *American Journal of Human Genetics* 27, 528-42.

Stellmacher, F. F., Nurse, G. T. and Jenkins, T. (1976). Some serogenetic markers and haematological parameters in the Basters of Rehoboth, SWA. *Central African Journal of Medicine* 22, 88-91.

Stern, J. T. and Singer R. (1967). Quantitative morphological distinctions between Bushman and Hottentot skulls: a preliminary report. *South African Archaeological Bulletin* 22, 103-11.

Stevenson, A. C., Johnson, H. A., Stewart, M. P. P. and Golding, D. R. (1966). Congenital malformations: a report of a series of consecutive births in 24 countries. *Bulletin of the World Health Organization* 34, supplement, 1-127.

Stoke, J. C. J. (1966). Blood groups in Rhodesia. *Central African Journal of Medicine* 12, 73-4.

Stokes, E. T. and Brown, R. (eds.) (1966). *The Zambesian past*. Manchester University Press, Manchester.

Strydom, N. B. (1974). *Endurance capacity of Bushmen*. Paper presented to the Bushman Symposium, Johannesburg.

Summers, R. (1958). *Inyanga*. Cambridge University Press, Cambridge.

Super, M. (1975). Cystic fibrosis in the South West African Afrikaner. An example of population drift possibly with heterozygote advantage. *South African Medical Journal* 49, 818-20.

Swadesh, M. (1948). *The time value of linguistic diversity*. Paper delivered at Viking Fund Supper-Conference, New York.

Sydow, W. (1967). The pre-European pottery of South West Africa. *Cimbebasia* 1, 1-74.

Szumowski, G. (1956). Fouilles de l'abri sous-roche de Kourounkoronkale (Soudan Français). *Bulletins de l'Institut Français de L'Afrique Noire* 18, sér B, 462-508.

Tanaka, J. (1969). The ecology and social structure of Central Kalahari Bushmen. *Kyoto University African Studies* 3, 1-26.

— (1976). Subsistence ecology of Central Kalahari San. In *Kalahari hunter-gatherers* (eds. I. DeVore and R. Lee), pp. 98-119. Harvard University Press, Cambridge, Mass.

— (1980). *The San, hunter-gatherers of the Kalahari.* University of Tokyo Press, Tokyo.

Theal, G. McC. (1898-1903). *Records of South-East Africa.* 9 vols. Government of the Cape Colony.

— (1908). *History of South Africa since 1795.* 5 Vols. Swann Sonnenschein, London.

— (1907-10). *History and ethnography of Africa south of the Zambesi before 1795.* 3 Vols. Swann Sonnenschein, London.

— (1910a). *Ethnography and condition of South Africa before A.D. 1505.* George Allen and Unwin, London.

— (1910b). *The yellow and dark-skinned people of Africa south of the Zambesi.* Swann Sonnenschein, London.

— (1919). *History of South Africa, 1873 to 1894.* 2 Vols. George Allen and Unwin, London.

Thompson, L. (ed.) (1969). *African societies in Southern Africa.* Heinemann, London, Ibadan and Nairobi.

Tills, D., Kopeć, A. C., Fox, R. F. and Mourant, A. E. (1979). The inherited blood factors of some Northern Nigerians. *Human Heredity* 29, 172-6.

Tippler, T. S., Dunn, D. S. and Jenkins, T. (1982). Phosphoglucomutase first locus polymorphism as revealed by isoelectric focusing in Southern Africa. *Human Heredity* 32, 80-93.

Tlou, T. (1977). Servility and political control: Botlhanka among the Batawana of Northwestern Botswana c.1750-1906. In *Slavery in Africa* (eds. S. Miers and I. Kopytoff), pp. 367-409. University of Wisconsin Press, Madigon.

Tobias, P. V. (1955a). Physical anthropology and the somatic origins of the Hottentots. *African Studies* 14, 1-22.

— (1955b). Les Bochimans Auen et Naron de Ghanzi. Contribution a l'étude des 'Anciens Jaunes' sud-africains. II. *Anthropologie* 59, 429-61.

— (1956). Les Bochimans Auen et Naron de Ghanzi. Contribution a l'étude des 'Anciens Jaunes' sud-africains. III. *Anthropologie* 60, 22-52.

— (1958). Skeletal remains from Inyanga. In *Inyanga* (ed. R. Summers), pp. 159-72. Cambridge University Press, Cambridge.

— (1961). Fingerprints and palmar prints of Kalahari Bushmen. *South African Journal of Science* 57, 333-45.

— (1962a). Early members of the genus *Homo* in Africa. In *Evolution and hominisation* (ed. G. Kruth), pp. 191-204. Fischer Verlag, Stuttgart.

— (1962b). On the increasing stature of the Bushman. *Anthropos* 57, 801-10.

— (1964). Bushman hunter-gatherers: a study in human ecology. In *Ecological studies in Southern Africa* (ed. D. H. S. Davies), pp. 67-86. Junk, The Hague.

— (1965). Early man in East Africa. *Science* 199, 22-33.

— (1966). The peoples of Africa south of the Sahara. In *The biology of human adaptability* (eds. P. Baker and J. S. Weiner), pp. 111–200. Oxford University Press, Oxford.

— (1971). *The brain in hominid evolution.* Columbia University Press, New York.

— (1972a). Recent human biological studies in Southern Africa, with special reference to Negroes and Khoisans. *Transactions of the Royal Society of South Africa* **40**, 109–33.

— (1972b). Growth and stature in Southern African populations. In *The human biology of environmental change* (ed. D. J. M. Vorster), pp. 96–104. Taylor and Francis, London.

— (1974). The biology of the Southern African Negro. In *The Bantu-speaking peoples of Southern Africa* (ed. W. D. Hammond-Tooke), pp. 3–45. Routledge and Kegan Paul, London and Boston.

— (1978). South African australopithecines in time and hominid phylogeny, with especial reference to dating and affinities of the Taung skull. In *Early hominids of Africa* (ed. C. J. Jolly), pp. 45–84. Duckworth, London.

— (1980). A survey and synthesis of the African hominids of the late Tertiary and early Quaternary periods. In *Current argument on early man* (ed. L. Konigsson), pp. 86–113. Pergamon Press, Oxford.

Toerien, M. J. (1958). The physical characters of the Lake Chrissie Bushmen. *South African Journal of Medical Science* **23**, 121–4.

— and Hughes, A. R. (1955). The limb bones of Springbok Flats man. *South African Journal of Science* **52**, 125–8.

Traill, A. (1973). 'N4 or S7': another Bushman language. *African Studies* **32**, 25–32.

Trevor, J. C. (1950). The physical characters of the Sandawe. *Journal of the Royal Anthropological Institute* **77**, 61–78.

— Mukherjee, R. and Rao, C. R. (1955). *The ancient inhabitants of Jebel Moya.* Cambridge University Press, Cambridge.

Trewartha, G. T. (1961). *The earth's problem climates.* University of Wisconsin Press, Madison.

— Robinson, A. H. and Hammon, E. H. (1957). *Elements of geography.* McGraw Hill, New York.

Truswell, A. S. and Hansen, J. D. L. (1968a). Serum lipids in Bushmen. *Lancet* **ii**, 684.

— — (1968b). Medical and nutritional studies of !Kung Bushmen in north-west Botswana: a preliminary report. *South African Medical Journal* **42**, 1338–9.

— — (1975). Biochemical research among the Bushmen. In *Kalahari hunter-gatherers* (eds. I. DeVore and R. Lee), pp. 166–94. Harvard University Press, Cambridge, Mass.

— and Mann, J. I. (1972). Epidemiology of serum lipids in Southern Africa. *Atherosclerosis* **16**, 15–29.

Turnbull, C. M. (1965). *Wayward servants.* Eyre and Spottiswoode, London.

Twisselmann, F. (1958). Les ossements humains du gîte mésolithique d'Ishango. *Explorations du Parc National Albert, par la Mission J. de Heinzelin de Brancourt* **5**, 3–123.

de Vaal, J. B. (1943). 'n Soutpansbergse Zimbabwe. *South African Journal of Science* **40**, 303–22.

Vansina, J. (1966). *Kingdoms of the savannah.* University of Wisconsin Press, Madison.

— Mauny, R. and Thomas, L. V. (eds.) (1964). *The historian in Central Africa.* Oxford University Press for International African Institute, London.

Vedder, H. (1923). *Die Bergdama.* Friedrichsen, Hamburg.

— (1928*a*). The Berg Damara. In *The native tribes of South West Africa,* pp. 39-78. Cape Times, Cape Town.

— (1928*b*). The Nama. In *The native tribes of South West Africa,* pp. 109-48. Cape Times, Cape Town.

— (1928*c*). The Herero. In *The native tribes of South West Africa,* pp. 155-208. Cape Times, Cape Town.

— (1934). *Das alte Südwestafrika.* Martin Warneck Verlag, Berlin.

— (1938). *South West Africa in early times.* Oxford University Press, Oxford.

— (1965). *Zur Vorgeschichte der eingeborenen Völkerschaften von Südwestafrika.* S.W.A. Wissenschaftlige Gesellschaft, Windhoek.

Velho, A. (1838). *Roteiro da viagem que em descobrimento da India pelo Cabo de Boa Esperanga fez Dom Vasco da Gama.* Edited by D. Kopke and A. da Costa Paiva. Academia Politécnica, Porto.

van Velsen, J. (1964). *The politics of kinship.* Manchester University Press, Manchester.

de Villiers, C. C. (1893-5). *Geslachtregister de oude Kaapsche familien.* 3 Vols. Van de Sandt de Villiers and Co., Cape Town.

de Villiers, H. (1961). The tablier and steatopygia in Kalahari Bushwomen. *South African Journal of Science* 57, 223-7.

— (1968). The morphology and incidence of the tablier in Bushman, Griqua and Negro females. *Proceedings of the Eighth Congress of Anthropological and Ethnological Sciences,* 48-51.

— (1970). A note on taste blindess in Venda males. *South African Journal of Science* 66, 26-8.

— (1972). A study of morphological variables in urban and rural Venda male populations. In *The human biology of environmental change* (ed. D. J. M. Vorster), pp. 110-13. Taylor and Francis, London.

— (1973). Human skeletal remains from Border Cave, Ingwavuma District, KwaZulu, South Africa. *Annals of the Transvaal Museum* 28, 229-56.

— (1977). Urban and rural South African Negro male populations: a study of morphological variables and genetic markers. In *Human adaptability* (eds. K. O. Collins and J. S. Weiner), pp. 251-3. Taylor and Francis, London.

— and Fatti, L. P. (1982). The antiquity of the Negro. *South African Journal of Science* 78, 321-32.

Vivelo, F. M. (1977). *The Herero of Western Botswana.* West Publishing, St. Paul.

Vogel, F. and Chakravartti, M. R. (1966). ABO blood groups and smallpox in a rural population of West Bengal and Bihar (India) *Humangenetik* 3, 166-80.

— Pettenkofer, H. J. and Helmbold, W. (1960). Über die Populationsgenetik der ABO-Blutgruppen. *Acta Genetica, Basel* 11, 267-94.

Vorster, D. J. M. (ed.) (1972). *The human biology of environmental change.* Taylor and Francis, London.

— (1977). Adaptation to urbanization in South Africa. In *Population structure and human variation* (ed. G. A. Harrison), pp. 313-32. Cambridge University Press, Cambridge.

Vrba, E. S. (1980). Evolution, species and fossils: how does life evolve? *South African Journal of Science* 76, 61-84.

Wade, P. T., Jenkins T. and Huehns, E. R. (1967). Haemoglobin variant in a

Bushman: Haemoglobin D$\beta$ Bushman $\alpha_2\beta_2$ Gly→Arg. *Nature* **216**, 688–90.

Walker, A. R. P. (1979). South African Black, Indian and Coloured Populations. In *Western diseases: their emergence and prevention* (eds. H. C. Trowell and D. P. Burkitt). Edward Arnold, London.

— and Walker, B. F. (1978). Blood pressure in South African Black school-children aged 10–12 years. *Journal of Tropical Medicine and Hygiene* **81**, 159–63.

Walker, E. A. (1934). *The Great Trek*. A. and C. Black, London.

— (1962). *A history of Southern Africa* (3rd edn corrected). Longman, London.

van der Walt, L. A., Wilmsen, E. N., and Jenkins, T. (1977). Unusual sex hormone patterns among desert-dwelling hunter-gatherers. *Journal of Clinical Endocrinology and Metabolism* **46**, 658–63.

— — Levin, J. and Jenkins, T. (1978). Endocrine studies on the San ('Bushmen') of Botswana. *South African Medical Journal* **52**, 230–2.

Walton, J. (1956). Early Bafokeng settlement in South Africa. *African Studies* **15**, 37–43.

van Warmelo, N. J. (1962). *Notes on the Kaokoveld (South West Africa) and its people*. Government Printer, Pretoria.

— (1966). Zur Spreche und Herkunft der Lemba. *Deutsches Institut für Afrika-Forschung* Beil. 5, 273–83.

Waterhouse, G. (1932). *Simon van der Stel's journal of his expedition to Namaqualand*. Longmans, Green and Co., London.

Weiner, J. S. (1954). Nose shape and climate. *American Journal of Physical Anthropology* **12**, 1–4.

— (1958). The pattern of evolutionary development of the genus *Homo*. *South African Journal of Medical Sciences* **23**, 111–20.

— (1962). The pattern of evolutionary development of the genus *Homo*. In *Ideas on human evolution* (ed. W. W. Howells), pp. 521–31. Harvard University Press, Cambridge, Mass.

— (1965). Pattern and process in the evolution of man. (Presidential address). *Proceedings of the Royal Anthropological Institute*, 5–11.

— (1971a). *Man's natural history*. Weidenfeld and Nicholson, London.

— (1971b). The human biology of populations undergoing environmental change. *Journal of Tropical Medicine and Hygiene* **74**, 207–10.

— (1972). Tropical ecology and population structure. In *The structure of human populations* (ed. G. A. Harrison), pp. 393–410. Clarendon Press, Oxford.

— (1974a). Adaptation and variation among hunter-gatherers. In *Man in adaptation: the biosocial background* (ed. Y. A. Cohen). 2nd Edition. pp. 248–58. Aldine Publishing Company, Chicago.

— (1974b). Human biology and the ecosystem concept. *Proceedings of the First Congress on Ecology*, The Hague, 378–80.

— (1980). Work and well-being: physiological considerations. In *Human ecology in savannah environments* (ed. D. Harris), pp. 417–37. Academic Press, London.

— and Campbell, B. G. (1964). The taxonomic status of the Swanscombe skull. In *The Swanscombe skull* (ed. C. D. Ovey), pp. 175–209. Royal Anthropological Institute, London.

— Harrison, G. A., Singer, R., Harris, R. and Jopp, W. (1964). Skin colour in Southern Africa. *Human biology* **36**, 294–307.

— and Huizinga, J. (eds.) (1972). *The assessment of population affinities in man*. Oxford University Press, Oxford.

von Weismann, J., Vollmer, M. and Pribilla, O. (1981). Phenotype distribution and gene frequencies of adenosine deaminase in Schleswig–Holstein compared with samples from Portugal, Brazil and South Africa (Bantu-Xhosa and Whites). *Zeitschrift für Rechtsmedizin* 86, 227–32.

— Zippel, J. and Pribilla, O. (1982). Evidence of the factors Gm(1, 2, 4, 10, 12, 21) and Km(1, 3) as well as of the systems Gc and Hp on a random sample of Xhosa–Bantus, South Africa. *Zeitschrift für Morphologie und Anthropologie* 73, 73–8.

Welch, S. G., McGregor, I. A. and Williams, K. (1977). The Duffy blood group and malaria prevalence in Gambian West Africans. *Transactions of the Royal Society of Tropical Medicine and Hygiene* 71, 295–6.

— — — (1978). A new variant of human erythrocyte G6PD occurring at a high frequency amongst the population of two villages in the Gambia, West Africa. *Human Genetics* 40, 305–9.

— Swindlehurst, C. A., McGregor, I. A. and Williams, K. (1978). Isoelectric focusing of human red cell PGM: the distribution of variant phenotypes in a village population from the Gambia (West Africa). *Human Genetics* 43, 307–13.

— — — — (1979). Serum protein polymorphisms in a village community from the Gambia, West Africa (Hp, Tf, and Gc). *Human Genetics* 48, 81–4.

Welch, S. R. (1946). *South Africa under King Manuel*. Juta, Cape Town.

Wellington, J. H. (1955). *Southern Africa: a geographical study*. Vol. 1. *Physical geography*. Vol. II. *Economic and human geography*. Cambridge University Press, Cambridge.

— (1967). *South West Africa and its human issues*. Oxford University Press, Oxford.

Wells, L. H. (1952*a*). Human crania of the Middle Stone Age in South Africa. In *Proceedings of the First Pan-African Congress of Prehistory* (ed. L. S. B. Leakey), pp. 125–33. Blackwell, Oxford.

— (1952*b*). Physical measurements of Northern Bushmen. *Man (o.s.)* 52, 53–6.

— (1957). Late Stone Age human types in Central Africa. In *Proceedings of the Third Pan-African Congress on Prehistory, Livingstone 1955* (ed. J. D. Clark), pp. 183–5. Chatto and Windus, London.

Wendorf, F. (ed.) (1968). *The prehistory of Nubia*. Vol. II. Southern Methodist University Press, Dallas.

Werbner, R. (1970). Land and chieftainship in the Tati Concession. *Botswana Notes and Records* 2, 6–13.

Westphal, E. O. J. (1963). The linguistic prehistory of Southern Africa: Bush, Kwadi, Hottentot and Bantu linguistic relationships. *Africa* 33, 237–65.

— (1971). The click languages of Southern and Eastern Africa. In *Current trends in linguistics* (eds. J. Barry and J. H. Greenberg), pp. 367–420. Mouton, The Hague.

White, M. J. D. (1978). *Modes of speciation*. Freeman, San Francisco.

White, T. D., Johanson, D. C. and Kimbel, W. H. (1981). *Australopithecus africanus*: its phyletic position reconsidered. *South African Journal of Science* 77, 445–70.

Willcox, M., Brohult, J., Sirleaf, V. and Bengtsson, E. (1979). Malaria and haemoglobin $A_2$ levels in norther Liberia. *Transactions of the Royal Society of Tropical Medicine and Hygiene* 73, 209–11.

Willcox, M. C., Weatherall, D. J. and Clegg, J. B. (1975). Homozygous β-thalas-saemia in Liberia. *Journal of Medical Genetics* 12, 165–73.

Wilmsen, E. N. (1973). Interaction, spacing behaviour, and the organizing of hunting bands. *Journal of Anthropological Research* 29, 1–31.

Wilson, M. and Thompson, L. (1969–71). *The Oxford history of South Africa.* Vols. I and II. Oxford University Press, London.

Zoutendyk, A., Kopeć, A. C. and Mourant, A. E. (1955). The blood groups of the Hottentots. *American Journal of Physical Anthropology* 13, 691–8.

# GLOSSARY

**adaptation** This term is used in two senses. It may mean either the *process* or the *trait* by means of which an organism or a population of organisms is enabled to function more efficiently in a given environment.

**affinal kin** Relatives by marriage.

**alleles** Alternative forms of a gene at a single locus.

**allopatry** The occupation of distinctly separate geographical ranges.

**amorph** A gene producing no detectable gene product.

**antibody** An immunoglobulin whose production is stimulated by the introduction into the body of an antigen, a foreign substance with which it combines and thereby neutralizes.

**antigen** A substance which the body recognizes as foreign and strives through its immune mechanisms to neutralize. Certain antigens may be present normally in some members of a species but provoke an immune response in others.

**association** The occurence together of two traits more frequently than might be expected by chance. This does *not necessarily* indicate linkage.

**assortative mating** The conjunction of male and female for reproduction, in which pairing is determined by non-random factors. An example of *negative assortative mating* would be the (experimentally unproven but widely credited) tendency of redheads not to marry one another.

**blood group** A gene marker expressed as the presence of specific antigens on the surface of the red cells. Most blood groups are polymorphic.

**chromosome** The structures in the cell nucleus which by arrangements of the chemical molecules of which they are built up store and transmit genetic information as genes. Chromosomes replicate during the process of cell division.

**cline** A geographical gradient in gene frequencies.

**codominance** The equal expression of alleles, each independently of the presence of the other.

**cognatic kin** Relatives on both the mother's and the father's side.

**congenital** Present at birth. Congenital traits are not necessarily inherited but may be environmentally determined.

**consanguinity** Being cognatically related through a common ancestor.

**coupling** This is said to occur when two particular genes at linked loci are found on the same chromosome. They are said to be in repulsion when on different chromosomes.

**cross-cousin** First cousin through parental sibs of opposite sexes; the offspring of father's sister or of mother's brother.

**crossing over** A process in which homologous chromosomes in the course of replication exchange segments. This occurs most commonly during the maturation of the germ cells.

**cytogenetics** The study of chromosomes.

**dermatoglyphics** The study of the dermal ridges of the extremities (e.g. fingerprints).

**deterministic** Non-random; occurring as part of a set process.

**dominance** The action of one allele which is expressed, in the presence at the equivalent locus on the homologous chromosome of another which is not expressed (i.e. is recessive).

**dual-descent** A kinship system in which two categories of kin are recognized, one matrilineal and the other patrilineal.

**endemic** Said of a disease of man of continuous prevalence in a defined population or geographical area.

**endogamy** The custom of marrying within the family or social group.

**enzoötic** Said of a disease of animals of continuous prevalence in a defined geographical area.

**enzyme** A protein which acts as a catalyst in a specific biochemical reaction.

**enzyme system** A gene marker system where the polymorphic protein is an enzyme. In most of those which are used for the study of populations, the enzymes investigated are those inside the red blood cells. Other enzyme systems are those of the serum and of the tissues.

**epidemic** Said of a disease of man periodically prevalent in a defined population or geographical area.

**epidemiology** The study of incidences and prevalences of diseases or traits over time and space among human populations.

**epistasis** Interaction between gene loci such that the genotype at one or both loci affects the expression of the genotype at the other.

**epizoötic** Said of a disease of animals periodically prevalent in a defined population or geographical area.

**exogamy** The custom of marrying outside the family or social group.

**expressivity** Variation in phenotype determined by a single genotype. Particularly associated with dominantly inherited diseases and traits.

**fitness, genetic** The ability of an organism to survive and attain its maximum reproductive capacity.

**fixation of a gene** Loss of polymorphism at a locus and the consequent establishment of a monomorphism of a particular allele.

**gamete** The male or female cell which at fertilization fuses with the cell of the other sex to form the zygote from which the foetus develops.

**gene** A unit of heredity, usually equated with a unit of inherited function such that it determines a particular trait inherited in a Mendelian fashion. May also be defined as a sequence of nucleotides on a chromosome, programmed to code for a particular polypeptide (protein) chain or a molecule of ribonucleic acid.

**gene flow** The exchange of genes between two populations, or the donation of genes from one gene pool to another. This usually occurs at such a low rate that both populations retain their identity.

**gene frequency** The proportion of a particular allele among all the alleles at a particular locus in a population.

**gene marker system** An identifiable range of alleles at a single locus. The system may determine a blood group or enzyme or serum protein or other polymorphism.

**gene pool** The total pool of genes in a population.

**genetic equilibrium** A state of genetic stability in a population, such that local temporal fluctuations are small in amplitude and over the generations obey the requirements of the Hardy–Weinberg Law.

**genome** The ensemble of genetic material in a cell. It must not be forgotten that all the cells of an organism carry an identical complement of genes, so that the genome may also be defined as the total genetic information contained in an organism.

**germ cell** Any cell of the cell line leading to the production of a gamete, and including the gamete itself.

**haplotype** A combination of alleles at closely linked loci found in a single chromosome.

**Hardy–Weinberg law** A rule for predicting genotype frequencies from gene frequencies, or deriving gene frequencies from phenotype arrays, on the assumption of random mating and the absence of selection in an infinitely large population.

**hemizygous** Having only one locus, instead of the two present with a pair of homologous chromosomes. Since human males normally possess only one X chromosome, they are hemizygous for all their X-linked genes.

**heterosis** See hybrid vigour.

**heterozygous** Having different alleles at a particular gene locus.

**histocompatibility** The capacity to accept a tissue graft from another individual. Mainly determined by an array of closely linked loci on chromosome no. 6 in man. These loci are also known as the HLA system or the Major Histocompatibility Complex (MHC).

**HLA** See Histocompatibility.

**hominid** Primate of the family Hominidae, of which man (*Homo sapiens sapiens*) is the only surviving member.

**homoeostasis** A state of physiological or similar balance free of any tendency to fluctuation. This represents the most efficient state of a particular trait in an organism, generally irrespective of the environment. An example is the normal body temperature, which remains the same no matter where the individual is.

**homologous** Applies to chromosomes which pair during cell division, and which consequently have the same morphology and carry the same array of gene loci.

**homozygous** Having similar alleles at a particular gene locus.

**hybrid vigour** The increased vigour of hybrids when compared with both parental lines. Hybrid vigour does not necessarily confer genetic fitness, and when it is the result of mating across species lines it is often associated with sterility.

**immunoglobulins** Antibody molecules. These fall into five classes and have widely varying antibody specificities. Certain regions of the molecules carry gene marker sequences which appear to be unrelated to their immune properties.

**inbreeding** The consequences of the mating of consanguineous individuals. See also Endogamy.

**incidence** The occurrence rate of an event, for instance, the frequency of a trait at birth.

**interbreeding** Mating between members of populations between which there has been little or no previous gene flow.

**isozymes** Enzymes determined at the same locus but distinguishable by some property such as difference in electrical charge.

**lexis** The corpus and form of words in a language.

**lineage** A group of individuals sharing a common ancestor.

**linkage** The presence of two or more loci so close together on a chromosome that crossing-over between them is infrequent and they tend to be inherited together.

**linkage disequilibrium** A tendency for particular alleles to associate together at linked loci at a higher frequency than could be expected by chance.

**locus** The position of a gene on a chromosome. May be used as a synonym for gene but not for allele.

**matrilineality** Descent or inheritance through the female line.

**Mendelian inheritance** Inheritance of a trait in a manner which follows Mendel's laws. In practice, this means monogenic inheritance.

**MHC** Major Histocompatibility Complex. See Histocompatibility.

**mimicry** The similarity in appearance of one race or species to another. Mimicry is not necessarily determined by similar alleles or even at similar loci.

**miscegenation** Mating between members of distinctly separate populations, most commonly different races.

**monogenic** Said of the determination of a trait at a single locus.

**monomorphism** The present at a particular locus of only one allele.

**morphology** Gross physical characters, generally multifactorially determined.

**mulierilocality** Residence of a married couple at the wife's parental place of residence.

**multifactorial** Applied to a trait whose phenotypic expression is determined by the cumulative effects of a number of factors, of which some or all may be genetic and some may be environmental.

**mutation** A heritable change in the genetic material at a locus.

**ortho-cousin** First cousin through parental sibs of the same sex; the offspring of father's brother or mother's sister.

**outbreeding** See Miscegenation; Exogamy.

**overdominance** See Hybrid Vigour.

**panmixis** Random mating; the opposite of assortative mating.

**patrilineality** Descent or inheritance through the male line.

**penetrance** The relative prevalence of a trait given a particular genotype.

**pericentric inversion** Reversal of the order of the genes around the central point of a chromosome, due to breaks occurring on opposite sides of the centre and the reinsertion of the central portion upside down.

**phenocopy** Either an environmentally produced phenotype closely resembling that produced by a particular genotype, or a phenotype produced by one genotype but closely resembling one produced by a different genotype, possibly at a different locus.

**phenotype** An observable trait or set of traits produced by the interaction of a particular genotype with the environment in which development takes place.

**phoneme** A sound or set of sounds in a language regarded by speakers of that language as a single phonetic item.

**plasma protein** A protein found in solution in the plasma, the fluid in which the blood cells are suspended.

**pleiotropy** The capacity of a gene to determine or influence the expression of more than one trait.

**polygenic** Applied to a trait whose phenotypic expression is determined by the interaction and/or cumulative effects of a number of genes at different loci.

**polymorphism** The occurrence at a particular locus of two or more alleles at frequencies higher than could be maintained by mutation alone.

**population** A group of individuals of both sexes associated together on any of a variety of grounds and capable of maintaining itself by natural reproduction or by a uniform and unvarying pattern of migration.

**population structure** The patterns by which a population either maintains stability or undergoes change.

**prevalence** The relative frequency of a trait or disease in a particular population.

**proband** The person through whom the presence of a trait in a population or a lineage is ascertained.

**race** A geographically and usually genetically distinct and self-perpetuating division of a species.

**random mating** A situation in which the genetic trait being studied has no bearing on the choice of mates.

**recessivity** The action of one allele which is not expressed in the presence at the equivalent locus on the homologous chromosome of another which is expressed (i.e. is dominant).

**recombination** The genetic results of crossing-over; the association of new sets of alleles on homologous chromosomes.

**segregation** The paths of descent of homologous alleles.

**selection** The process which determines the relative part which individuals of particular genotypes play in the propagation of a population. The selective effect of a gene is the probability that a carrier of a gene will reproduce.

**semantic** Appertaining to the domain of meaning of words or phrases.

**sex chromosomes** Those chromosomes on which the principal determinants of sex are situated. There is a morphological difference between the sexes in mammals, such that male posseses an X and a Y chromosome and females two X chromosomes.

**sex ratio** The relative proportion of males and females in a population.

**sibling** Brother or sister.

**sibship** A group of brothers, sisters, or brothers and sisters.

**significance** An arbitrary probability level chosen for the appropriate judging of contrasting findings or results.

**stochastic** Applied to a process consisting of a sequence of steps the transition from one to the next of which happens at random.

**stock** A lineage or closely related group of lineages.

**stratification** The coexistence of two or more sub-populations within a larger population.

**structural gene** A gene which codes directly for the amino acid sequence of a particular polypeptide.

**sympatry** The occupation of the same geographical range.

**synteny** The location of loci on the same chromosome.

**tissue antigens** See Histocompatibility.

**virilocality** Residence of a married couple at the husband's parents' place of residence.

**wild type** The normal type of an organism; the normal allele as compared to a mutant allele.

**X-linked** A trait determined by a gene situated on the X chromosome is known as an X-linked trait.

# AUTHOR INDEX

# SUBJECT INDEX